Homogeneous Groups: Hardy Inequalities (Volume 1)

Homogeneous Groups: Hardy Inequalities (Volume 1)

Hart Scott

MURPHY & MOORE
www.murphy-moorepublishing.com

Published by Murphy & Moore Publishing,
1 Rockefeller Plaza,
New York City, NY 10020, USA

ISBN: 978-1-63987-307-4

Cataloging-in-Publication Data

Homogeneous groups : Hardy inequalities. Volume 1 / Hart Scott.
p. cm.
Includes bibliographical references and index.
ISBN 978-1-63987-307-4
1. Inequalities (Mathematics). 2. Integral inequalities. 3. Mathematics. I. Scott, Hart.

QA295 .H66 2022
515.243--dc23

For information on all Murphy & Moore Publications
visit our website at www.murphy-moorepublishing.com

MURPHY & MOORE

Contents

Preface

Homogeneous groups are a part of the theories of Lie groups, algebraic groups and topological groups. A homogeneous space for a group G is a non-empty manifold or topological space X on which G acts transitively. The elements of G are known as the symmetries of X. When the group G in question is the automorphism group of the space X, a special case arises. An isometry group, a diffeomorphism group or a homeomorphism group can be called an automorphism group. In this case, X is homogeneous if naturally X looks locally identical at each point, either in the sense of isometry, diffeomorphism or homeomorphism. This book outlines the processes and applications of homogenous groups in detail. It presents this complex subject in the most comprehensible and easy to understand language. This textbook will serve as a valuable source of reference for graduate and post graduate students.

This book is a comprehensive compilation of works of different researchers from varied parts of the world. It includes valuable experiences of the researchers with the sole objective of providing the readers (learners) with a proper knowledge of the concerned field. This book will be beneficial in evoking inspiration and enhancing the knowledge of the interested readers.

In the end, I would like to extend my heartiest thanks to the authors who worked with great determination on their chapters. I also appreciate the publisher's support in the course of the book. I would also like to deeply acknowledge my family who stood by me as a source of inspiration during the project.

Hart Scott

Introduction

The present book is devoted to the exposition of the research developments at the intersection of two active fields of mathematics: Hardy inequalities and related analysis, and the noncommutative analysis in the setting of nilpotent Lie groups of different types. Both subjects are very broad and deserve separate monograph presentations on their own, and many good books are already available.

However, the recent active research in the area does allow one to make a consistent treatment of 'anisotropic' Hardy inequalities, their numerous features, and a number of related topics. This brings many new insights to the subject, also allowing to underline the interesting character of its subelliptic features. The progress in this field is facilitated by the rapid developments in both areas of Hardy inequalities and related topics, and in the noncommutative analysis on Folland and Stein's homogeneous groups.

We will now give some short insights into both fields and into the scope of this book. Here we only give a general overview, with more detailed references and explanations of different features presented throughout the monograph.

Hardy inequalities and related topics

The classical L^2-Hardy inequality in the modern literature in the Euclidean space $\mathbb{R}^n$ with $n \geq 3$ can be written in the form

$$\left\| \frac{f}{|x|_E} \right\|_{L^2(\mathbb{R}^n)} \leq \frac{2}{n-2} \|\nabla f\|_{L^2(\mathbb{R}^n)}, \tag{1}$$

where ∇ is the standard gradient in $\mathbb{R}^n$,

$$|x|_E = \sqrt{x_1^2 + \cdots + x_n^2}$$

is the Euclidean norm, $f \in C_0^\infty(\mathbb{R}^n)$, and where the constant $\frac{2}{n-2}$ is known to be sharp. In addition to references in the preface, the multidimensional version of the Hardy inequality was proved by J. Leray [Ler33].

It has numerous applications in different fields, for example in the spectral theory, leading to the lower bounds for the quadratic form associated to the Laplacian operator. It is also related to many other areas and fields, notably to the uncertainty principles. The uncertainty principle in physics is a fundamental concept going back to Heisenberg's work on quantum mechanics [Hei27, Hei85], as well as to its mathematical justification by Hermann Weyl [Wey50]. In the simplest Euclidean setting it can be stated as the inequality

$$\left(\int_{\mathbb{R}^n} |\nabla\phi|^2 dx\right)\left(\int_{\mathbb{R}^n} |x|_E^2\phi^2 dx\right) \geq \frac{n^2}{4}\left(\int_{\mathbb{R}^n} \phi^2 dx\right)^2, \tag{2}$$

for all real-valued functions $\phi \in C_0^\infty(\mathbb{R}^n)$, where the constant $\frac{n^2}{4}$ is sharp. It can be shown to be a consequence of (1). There are good surveys on the mathematical aspects of uncertainty principles by Fefferman [Fef83] and by Folland and Sitaram [FS97]. We also note that the uncertainty principle can be also obtained without Hardy inequalities, see, e.g., Ciatti, Ricci and Sundari [CRS07].

The inequality (1) can be extended to L^p-spaces, taking the form

$$\left\|\frac{f}{|x|_E}\right\|_{L^p(\mathbb{R}^n)} \leq \frac{p}{n-p}\|\nabla f\|_{L^p(\mathbb{R}^n)}, \quad n \geq 2, \quad 1 \leq p < n, \tag{3}$$

where $f \in C_0^\infty(\mathbb{R}^n)$, and where the constant $\frac{p}{n-p}$ is known to be sharp.

As mentioned in the preface, such inequalities go back to Hardy [Har19], and have been evolving and growing over the years. In fact, the subject is so deep and broad at the same time that it would be impossible to give justice to all the authors who have made their contributions. To this end, we can refer to several extensive presentations of the subject in the books and surveys and the references therein: Opic and Kufner [OK90] in 1990, Davies [Dav99] in 1999, Edmunds and Evans [EE04] in 2004, part of Mazya's books [Maz85, Maz11], Ghoussoub and Moradifam [GM13] in 2013, and the recent book by Balinsky, Evans and Lewis [BEL15]. Hardy type inequalities have been very intensively studied, see, e.g., also Davies and Hinz [DH98], Davies [Dav99] as well as Ghoussoub and Moradifam [GM11] for reviews and applications.

One further extension of the Hardy inequality is the now classical result by Rellich appearing at the 1954 ICM in Amsterdam [Rel56] with the inequality

$$\left\|\frac{f}{|x|_E^2}\right\|_{L^2(\mathbb{R}^n)} \leq \frac{4}{n(n-4)}\|\Delta f\|_{L^2(\mathbb{R}^n)}, \quad n \geq 5, \tag{4}$$

with the sharp constant. We can refer, for example, to Davies and Hinz [DH98] (see also Brézis and Vázquez [BV97]) for further history and later extensions, including the derivation of sharp constants.

Higher-order Hardy inequalities have been also intensively investigated. Some of such results go back to 1961 to Birman [Bir61, p. 48] who has shown, for functions $f \in C_0^k(0,\infty)$, the family of inequalities

$$\left\|\frac{f}{x^k}\right\|_{L^2(0,\infty)} \leq \frac{2^k}{(2k-1)!!}\left\|f^{(k)}\right\|_{L^2(0,\infty)}, \quad k \in \mathbb{N}, \tag{5}$$

where $(2k-1)!! = (2k-1)\cdot(2k-3)\cdots 3\cdot 1$. For $k=1$ and $k=2$ this reduces to one-dimensional Hardy and Rellich inequalities, respectively. Such one-dimensional inequalities have recently found new life and one can find their historical discussion by Gesztesy, Littlejohn, Michael and Wellman in [GLMW17].

There is now a whole scope of related inequalities playing fundamental roles in different branches of mathematics, in particular, in the theory of linear and nonlinear partial differential equations. For example, the analysis of more general weighted Hardy–Sobolev type inequalities has also a long history, initiated by Caffarelli, Kohn and Nirenberg [CKN84] as well as by Brézis and Nirenberg in [BN83], and then Brézis and Lieb [BL85] with a mixture with Sobolev inequalities, Brézis and Vázquez in [BV97, Section 4], also [BM97], with many subsequent works in this direction. We also refer to more recent paper of Hoffmann-Ostenhof and Laptev [HOL15] on this subject and to further references therein. Many of these inequalities will be also appearing in the present book.

Of course, there are many more aspects to Hardy inequalities. In particular, working in domains, one can establish inequalities under certain boundary conditions. For example, for Hardy inequalities for Robin Laplacians and p-Laplacians see [KL12] and [EKL15], respectively, or [LW99, BLS04] for magnetic versions, or [BM97, HOHOL02] for versions involving the distance to the boundary. For Hardy inequalities for discrete Laplacians see, e.g., [KL16], or [HOHOLT08] for many-particle versions.

Homogeneous groups of different types

The harmonic analysis on homogeneous groups goes back to 1982 to Folland and Stein who in their book [FS82] laid down the foundations of 'anisotropic' harmonic analysis, that is, the harmonic analysis that depends only on the group and dilation structures of the group.

Such homogeneous groups are necessarily nilpotent, and provide a unified framework including many well-known classes of (nilpotent) Lie groups: *the Euclidean space, the Heisenberg group, H-type groups, polarizable Carnot groups, stratified groups (homogeneous Carnot groups), graded groups.* All of these groups are homogeneous and have the rational weights for their dilations.

The class of homogeneous groups is closer to the classical analysis than one might first think: in fact, any homogeneous group can be identified with some space $\mathbb{R}^n$ with a polynomial group law. The simplest examples are $\mathbb{R}^n$ itself, where the group law is linear, or the Heisenberg group, where the group law is quadratic in the last variable.

An important feature of homogeneous groups is that they do not have to allow for homogeneous hypoelliptic left invariant partial differential operators. In fact, if such an operator exists, the group has to be *graded* and its weights of dilations are rational. The class of stratified groups is a particularly important class of graded groups allowing for a homogeneous second-order sub-Laplacian. In general, nilpotent Lie groups provide local models for many questions in subelliptic analysis and sub-Riemannian geometry, their importance widely recognized since the essential role they played in deriving sharp subelliptic estimates for differential operators on manifolds, starting from the seminal paper by Rothschild and Stein [RS76] (see also [Fol77, Rot83]).

In order to facilitate the exposition in the sequel, in Chapter 1 we will recall all the necessary facts needed for the analysis in this book.

We note, however, that the general scope of techniques available on such groups is much more extensive than presented in Chapter 1. The fundamental paper by Folland [Fol75] developed the rich functional analysis on stratified groups. Further functional spaces (e.g., of Besov type) on stratified groups have been analysed by Saka [Sak79]. There are many sources with rather comprehensive and deep treatments of general nilpotent Lie groups, for example, the books by Goodman [Goo76] or Corwin and Greanleaf [CG90]. Good sources of information are the notes by Fulvio Ricci [Ric] and Folland's books [Fol89, Fol95, Fol16].

As a side remark we can note that there is also a number of recent works developing function spaces on graded groups extending Folland and Saka's constructions in the stratified case, see [FR17] and [FR16] for Sobolev, and [CR16] and [CR17] for Besov spaces, respectively.

In our presentation and approach to the basic analysis on homogeneous groups of different types we mostly rely on the recent *open access* book [FR16]. Moreover, the exposition of the topics in this book is done more in the spirit of the classical potential theory, without much reference to the Fourier analysis. However, here we should mention that the noncommutative Fourier analysis on nilpotent Lie groups is extremely rich, with many powerful approaches available, such as Kirillov's orbit method [Kir04], Mackey general description of the unitary dual, or the von Neumann algebra approaches of Dixmier [Dix77, Dix81]. We can refer to [FR16, Appendix B] for a workable summary of these methods.

The recently developed noncommutative quantization theories on nilpotent Lie groups, in particular, the global theory of pseudo-differential operators on graded groups, indeed heavily rely on such Fourier analysis. We refer the interested reader to [FR16] for the thorough exposition and application of such methods. A good exposition of the analysis of questions not requiring the Fourier analysis, in the setting of stratified groups, can be found in the book [BLU07] by Bonfiglioli, Lanconelli and Uguzzoni.

Hardy inequalities and potential theory on stratified groups

The study of the subelliptic Hardy inequalities has also begun more than 40 years ago due to their importance for many questions involving subelliptic partial differential equations, unique continuation, sub-Riemannian geometry, subelliptic spectral theory, etc. Not surprisingly, here the work started with the most important example of the Heisenberg group, where we can mention a fundamental contribution by Garofalo and Lanconelli [GL90].

There is a deep link with the properties of the fundamental solutions for the sub-Laplacian on stratified groups. In general, the understanding of the fundamental solutions of differential operators is one of the keys for solving boundary value problems for differential equations in a domain, and this idea has a long history dating back to the works of mathematicians such as Gauss [Gau77, Gau29] and Green [Gre28].

In general, the sub-Laplacians on stratified (and on more general graded) groups play important roles not only in theoretical settings (see, e.g., Gromov [Gro96] or Danielli, Garofalo and Nhieu [DGN07] for general expositions from different points of view), but also in applications of mathematics, for example in mathematical models of crystal material and human vision (see, for example, [Chr98] and [CMS04]).

The fundamental solution for the sub-Laplacian on stratified groups behaves well and was already understood by Folland [Fol75]. In particular, one always has its existence and uniqueness, an advantageous feature when compared to higher-order operators on stratified groups, or more general hypoelliptic operators on graded groups, see Geller [Gel83], and an exposition in [FR16, Section 3.2.7].

Roughly speaking, there are three versions of Hardy type inequalities on stratified groups available in the literature:

(A) Using the homogeneous quasi-norm, sometimes called the $\mathcal{L}$-gauge, given by the appropriate power of the fundamental solution of the sub-Laplacian $\mathcal{L}$. Thus, if $d(x)$ is the $\mathcal{L}$-gauge, then $d(x)^{2-Q}$ is a constant multiple of Folland's [Fol75] fundamental solution of the sub-Laplacian $\mathcal{L}$, with Q being the homogeneous dimension of the stratified group $\mathbb{G}$; these will be discussed in Chapter 7.

(B) Using the Carnot–Carathéodory distance, i.e., the control distance associated to the sub-Laplacian.

(C) Using the Euclidean distance on the first stratum of the group.

One can note that if one is not interested in best constants in such inequalities one can work with any of these equivalent quasi-norms. In fact, in such a case one can also work with fractional-order derivatives expressed as arbitrary powers of the sub-Laplacian, see, e.g., Ciatti, Cowling and Ricci [CCR15] for such an analysis on stratified groups, as well as Yafaev [Yaf99] for some Euclidean considerations also with best constants, or Hoffmann-Ostenhof and Laptev [HOL15] and references therein.

However, the best constants in the corresponding inequalities in cases (A)–(C) above may depend on the quasi-norm that one is using.

Thus, in the case (A) there is an extensive literature on Hardy inequalities and related topics on stratified groups relating them to the fundamental solution to the sub-Laplacian. Here we can briefly mention some papers [GL90, GK08, Gri03, NZW01, HN03, D'A04b, WN08, DGP11, Kom10, JS11, Lia13, Yan13, CCR15,

Yen16, GKY17], with more details and acknowledgements given throughout the book. Here, the Hardy inequality typically takes the form

$$\left\| \frac{f}{d(x)} \right\|_{L^p(\mathbb{G})} \leq \frac{p}{Q-p} \|\nabla_H f\|_{L^p(\mathbb{G})}, \quad Q \geq 3,\ 1 < p < Q, \tag{6}$$

where Q is the homogeneous dimension of the stratified group $\mathbb{G}$, ∇_H is the horizontal gradient, and $d(x)$ is the so-called $\mathcal{L}$-gauge related to the fundamental solution of the sub-Laplacian, and the constant is sharp. The analysis in the case (A) in terms of the fundamental solution of the sub-Laplacian will be the subject of Chapter 11 of this book.

The results on Hardy and other inequalities for the case (B) are less extensive, mostly devoted to the case of the Heisenberg group. However, the case (C) has recently attracted a lot of attention due to its geometrically clear nature and importance for questions in partial differential equations, see, e.g., [BT02a] and [D'A04b]. A typical horizontal Hardy inequality would take the form

$$\left\| \frac{f}{|x'|} \right\|_{L^p(\mathbb{G})} \leq \frac{p}{N-p} \|\nabla_H f\|_{L^p(\mathbb{G})}, \quad 1 < p < N, \tag{7}$$

where N is the dimension of the first stratum, x' denotes the variables in the first stratum of $\mathbb{G}$, and

$$|x'| = \sqrt{x_1'^2 + \cdots + x_N'^2}$$

is the Euclidean norm on the first stratum of $\mathbb{G}$, which can be identified with $\mathbb{R}^N$. The constant $\frac{p}{N-p}$ in (7) is also sharp.

In Chapter 6 we aim at giving a comprehensive treatment of such horizontal estimates based on the divergence relations and on the potential theory on the stratified groups.

Another ingredient that we find to be missing in the literature in the setting of stratified groups is the *classical style potential theory* working with layer potential operators. Indeed, nowadays the appearing boundary layer operators and elements of the potential theory serve as the main apparatus for the analysis and construction of solutions to boundary value problems. That have led to a vast literature concerning modern theory of boundary layer operators and potential theory in $\mathbb{R}^n$ as well as their important applications. In the subelliptic setting we can mention the works of Jerison [Jer81] and Romero [Rom91] on the Heisenberg group. We refer to [GL03], [GN88], [GW92], [LU97], [RS17d] and [WN16] as well as to references therein for more general Green function analysis of second order subelliptic (and weighted degenerate) operators. In this book, we follow a more geometric approach of our recent paper [RS17c] to present such a subject in the setting of general stratified groups, and to give its applications to several questions, such as boundary value problems for the sub-Laplacian, traces of Newton potential operators, and Hardy inequalities with boundary terms. All these topics will be the subject of discussions in Chapter 11.

The boundary value problems in the subelliptic settings are substantially more complicated than in the elliptic case due to the appearance of so-called *characteristic points* at the boundary – some problems of such a type are well explained, e.g., in [DGN06]. However, there is still a particular type of boundary conditions (Kac' boundary value problem) which can be viewed as a subelliptic version of M. Kac's question: *is there any boundary value problem for the Laplacian which is explicitly solvable in the classical sense for any smooth domain?*

An answer to M. Kac's question was given in [RS16c] for the Heisenberg group and in [RS17c] for general stratified groups. The appearing boundary conditions are, however, nonlocal and the corresponding boundary value problem can be called Kac's boundary value problem. One interesting fact is that the explicit solutions that one constructs for Kac's boundary value problem for the sub-Laplacian work also well in the presence of characteristic points on the boundary.

In Section 11.5 we also discuss another version of such a question: *is there a class of domains in which the Dirichlet boundary value problem for the sub-Laplacian is explicitly solvable in the classical sense?* This is discussed in the setting of H-type groups following our recent paper [GRS17] with Nicola Garofalo.

Furthermore, in Chapter 12 we will give an exposition of the potential theory and related Hardy–Rellich inequalities for more general Hörmander's sums of squares, based on the properties of the fundamental solutions rather than those of the $\mathcal{L}$-gauge. This has a definitive advantage of eliminating the need to use Folland's formula relating the $\mathcal{L}$-gauge with the fundamental solution. As a result, we can extend the analysis to more general settings, also those without any group structure, dealing with Hörmander's sums of squares beyond the setting of the stratified groups.

In general, there are several ways to obtain improvements of Hardy inequalities by including boundary terms. For the Laplacians such problems have been considered in [ACR02] by using variational method and in [WZ03] by using conformal transformation method. The methods described in this book are based on the potential theory and, compared to other approaches, do not rely so much on the particular structure of the Euclidean space. Certain Hardy and Rellich inequalities for sums of squares have been considered by Grillo [Gri03], compared to which our approach provides refinements from several points of view, based on the generalized representation formulae (Green's formulae) of non-subharmonic functions for improved Hardy and Rellich type inequalities with boundary terms.

Hardy inequalities and related topics on homogeneous groups

The lack of homogeneous hypoelliptic left invariant differential operators on general homogeneous groups is compensated by several other advantageous properties, such as a good polar decomposition which, combined with dilations, still allows one to explore the radial structure of the group. For this purpose, we extensively work with the radial derivative operator $\mathcal{R}$ and the Euler operator $\mathbb{E}$ which we describe in Section 1.3.

This fits well with the structure of Hardy and other inequalities as their maximisers are often achieved on radial functions. An advantage of such an approach is that one can also work with arbitrary quasi-norms and anisotropic structures still yielding similar properties and best constants in the inequalities. In addition, in Section 1.3.3 we demonstrate how the Hardy type inequalities for radial functions often imply similar inequalities for functions of general (non-radial) type.

Thus, Chapter 2 is devoted to **Homogeneous Groups: Hardy Inequalities**, their weighted and critical versions, stability and remainder estimates. Further-more, Chapter 3 is devoted to Rellich, Caffarelli–Kohn–Nirenberg and Sobolev type inequalities on homogeneous groups. In Chapter 9 we present different ver-sions of uncertainty principles on homogeneous groups. We follow an abstract approach by defining abstract position and momentum operators satisfying minimal structural properties, already allowing one to establish a number of uncertainty type relations. Consequently, different choices of such abstract position and momentum operators are possible based on the additional structural properties available on the group.

In Chapter 10 we discuss different **Homogeneous Groups: Function Spaces**, with or without differentiability properties. The spaces involving radial derivatives are the Euler–Hilbert–Sobolev and Sobolev–Lorentz–Zygmund spaces. We investigate their basic properties and embeddings. The spaces not involving the differentiable structure are the Morrey and Campanato spaces. There, we discuss the boundedness of integral operators, namely, the Hardy–Littlewood maximal operator, Bessel–Riesz operators, generalized Bessel–Riesz operators, generalized fractional integral operators and Olsen type inequalities in generalized Morrey spaces on homogeneous groups.

Incidentally, all these and other results on homogeneous groups in this book give new statements already in the Euclidean setting of $\mathbb{R}^n$ when we are working with anisotropic differential structure in view of the arbitrariness of the choice of any homogeneous quasi-norm.

Let us consider, for instance, the following Bessel–Riesz operators

$$I_{\alpha,\gamma}f(x) = \int_{\mathbb{R}^n} \frac{|x-y|_E^{\alpha-n}}{(1+|x-y|_E)^\gamma} f(y)dy, \tag{8}$$

where $f \in L^p_{\text{loc}}(\mathbb{R}^n)$, $p \geq 1$, $\gamma \geq 0$ and $0 < \alpha < n$. Classical results on the Bessel–Riesz operators are due to Hardy, Littlewood and Sobolev, precisely, the boundedness of the Bessel–Riesz operators on Lebesgue spaces was shown by Hardy and Littlewood in [HL27], [HL32] and by Sobolev in [Sob38]. In the case of $\mathbb{R}^n$, the Hardy–Littlewood maximal operator, the Riesz potential $I_{\alpha,0} = I_\alpha$, the generalized fractional integral operators, which are a generalized form of the Riesz potential $I_{\alpha,0} = I_\alpha$, Bessel–Riesz operators and Olsen type inequalities are widely analysed on Lebesgue spaces, Morrey spaces and generalized Morrey spaces. For further discussions in this direction we refer to [Ada75, CF87, BNC14, Nak94, EN04, Eri02, KNS99, Nak01, Nak02, GE09, SST12, IGLE15, IGE16], as well as to

[Bur13] for a recent survey. Morrey spaces for non-Euclidean distances find their applications in many problems, see, e.g., [GS15a, GS15b] and [GS16]. A natural analogue of the Bessel–Riesz operator (8) on homogeneous groups is the operator

$$I_{\alpha,\gamma}f(x) := \int_{\mathbb{G}} \frac{|xy^{-1}|^{\alpha-Q}}{(1+|xy^{-1}|)^{\gamma}} f(y)dy.$$

In the setting of graded Lie groups the connections between these operators and the Sobolev spaces have been investigated in [FR16, Chapter 4] using the heat kernel methods. Here, in Chapter 10 we concentrate on the harmonic analysis aspects in the framework of Morrey and Campanato spaces on homogeneous groups.

In addition, we can mention an overview of constructions for Morrey–Campanato spaces in [RSS13] by Rafeiro, N. Samko and S. Samko, or in [RT15] by Rosenthal and Triebel. It is worth noting that Morrey–Campanato spaces can be interpolated [VS14]. One also considered variable exponent versions of Morrey spaces and maximal and singular operators there, see [GS13, GS16] and references therein.

In Chapter 4 we look at the Hardy inequalities from the point of view of operators of fractional orders. Certainly, fractional powers of Laplacians and sub-Laplacians can be defined in different ways, e.g., using the Fourier or spectral analysis. However, here, we first adopt the integral representation that turns out to make perfect sense on general homogeneous groups. More specifically, for $p > 1$ and $s \in (0,1)$, we consider the *fractional p-sub-Laplacian* $(-\Delta_p)^s$ on a general homogeneous group $\mathbb{G}$ defined by the formula

$$(-\Delta_p)^s u(x) := 2 \lim_{\delta \searrow 0} \int_{\mathbb{G}\setminus B(x,\delta)} \frac{|u(x)-u(y)|^{p-2}(u(x)-u(y))}{|y^{-1}x|^{Q+sp}} dy, \quad x \in \mathbb{G},$$

where $B(x,\delta) = B_{|\cdot|}(x,\delta)$ is a quasi-ball with respect to the quasi-norm $|\cdot|$, with radius δ centred at $x \in \mathbb{G}$. It turns out that this operator has many advantageous properties similar to those exhibited by the usual p-Laplacians on the Euclidean spaces, and in Chapter 4 we present their analysis and some applications to 'partial differential' functional equations and related spectral questions. Consequently, we look at operators of fractional orders from a different point of view, and in Section 4.7 we discuss the boundedness of the operator

$$T_\alpha f(x) := |x|^{-\alpha}\mathcal{L}^{-\alpha/2} f(x),$$

on L^p-spaces on stratified Lie groups, where $\mathcal{L}$ is a sub-Laplacian. The analysis is based on the Riesz kernel representation of such operators, and we also supplement it with several versions of the Landau–Kolmogorov inequalities.

In Chapter 5 we discuss integral versions of Hardy inequalities. In fact, such a point of view goes back to one of the original versions of such inequalities by Hardy [Har20], where he has shown the inequality

$$\int_b^\infty \left(\frac{\int_b^x f(t)dt}{x}\right)^p dx \le \left(\frac{p}{p-1}\right)^p \int_b^\infty f(x)^p dx,$$

where $p > 1$, $b > 0$, and $f \geq 0$ is a non-negative function. It turns out that such an inequality can be also put in the framework of general homogeneous groups, especially since it does not involve derivatives, so that one does not need to specify one's analysis to a particular choice of the gradient. Thus, in Chapter 5 we present inequalities of such a type in weighted and unweighted settings, actually providing characterizations of weights for which integral Hardy inequalities hold true. We also present inequalities in the convolution form which, in turn, can be used for the derivation of Hardy–Littlewood–Sobolev and Stein–Weiss inequalities. The latter can be then established both on general homogeneous groups as well as on stratified/graded groups using Riesz kernels of hypoelliptic differential operators.

In Chapter 8 we discuss the so-called geometric versions of Hardy inequalities. By this one usually means Hardy inequalities on domains when the distance to the boundary enters the inequality as a weight. For example, if Ω is a convex open set of the Euclidean space $\mathbb{R}^n$, then a geometric version of the Hardy inequality can take a form

$$\int_\Omega |\nabla u|^2 dx \geq \frac{1}{4} \int_\Omega \frac{|u|^2}{\operatorname{dist}(x, \partial\Omega)^2} dx,$$

for all $u \in C_0^\infty(\Omega)$, with the sharp constant $1/4$. In the case of half-spaces and, more generally, convex domains, in Chapter 8 we present such inequalities in the setting of stratified groups.

1

Homogeneous Groups: An Analytical Approach

In this chapter we provide preliminaries for the analysis on homogeneous groups to make the use of the monograph more self-sufficient. We make a selection of topics which will be playing a role in the subsequent analysis. Thus, we first discuss relevant properties of general Lie groups and algebras and then concentrate on properties of homogeneous groups required for our further analysis. Lastly, we introduce the notion of the Euler operator on homogeneous groups and establish its main properties.

This chapter is not intended to be a comprehensive treatise of homogeneous groups but rather a description of a collection of tools used throughout the book. The theory of homogeneous groups for their use in analysis was developed by Folland and Stein [FS82]. A recent rather comprehensive description of homogeneous groups and their place among nilpotent Lie groups have appeared in [FR16]. We refer to both books for the expositions devoted specifically to homogeneous groups. For some related information we may also refer to Ricci's notes [Ric].

There are many sources with rather comprehensive and deep treatments of nilpotent Lie groups, for example the books by Goodman [Goo76] or Corwin and Greanleaf [CG90]. There are also many books on groups or Lie groups, we can refer for example to [RT10, Part III] for a basic introduction. Therefore, we assume the reader to have some familiarity with the concepts of the Lie groups and Lie algebras.

1.1 Homogeneous groups

In this section we discuss nilpotent Lie algebras and groups in the spirit of Folland and Stein's book [FS82] as well as introduce homogeneous (Lie) groups. For more analysis and details in this direction we refer to the recent open access book [FR16].

Let $\mathfrak{g}$ be a Lie algebra (always assumed real and finite-dimensional), and let $\mathbb{G}$ be the corresponding connected and simply-connected Lie group. The lower central series of $\mathfrak{g}$ is defined inductively by

$$\mathfrak{g}_{(1)} :=\mathfrak{g}, \quad \mathfrak{g}_{(j)} := [\mathfrak{g}, \mathfrak{g}_{(j-1)}].$$

If the lower central series of a Lie algebra $\mathfrak{g}$ terminates at 0 in a finite number of steps then this Lie algebra is called *nilpotent.* Moreover, if $\mathfrak{g}_{(s+1)} = \{0\}$ and $\mathfrak{g}_{(s)} \neq \{0\}$, then $\mathfrak{g}$ is said to be *nilpotent of step s*. A Lie group $\mathbb{G}$ is *nilpotent* (of step s) whenever its Lie algebra is nilpotent (of step s). If $\exp : \mathfrak{g} \to \mathbb{G}$ is the exponential map, by the Campbell–Hausdorff formula for $X, Y \in \mathfrak{g}$ sufficiently close to 0 we have

$$\exp X \exp Y = \exp H(X,Y),$$

where $H(X,Y)$, the Campbell–Hausdorff series, is an infinite linear combination of X and Y and their iterated commutators and H is universal, i.e., independent of $\mathfrak{g}$, and that

$$H(X,Y) = X + Y + \frac{1}{2}[X,Y] + \cdots,$$

where the dots indicate terms of order ≥ 3.

If $\mathfrak{g}$ is nilpotent, the Campbell–Hausdorff series terminates after finitely many terms and defines a polynomial map from $V \times V$ to V, where V is the underlying vector space of $\mathfrak{g}$.

Altogether, we have the following useful properties:

Proposition 1.1.1 (Exponential mapping and Haar measure). *Let* $\mathbb{G}$ *be a connected and simply-connected nilpotent Lie group with Lie algebra* $\mathfrak{g}$. *Then:*

(i) *The exponential map* exp *is a diffeomorphism from* $\mathfrak{g}$ *to* $\mathbb{G}$. *Moreover, if* $\mathbb{G}$ *is identified with* $\mathfrak{g}$ *via* exp, *then the group law* $(x, y) \mapsto xy$ *is a polynomial map.*

(ii) *If* λ *denotes a Lebesgue measure on* $\mathfrak{g}$, *then* $\lambda \circ \exp^{-1}$ *is a bi-invariant Haar measure on* $\mathbb{G}$.

Proof of Proposition 1.1.1. Part (i) is a direct consequence of the fact that $\mathbb{G}$ is uniquely (up to isomorphism) determined by $\mathfrak{g}$, see, e.g., [FS82, Proposition 1.2], [CG90, Section 1.2] or [FR16, Proposition 1.6.6].

Let us give an argument for Part (ii). Let us denote the lower central series for $\mathfrak{g}$ by

$$\mathfrak{g}_{(1)}, \ldots, \mathfrak{g}_{(m)}, \mathfrak{g}_{(m+1)} = \{0\}$$

and denote

$$n := \dim \mathfrak{g} \quad \text{and} \quad n_j := \dim \mathfrak{g}_{(j)}.$$

Let $X_{n-n_m+1}, \ldots, X_n$ be a basis for $\mathfrak{g}_{(m)}$, and we extend it to a basis

$$X_{n-n_{m-1}+1}, \ldots, X_n$$

for $\mathfrak{g}_{(m-1)}$, and so forth obtaining eventually a basis $X_1, \ldots, X_n$ for $\mathfrak{g}$. Let $\xi_1, \ldots, \xi_n$ be the dual basis for $\mathfrak{g}^*$, and let $\eta_k := \xi_k \circ \exp^{-1}$. These $\eta_1, \ldots, \eta_n$ are a system of global coordinates on $\mathbb{G}$. By using the Campbell–Hausdorff formula and the construction of the η_k's we obtain

$$\eta_k(xy) = \eta_k(x) + \eta_k(y) + P_k(x, y),$$

where $P_k(x, y)$ depends only on the coordinates $\eta_i(x), \eta_i(y)$ with $i < k$. Thus, with respect to the coordinates η_k, the differentials of the maps $x \mapsto xy$ with fixed y and $y \mapsto xy$ with fixed x are given by lower triangular matrices with only 1 elements on the diagonal, and therefore, each of the determinants is equal to one. This implies that the volume form $d\eta_1 \cdots d\eta_n$ on $\mathbb{G}$, which corresponds to the Lebesgue measure on $\mathfrak{g}$, is left and right invariant. □

Definition 1.1.2 (Dilations on a Lie algebra). A family of *dilations* of a Lie algebra $\mathfrak{g}$ is a family of linear mappings

$$\{\delta_r : r > 0\}$$

from $\mathfrak{g}$ to itself which satisfies:

- the mappings are of the form

  $$\delta_r = \exp(A \log r),$$

  where A is a diagonalisable linear operator on $\mathfrak{g}$ with positive eigenvalues.
- In particular, $\delta_{rs} = \delta_r \delta_s$ for all $r, s > 0$. If $\alpha > 0$ and $\{\delta_r\}$ is a family of dilations on $\mathfrak{g}$, then so is $\{\widetilde{\delta}_r\}$, where

  $$\widetilde{\delta}_r := \delta_{r^\alpha} = \exp(\alpha A \log r).$$

  By adjusting α we can always assume that the minimum eigenvalue of A is equal to 1.

Let $\mathcal{A}$ be the set of eigenvalues of A and denote by $W_a \subset \mathfrak{g}$ the corresponding eigenfunction space of A, where $a \in \mathcal{A}$. Then we have

$$\delta_r X = r^a X \quad \text{for} \quad X \in W_a.$$

If $X \in W_a$ and $Y \in W_b$, then

$$\delta_r[X, Y] = [\delta_r X, \delta_r Y] = r^{a+b}[X, Y]$$

and thus $[W_a, W_b] \subset W_{a+b}$. In particular, since $a \geq 1$ for $a \in \mathcal{A}$, we see that $\mathfrak{g}_{(j)} \subset \bigoplus_{a \geq j} W_a$. Since the set $\mathcal{A}$ is finite, it follows that $\mathfrak{g}_{(j)} = \{0\}$ for j sufficiently large. Thus, we obtain:

Proposition 1.1.3 (Lie algebras with dilations are nilpotent). *If a Lie algebra $\mathfrak{g}$ admits a family of dilations then it is nilpotent.*

However, not all nilpotent Lie algebras admit a dilation structure: an example of a (nine-dimensional) nilpotent Lie algebra that does not allow any compatible family of dilations was constructed by Dyer [Dye70].

Definition 1.1.4 (Graded Lie algebras and groups). A Lie algebra $\mathfrak{g}$ is called *graded* if it is endowed with a vector space decomposition (where all but finitely many of the V_k's are 0)

$$\mathfrak{g} = \oplus_{j=1}^{\infty} V_j \quad \text{such that} \quad [V_i, V_j] \subset V_{i+j}.$$

Consequently, a Lie group is called *graded* if it is a connected simply-connected Lie group whose Lie algebra is graded.

Definition 1.1.5 (Stratified Lie algebras and groups). A graded Lie algebra $\mathfrak{g}$ is called *stratified* if V_1 generates $\mathfrak{g}$ as an algebra. In this case, if $\mathfrak{g}$ is nilpotent of step m we have

$$\mathfrak{g} = \oplus_{j=1}^{m} V_j, \quad [V_j, V_1] = V_{j+1},$$

and the natural dilations of $\mathfrak{g}$ are given by

$$\delta_r \left(\sum_{k=1}^{m} X_k \right) = \sum_{k=1}^{m} r^k X_k, \quad (X_k \in V_k).$$

Consequently, a Lie group is called *stratified* if it is a connected simply-connected Lie group whose Lie algebra is stratified.

Definition 1.1.6 (Homogeneous groups). Let δ_r be dilations on $\mathbb{G}$. We say that a Lie group $\mathbb{G}$ is a *homogeneous group* if:

a. It is a connected and simply-connected nilpotent Lie group $\mathbb{G}$ whose Lie algebra $\mathfrak{g}$ is endowed with a family of dilations $\{\delta_r\}$.
b. The maps $\exp \circ \delta_r \circ \exp^{-1}$ are group automorphism of $\mathbb{G}$.

Since the exponential mapping exp is a global diffeomorphism from $\mathfrak{g}$ to $\mathbb{G}$ by Proposition 1.1.1, (i), it induces the corresponding family on $\mathbb{G}$ which we may still call the dilations on $\mathbb{G}$ and denote by δ_r. Thus, for $x \in \mathbb{G}$ we will write $\delta_r(x)$ or abbreviate it writing simply rx.

The origin of $\mathbb{G}$ will be usually denoted by 0.

Now let us give some well-known examples of homogeneous groups.

Example 1.1.7 (Abelian groups). The Euclidean space $\mathbb{R}^n$ is a homogeneous group with dilation given by the scalar multiplication.

Example 1.1.8 (Heisenberg groups). If n is a positive integer, the Heisenberg group $\mathbb{H}^n$ is the group whose underlying manifold is $\mathbb{C}^n \times \mathbb{R}$ and whose multiplication is given by

$$(z_1, \ldots, z_n, t)(z_1', \ldots, z_n', t') = \left(z_1 + z_1', \ldots, z_n + z_n', t + t' + 2\mathrm{Im} \sum_{k=1}^{n} z_k \overline{z}_k' \right).$$

The Heisenberg group $\mathbb{H}^n$ is a homogeneous group with dilations

$$\delta_r(z_1, \ldots, z_n, t) = (rz_1, \ldots, rz_n, r^2 t).$$

Example 1.1.9 (Upper triangular groups). Let $\mathbb{G}$ be the group of all $n \times n$ real matrices (a_{ij}) such that $a_{ii} = 1$ for $1 \leq i \leq n$ and $a_{ij} = 0$ when $i > j$. Then $\mathbb{G}$ is a homogeneous group with dilations

$$\delta_r(a_{ij}) = r^{j-i} a_{ij}.$$

These Examples 1.1.7, 1.1.8 and 1.1.9 are all examples of the stratified groups. It is also possible to define other families of dilations on these groups. For instance, on $\mathbb{R}^n$ we can define

$$\delta_r(x_1, \ldots, x_n) = (r^{d_1} x_1, \ldots, r^{d_n} x_n),$$

where $1 = d_1 \leq d_2 \leq \cdots \leq d_n$, and on $\mathbb{H}^n$ we can define

$$\delta_r(x_1 + iy_1, \ldots, x_n + iy_n, t) = (r^{a_1} x_1 + ir^{b_1} y_1, \ldots, r^{a_n} x_n + ir^{b_n} y_n, r^c t),$$

where $\min\{a_1, \ldots, a_n, b_1, \ldots, b_n\} = 1$ and $a_j + b_j = c$ for all j. In general, these dilations do not have to be stratified. However, when we refer to $\mathbb{R}^n$ or $\mathbb{H}^n$ we shall assume that they are equipped with the natural dilations defined in Examples 1.1.7, 1.1.8 unless we state otherwise.

Let $d_1, \ldots, d_n$ be the eigenvalues of A, enumerated in nondecreasing order according to their multiplicity, and let $\overline{d} = \max d_k$. The mappings $\{\delta_r = \exp(A \log r)\}$ give the dilation structure to an n-dimensional homogeneous group $\mathbb{G}$, with

$$1 = d_1 \leq d_2 \leq \cdots \leq d_n = \overline{d}. \tag{1.1}$$

Let us fix a basis $\{X_k\}_{k=1}^n$ of the Lie algebra $\mathfrak{g}$ of the Lie group $\mathbb{G}$ such that

$$AX_k = d_k X_k$$

for each k. Then one can define a standard Euclidean norm $\|\cdot\|$ on $\mathfrak{g}$ by declaring the X_k's to be orthonormal. This norm can be also considered as a function on $\mathbb{G}$ by the formula

$$\|x\| = \|\exp^{-1} x\|. \tag{1.2}$$

The number

$$Q := \sum_{k=1}^{n} d_k = \operatorname{Tr}(A) \tag{1.3}$$

is called the *homogeneous dimension* of $\mathbb{G}$. From now on Q will always denote the homogeneous dimension of $\mathbb{G}$.

1.2 Properties of homogeneous groups

In this section we discuss properties of homogeneous groups that are important for their understanding and that will be also useful for our further analysis. For different further properties of homogeneous and graded groups we can refer the reader to the open access book [FR16, Chapter 3].

1.2.1 Homogeneous quasi-norms

We start by the definition of a quasi-norm.

Definition 1.2.1 (Quasi-norms). Let us define a *homogeneous quasi-norm* on a homogeneous group $\mathbb{G}$ to be a continuous function $x \mapsto |x|$ from $\mathbb{G}$ to $[0, \infty)$ that satisfies

(a) for all $x \in \mathbb{G}$ and $r > 0$:

$$|x^{-1}| = |x| \quad \text{and} \quad |rx| = r|x|.$$

(b) The non-degeneracy:

$$|x| = 0 \quad \text{if and only if} \quad x = 0.$$

Here and elsewhere we denote by $rx = \delta_r x$ the dilation of x induced by the dilations on the Lie algebra through the exponential mapping.

There always exist homogeneous quasi-norms on homogeneous groups. Moreover, there always exist quasi-norms that are C^∞-smooth on $\mathbb{G}\backslash\{0\}$. Let us give such an example. Observe that

$$X = \sum_{k=1}^{n} c_k X_k \in \mathfrak{g} \quad \text{implies} \quad \|\delta_r X\| = \left(\sum_{k=1}^{n} c_k^2 r^{2d_k}\right)^{1/2},$$

where $\|\cdot\|$ is the Euclidean norm from (1.2). We can notice that for $X \neq 0$ the function $\|\delta_r X\|$ is a strictly increasing function of r, and it tends to 0 and ∞ as $r \to 0$ and $r \to \infty$, respectively. Now, for $x = \exp X$, we can define a homogeneous quasi-norm on $\mathbb{G}$ by setting

$$|0| := 0 \quad \text{and} \quad |x| := 1/r \quad \text{for} \quad x \neq 0,$$

where $r = r(X) > 0$ is the unique number such that

$$\|\delta_{r(X)} X\| = 1.$$

By the implicit function theorem and the fact that the Euclidean unit sphere is a C^∞ manifold we see that this function is C^∞ on $\mathbb{G}\backslash\{0\}$.

If $x \in \mathbb{G}$ and $r > 0$ we define the ball of radius r about x by

$$B(x,r) := \{y \in \mathbb{G} : |x^{-1}y| < r\}.$$

It can be noticed that $B(x,r)$ is the left translate by x of $B(0,r)$, which in turn is the image under δ_r of $B(0,1)$.

Lemma 1.2.2 (Closed quasi-balls are compact). *$\overline{B(x,r)}$ is compact for any $x \in \mathbb{G}$ and $r > 0$.*

Proof. Let us define

$$\rho(x) := \sum_{k=1}^{n} \frac{|c_k|}{d_k} \quad \text{for} \quad x = \exp\left(\sum_{k=1}^{n} c_k X_k\right),$$

where d_k are as in (1.1). Then ρ satisfies all the properties of a homogeneous quasi-norm. Obviously $\{x : \rho(x) = 1\}$ is compact and does not contain 0, so the function $x \mapsto |x|$ attains a positive minimum η on it. Since $|rx| = r|x|$ and $\rho(rx) = r\rho(x)$, it follows that $|x| \geq \eta\rho(x)$ for all x and for some $\eta > 0$, and hence that

$$\overline{B(0,\eta)} \subset \{x : \rho(x) \leq 1\}.$$

Thus, $\overline{B(0,\eta)}$ is compact, and it follows by dilation and translation that $\overline{B(x,r)}$ is compact for all $r > 0$, $x \in \mathbb{G}$. □

We can compare the quasi-norms with each other and with the Euclidean norm (1.2).

Proposition 1.2.3 (Quasi-norms and the Euclidean norm). *We have the following properties:*

(1) *Any two homogeneous quasi-norms on a homogeneous group are equivalent.*

(2) *There are the constants $C_1, C_2 > 0$ such that*

$$C_1\|x\| \leq |x| \leq C_2\|x\|^{1/\overline{d}} \quad \textit{for all} \quad |x| \leq 1.$$

Proof. Proof of Part (2). When $y = \exp(\sum c_k X_k)$ we have $\|ry\| = (\sum c_k^2 r^{2d_k})^{1/2}$ and hence

$$r^{\overline{d}}\|y\| \leq \|ry\| \leq r\|y\|$$

for $r \leq 1$. A positive maximum C_1^{-1} and a positive minimum $C_2^{-\overline{d}}$ on $\{y : |y| = 1\}$ are attained by the Euclidean norm $\|y\|$ in view of the compactness in Lemma 1.2.2. Any $x \neq 0$ can be written as $x = |x|y$ where $|y| = 1$, so that for $|x| \leq 1$,

$$\|x\| \leq |x|\|y\| \leq C_1^{-1}|x|, \quad \|x\| \geq |x|^{\overline{d}}\|y\| \geq C_2^{-\overline{d}}|x|^{\overline{d}},$$

completing the proof of Part (2).

Proof of Part (1). Let $|\cdot|_1$ and $|\cdot|_2$ be two homogeneous quasi-norms. By a similar argument to Part (2), we observe that since the ball $\overline{B(0,1)}$ with respect to $|\cdot|_1$ is compact by Lemma 1.2.2, and $|\cdot|_2$ is continuous, we have

$$\left|\frac{x}{|x|_1}\right|_2 \leq C < \infty$$

for all $x \neq 0$. By homogeneity it follows that $|x|_2 \leq C|x|_1$ for all $x \in \mathbb{G}$. Switching the roles of $|\cdot|_1$ and $|\cdot|_2$ we obtain the statement. ☐

The reason why the homogeneous quasi-norms have the prefix 'quasi' becomes clear from the following proposition that shows that in general the triangle inequality is satisfied only with some constant:

Proposition 1.2.4 (Triangle inequality with constant). *Let $\mathbb{G}$ be a homogeneous group. Then we have the following properties:*

(1) *If $|\cdot|$ is a homogeneous quasi-norm on $\mathbb{G}$, there exists $C > 0$ such that for every $x, y \in \mathbb{G}$, we have*
$$|xy| \leq C(|x| + |y|).$$

(2) *There always exists a homogeneous quasi-norm $|\cdot|$ on $\mathbb{G}$ which satisfies the triangle inequality (with constant $C = 1$):*
$$|xy| \leq |x| + |y| \tag{1.4}$$
for all $x, y \in \mathbb{G}$.

Proof. Let us prove Part (1). The function $(x, y) \mapsto |xy|$ attains a finite maximum $C > 0$ on the set $\{(x, y) \in \mathbb{G} \times \mathbb{G} : |x| + |y| = 1\}$ which is compact by Lemma 1.2.2. Then, given any $x, y \in \mathbb{G}$, set $r = |x| + |y|$. It follows that

$$|xy| = r|r^{-1}(xy)| = r|(r^{-1}x)(r^{-1}y)| \leq Cr = C(|x| + |y|),$$

completing the proof.

We leave Part (2) without proof, referring to [FR16, Proposition 3.1.38 and Theorem 3.1.39] for the complete argument. ☐

Proposition 1.2.5. *There exists a constant $C > 0$ such that for every $x \in \mathbb{G}$ and $s \in [0, 1]$, we have*
$$|\exp(s\log(x))| \leq C|x|. \tag{1.5}$$

Proof. Let $x \neq 0$, otherwise (1.5) is trivial. Using the fact that $|\cdot|$ is homogeneous of degree 1, we have

$$\frac{|\exp(s\log(x))|}{|x|} = |\delta_{1/|x|}(\exp(s\log(x))| = |\exp(s\log(\delta_{1/|x|}(x))|.$$

With $|\delta_{1/|x|}(x)| = 1$, it follows that

$$\frac{|\exp(s\log(x))|}{|x|} \le \max_{\xi\in\mathbb{G}:|\xi|=1, s\in[0,1]} \rho(\exp(s\log(\xi))) =: C.$$

Note that C is finite, since the set $\{\xi : |\xi| = 1\}$ is compact (see Lemma 1.2.2) as well as $|\cdot|$, exp and log are all continuous functions. $\square$

The bi-invariant Haar measure on $\mathbb{G}$ comes from the Lebesgue measure on $\mathfrak{g}$ by Proposition 1.1.1. Fixing the normalisation of the Haar measure on $\mathbb{G}$ we require that the Haar measure of $B(0,1)$ is 1. (Thus, if $\mathbb{G} = \mathbb{R}^n$ with the usual Lebesgue measure, our Haar measure is $\Gamma((n+2)/2)/\pi^{n/2}$ times the Lebesgue measure.) The measure of any measurable set $E \subset \mathbb{G}$ will be denoted by $|E|$, and we shall denote the integral of a function f with respect to this measure by $\int_{\mathbb{G}} f dx$ or by $\int_{\mathbb{G}} f(x)dx$, or simply by $\int f$ or by $\int f(x)dx$.

Recalling (1.3), the homogeneous dimension of $\mathbb{G}$ is

$$Q = \sum_{k=1}^{n} d_k = \mathrm{Tr}(A),$$

and we have

$$|\delta_r(E)| = r^Q|E|, \quad d(rx) = r^Q dx. \tag{1.6}$$

In particular, we have $|B(x,r)| = r^Q$ for all $r > 0$ and $x \in \mathbb{G}$.

Definition 1.2.6 (Homogeneous functions and operators). A function f on $\mathbb{G}\backslash\{0\}$ is said to be *homogeneous of degree* λ if it satisfies

$$f \circ \delta_r = r^\lambda f \quad \text{for all} \quad r > 0.$$

We note that for f and g, we have the formula

$$\int_{\mathbb{G}} f(x)\,(g\circ\delta_r)(x)dx = r^{-Q}\int_{\mathbb{G}}(f\circ\delta_{1/r})(x)g(x)dx,$$

given that the integrals exist. Hence we can extend the mapping $f \mapsto f\circ\delta_r$ to distributions by defining, for any distribution f and any test function ϕ, the distribution $f\circ\delta_r$ by

$$\langle f\circ\delta_r, \phi\rangle = r^{-Q}\langle f, \phi\circ\delta_{1/r}\rangle,$$

where $\langle\cdot,\cdot\rangle$ denotes the usual duality between functions and distributions. The distribution f is called *homogeneous of degree* λ if it satisfies

$$f \circ \delta_r = r^\lambda f \quad \text{for all} \quad r > 0.$$

Also, a linear operator D on $\mathbb{G}$ is called *homogeneous of degree* λ if it satisfies

$$D(f\circ\delta_r) = r^\lambda (Df)\circ\delta_r \quad \text{for all} \quad r > 0,$$

for any f. If D is a linear operator homogeneous of degree λ and f is a homogeneous function of degree μ, then Df is homogeneous of degree $\mu-\lambda$.

The following extension of the reverse triangle inequality is often useful:

Proposition 1.2.7 (Reverse triangle inequality). *Let f be a homogeneous function of degree λ and of class C^1 on $\mathbb{G}\backslash\{0\}$. Then there is a constant $C > 0$ such that we have*

$$|f(xy) - f(x)| \leq C|y||x|^{\lambda-1} \quad \text{for all} \quad |y| \leq |x|/2.$$

Proof. Suppose that $|x| = 1$ and $|y| \leq 1/2$, and we use the fact that both sides of the desired inequality are homogeneous of degree λ. In this case x and xy are bounded, and also bounded away from zero, and the map $y \mapsto xy$ is C^1, so by the usual mean value theorem and Proposition 1.2.3, we obtain

$$|f(xy) - f(x)| \leq C\|y\| \leq C'|y| = C'|y||x|^{\lambda-1},$$

using that both sides of the desired inequality are homogeneous functions of the same degree λ. □

In particular, this proposition can be applied to C^1 homogeneous quasi-norms. Specifically, the combination of Proposition 1.2.4 and Proposition 1.2.7 leads to a constant $\gamma > 0$ such that we have

$$|xy| \leq \gamma(|x| + |y|) \quad \text{for all} \quad x, y \in \mathbb{G}, \tag{1.7}$$

$$||xy| - |x|| \leq \gamma|y| \quad \text{for all} \quad x, y \in \mathbb{G} \quad \text{with} \quad |y| \leq |x|/2. \tag{1.8}$$

Henceforth, γ will always be called the minimal constant satisfying (1.7) and (1.8). Obviously, $\gamma \geq 1$. We will be using (1.7) and (1.8) without comment in the sequel. The following simple fact will also be useful later:

Lemma 1.2.8 (Peetre type inequality). *For every $x, y \in \mathbb{G}$ and $s > 0$, we have*

$$(1 + |x|)^s(1 + |y|)^{-s} \leq \gamma^s(1 + |xy^{-1}|)^s.$$

Proof. Because of $|x| \leq \gamma(|xy^{-1}| + |y|)$ we have

$$1 + |x| \leq \gamma(1 + |xy^{-1}|)(1 + |y|),$$

and we obtain the needed inequality by raising both sides to the sth power. □

Let us now fix the notation for some common function spaces on $\mathbb{G}$. Let $\Omega \subset \mathbb{G}$, and let $C(\Omega)$ ($C_0(\Omega)$) be the space of continuous functions on $\mathbb{G}$ (continuous functions with compact support, respectively). If Ω is open, then $C^{(k)}(\Omega)$ is called the class of k times continuously differentiable functions on Ω,

$$C^\infty(\Omega) = \bigcap_{k=1}^{\infty} C^{(k)}(\Omega) \quad \text{and} \quad C_0^\infty(\Omega) = C^\infty(\Omega) \cap C_0(\Omega).$$

When $\Omega = \mathbb{G}$ we shall usually omit mentioning it. If $0 < p \leq \infty$, then L^p will denote the usual Lebesgue space on $\mathbb{G}$. For $0 < p < \infty$ we write

$$\|f\|_p := \left(\int_{\mathbb{G}} |f(x)|^p dx\right)^{1/p},$$

despite the fact that this is not a norm for $p < 1$. However, the map $(f, g) \mapsto \|f - g\|_p^p$ is a metric on L^p for $p < 1$. We recall that if f is a measurable function on $\mathbb{G}$, its *distribution function* $\lambda_f : [0, \infty] \to [0, \infty]$ is defined by

$$\lambda_f(\alpha) := |\{x : |f(x)| > \alpha\}|, \tag{1.9}$$

and its *nonincreasing rearrangement* $f^* : [0, \infty) \to [0, \infty)$ is defined by

$$f^*(t) = \inf\{\alpha : \lambda_f(\alpha) \leq t\}. \tag{1.10}$$

Moreover,

$$\int_{\mathbb{G}} |f(x)|^p dx = -\int_0^\infty \alpha^p d\lambda_f(\alpha) = p\int_0^\infty \alpha^{p-1}\lambda_f(\alpha)d\alpha = \int_0^\infty f^*(t)^p dt.$$

For $0 < p < \infty$, the weak-L^p is the space of functions f such that

$$[f]_p := \sup_{\alpha>0} \alpha^p \lambda_f(\alpha) = \sup_{t>0} t^{1/p} f^*(t) < \infty.$$

This $[\cdot]_p$ is not a norm but it defines a topology on the weak-L^p space. A subadditive operator which is bounded from L^p to weak L^q is said to be *weak type* (p, q).

1.2.2 Polar coordinates

There is an analogue of polar coordinates on homogeneous groups. We start with the following observation:

Proposition 1.2.9 (Polar decomposition: a special case). *Let f be a locally integrable function on $\mathbb{G}\backslash\{0\}$ and assume that it is homogeneous of degree $-Q$. Then there is a constant μ_f (the 'average value' of f) such that for every $g \in L^1((0, \infty), r^{-1}dr)$, we have*

$$\int_{\mathbb{G}} f(x)g(|x|)dx = \mu_f \int_0^\infty g(r)r^{-1}dr. \tag{1.11}$$

Proof. Define $L_f : (0, \infty) \to \mathbb{C}$ by

$$L_f(r) := \begin{cases} \int_{1\leq|x|\leq r} f(x)dx & \text{if } \ r \geq 1, \\ -\int_{r\leq|x|\leq 1} f(x)dx & \text{if } \ r < 1. \end{cases}$$

By changing the variables $x \mapsto sx$ and using the homogeneity of f, it can be verified that

$$L_f(rs) = L_f(r) + L_f(s)$$

for all $r, s > 0$. From the continuity of L_f, it then follows that

$$L_f = L_f(e)\log r,$$

and we set $\mu_f := L_f(e)$. Then equality (1.11) is obvious when g is the characteristic function of an interval, and it follows in general by taking linear combinations and limits of such functions. □

Proposition 1.2.10 (Polar decomposition). *Let*

$$\wp := \{x \in \mathbb{G} : |x| = 1\} \tag{1.12}$$

be the unit sphere with respect to the homogeneous quasi-norm $|\cdot|$. *Then there is a unique Radon measure* σ *on* $\wp$ *such that for all* $f \in L^1(\mathbb{G})$,

$$\int_{\mathbb{G}} f(x)dx = \int_0^\infty \int_\wp f(ry)r^{Q-1}d\sigma(y)dr. \tag{1.13}$$

Proof. Let $\widetilde{f} \in C(\mathbb{G}\backslash\{0\})$ be the homogeneous extension of $f \in C(\wp)$ defined by

$$\widetilde{f}(x) := |x|^{-Q}f(|x|^{-1}x).$$

Then $\widetilde{f}$ satisfies the hypotheses of Proposition 1.2.9. The map $f \mapsto \mu_{\widetilde{f}}$ is clearly a positive linear functional on $C(\wp)$, so it is given by the integration against some Radon measure σ on $\wp$. If $g \in C_0(0,\infty)$ then we have

$$\begin{aligned}\int_{\mathbb{G}} f(|x|^{-1}x)g(|x|)dx &= \int_{\mathbb{G}} \widetilde{f}(x)|x|^Q g(|x|)dx = \mu_{\widetilde{f}} \int_0^\infty r^{Q-1}g(r)dr \\ &= \int_0^\infty \int_\wp f(y)g(r)r^{Q-1}d\sigma(y)dr.\end{aligned}$$

Since linear combination of functions of the form $f(|x|^{-1}x)g(|x|)$ are dense in $L^1(\mathbb{G})$, this completes the existence proof, and from the decomposition it follows that such a measure is necessarily unique. □

Corollary 1.2.11. *Let* $C := \sigma(\wp)$. *Then if* $0 < a < b < \infty$ *and* $\alpha \in \mathbb{C}$, *we have*

$$\int_{a<|x|<b} |x|^{\alpha-Q}dx = \begin{cases} C\alpha^{-1}(b^\alpha - a^\alpha) & \text{if } \alpha \neq 0, \\ C\log(b/a) & \text{if } \alpha = 0. \end{cases}$$

Corollary 1.2.12. *Let* f *be a measurable function on* $\mathbb{G}$ *such that*

$$f(x) = O(|x|^{\alpha-Q})$$

for some $\alpha \in \mathbb{R}$. *If* $\alpha > 0$ *then* f *is integrable near* 0, *and if* $\alpha < 0$ *then* f *is integrable near* ∞.

These two corollaries will be frequently used without comment in the sequel.

1.2.3 Convolutions

Let f and g be two integrable function on $\mathbb{G}$. Then their *convolution* $f * g$ is well defined by

$$(f * g)(x) := \int_{\mathbb{G}} f(y)g(y^{-1}x)dy = \int_{\mathbb{G}} f(xy^{-1})g(y)dy.$$

The basic facts about convolution of L^p and weak-L^p functions can be formulated in two propositions. For other properties of convolutions on groups we can refer to [FR16, Sections 1.5 and 3.1.10].

Proposition 1.2.13 (Young's inequality). *Suppose*

$$1 \leq p, q, r \leq \infty \quad \textit{and} \quad \frac{1}{p} + \frac{1}{q} = \frac{1}{r} + 1.$$

If $f \in L^p$ and $g \in L^q$, then $f * g \in L^r$ and

$$\|f * g\|_{L^r(\mathbb{G})} \leq \|f\|_{L^p(\mathbb{G})} \|g\|_{L^q(\mathbb{G})}.$$

Proof. First assuming $r = \infty$, in this case p and q are conjugate exponents and the result follows from Hölder's inequality.

Second assuming $r = q$, $p = 1$, let q' be the conjugate exponent to q. By Hölder's inequality,

$$\begin{aligned} |f * g(x)| &\leq \int_{\mathbb{G}} |f(xy^{-1})|^{(1/q)+(1/q')} |g(y)| dy \\ &\leq \left(\int_{\mathbb{G}} |f(xy^{-1})| dy \right)^{1/q'} \left(\int_{\mathbb{G}} |f(xy^{-1})| |g(y)|^q dy \right)^{1/q} \\ &= \|f\|_1^{1/q'} \left(\int_{\mathbb{G}} |f(xy^{-1})| |g(y)|^q dy \right)^{1/q}. \end{aligned}$$

Thus, by Fubini's theorem, we obtain

$$\begin{aligned} \int_{\mathbb{G}} |f * g(x)|^q dx &\leq \|f\|_1^{q/q'} \int_{\mathbb{G}} \int_{\mathbb{G}} |f(xy^{-1})| |g(y)|^q dy dx \\ &= \|f\|_1^{(q/q')+1} \|g\|_q^q, \end{aligned}$$

so that $\|f * g\|_q \leq \|f\|_1 \|g\|_q$. The rest follows by interpolation. ∎

Proposition 1.2.14 (Young's inequality for weak-L^p spaces). *Suppose*

$$q \leq p < \infty, \quad 1 < q, r < \infty, \quad \textit{and} \quad \frac{1}{p} + \frac{1}{q} = \frac{1}{r} + 1.$$

If $f \in L^p$ and $g \in L^q$ then $f * g \in$ weak-L^r and there exists $C_1 = C_1(p, q)$ such that

$$[f * g]_r \leq C_1 \|f\|_p [g]_q.$$

Moreover, if $p > 1$ then $f * g \in L^r$ and there exists $C_2 = C_2(p, q)$ such that

$$\|f * g\|_r \leq C_2 \|f\|_p [g]_q.$$

Proof. By the Marcinkiewicz interpolation theorem one can notice that the strong result for $p > 1$ follows from the weak result. Suppose then that $f \in L^p$ and $g \in L^q$, and (without loss of generality) that $\|f\|_p = [q]_q = 1$. Given $\alpha > 0$ set

$$M := (\alpha/2)^{r/q}(q/r)^{r/qp'},$$

where p' is the conjugate exponent to p. Define $g_1(x) := g(x)$ if $|g(x)| \leq M$ and $g_1(x) := 0$ otherwise, and set $g_2 := g - g_1$. Since

$$\lambda_{f*g}(\alpha) \leq \lambda_{f*g_1}(\alpha/2) + \lambda_{f*g_2}(\alpha/2),$$

it is enough to show that each term on the right side is bounded by $C\alpha^{-r}$, where C depends only on p and q. On the one hand, since $q^{-1} - (p')^{-1} = r^{-1} > 0$ we have $p'q > 0$ and therefore

$$\begin{aligned}\int_{\mathbb{G}} |g_1(x)|^{p'} dx &= p' \int_0^\infty \alpha^{p'-1} \lambda_{g_1}(\alpha) d\alpha \leq p' \int_0^M \alpha^{p'-1} \lambda_g(\alpha) d\alpha \\ &\leq p' \int_0^M \alpha^{p'-1-q} d\alpha = \frac{p'}{p'-q} M^{p'-q} = \frac{r}{q} M^{qp'/r} = (\alpha/2)^{p'}.\end{aligned}$$

Thus, for every $x \in \mathbb{G}$, by Hölder's inequality (or by Proposition 1.2.13) we have

$$|f * g_1(x)| \leq \|f\|_p \|g_1\|_{p'} \leq \alpha/2,$$

which implies that $\lambda_{f*g_1}(\alpha/2) = 0$. On the other hand, since $q > 1$, we have

$$\begin{aligned}\int_{\mathbb{G}} |g_2(x)| dx &= \int_0^\infty \lambda_{g_2}(\alpha) d\alpha = \int_0^M \lambda_g(M) d\alpha + \int_M^\infty \lambda_g(\alpha) d\alpha \\ &\leq M \cdot M^{-q} + \int_M^\infty \alpha^{-q} d\alpha = \frac{q}{q-1} M^{1-q},\end{aligned}$$

and therefore by Proposition 1.2.13,

$$\|f * g_2\|_p \leq \|f\|_p \|g_2\|_1 \leq q(q-1)^{-1} M^{1-q}.$$

But then

$$\begin{aligned}\lambda_{f*g_2}(\alpha/2) &\leq [2\|f * g_2\|_p/\alpha]^p \\ &\leq \left(\frac{2}{\alpha}\right)^p \left(\frac{q}{q-1}\right)^p M^{(1-q)p} \\ &= C(p, q)\alpha^{-r}.\end{aligned}$$

This completes the proof. □

Now let us summarize some properties of approximations to the identity in terms of the convolution. The following notation will be used throughout this monograph: if ϕ is a function on $\mathbb{G}$ and $t > 0$, we define ϕ_t by

$$\phi_t := t^{-Q}\phi \circ \delta_{1/t}, \quad \text{that is,} \quad \phi_t(x) := t^{-Q}\phi(x/t). \tag{1.14}$$

We notice that if $\phi \in L^1(\mathbb{G})$ then $\int_{\mathbb{G}} \phi_t(x)dx$ is independent of t.

Proposition 1.2.15 (Approximation of identity). *Let $\phi \in L^1(\mathbb{G})$ and let $a := \int_{\mathbb{G}} \phi(x)dx$. Then we have the following properties:*

(i) *If $f \in L^p(\mathbb{G})$ for $1 \leq p < \infty$, then $\|f * \phi_t - af\|_p \to 0$ as $t \to 0$.*

(ii) *If f is bounded and right uniformly continuous, then $\|f * \phi_t - af\|_\infty \to 0$ as $t \to 0$.*

(iii) *If f is bounded on $\mathbb{G}$ and continuous on an open set $\Omega \subset \mathbb{G}$, then $f*\phi_t - af \to 0$ uniformly on compact subsets of Ω as $t \to 0$.*

Proof. For a function f on $\mathbb{G}$ and $y \in \mathbb{G}$, let us define

$$f^y(x) := f(xy^{-1}).$$

If $f \in L^p$ for $1 \leq p < \infty$, then it can be shown that

$$\|f^y - f\|_p \to 0 \quad \text{as } y \to 0, \tag{1.15}$$

for example, using the fact that C_0 is dense in L^p. If $p = \infty$, property (1.15) holds if and only if f is (almost everywhere equal to) a right uniformly continuous function. We now observe that

$$\begin{aligned} f * \phi_t(x) - af(x) &= \int_{\mathbb{G}} f(xy^{-1})t^{-Q}\phi(y/t)dy - af(x) \\ &= \int_{\mathbb{G}} f(x(tz)^{-1})\phi(z)dz - af(x) \\ &= \int_{\mathbb{G}} [f(x(tz)^{-1}) - f(x)]\phi(z)dz. \end{aligned}$$

Hence by Minkowski's inequality,

$$\|f * \phi_t - af\|_p \leq \int_{\mathbb{G}} \|f^{tz} - f\|_p |\phi(z)|dz.$$

Since $\|f^{tz} - f\|_p \leq 2\|f\|_p$, under the hypothesis of (i) or (ii) it follows from (1.15) and the dominated convergence theorem that $\|f * \phi_t \ - af\| \to 0$. The routine modification of this argument (with $p = \infty$) needed to establish (iii) is left to the reader. □

1.2.4 Polynomials

The Lie algebra $\mathfrak{g}$ of $\mathbb{G}$ can be understood from different prospectives:

- as tangent vectors at the origin,
- as left invariant (and right invariant) vector fields.

While in this book we will not make much use of the first interpretation, we will need the second. Consequently, let us denote by $\mathfrak{g}_L$ and $\mathfrak{g}_R$ the spaces of left invariant and right invariant vector fields on $\mathbb{G}$.

Let us fix a basis $X_1, \ldots, X_n$ for $\mathfrak{g}$ consisting of eigenvectors for the dilations δ_r with eigenvalues $r^{d_1}, \ldots, r^{d_n}$, i.e., such that

$$\delta_r X_k = r^{d_k} X_k.$$

In other words, first, we consider X_k as left invariant differential operators on $\mathbb{G}$ and we denote by $Y_1, \ldots, Y_n$ the corresponding basis for $\mathfrak{g}_R$: that is, Y_k is the element of $\mathfrak{g}_R$ such that $Y_k|_0 = X_k|_0$ and for $f \in C^1$ we have

$$X_k f(y) = \frac{d}{dt} f(y \cdot \exp(tX_k))|_{t=0},$$
$$Y_k f(y) = \frac{d}{dt} f(\exp(tX_k) \cdot y)|_{t=0}.$$

Then X_k and Y_k are the differential operators homogeneous of degree d_k since

$$\begin{aligned} X_k(f \circ \delta_r)(y) &= \frac{d}{dt} f((ry)\exp(r^{d_k} tX_k))|_{t=0} \\ &= r^{d_k} \frac{d}{dt} f((ry)\exp(tX_k))|_{t=0} \\ &= r^{d_k} (X_k f \circ \delta_r)(y), \end{aligned}$$

and similarly for Y_k. For $I = (i_1, \ldots, i_n) \in \mathbb{N}^n$, we use the notation

$$X^I = X_1^{i_1} X_2^{i_2} \cdots X_n^{i_n}, \quad Y^I = Y_1^{i_1} Y_2^{i_2} \cdots Y_n^{i_n}.$$

According to the Poincaré–Birkhoff–Witt theorem, the operators X^I give a basis for the algebra of left invariant differential operators on the Lie group $\mathbb{G}$. In addition, we also use the notations

$$|I| := i_1 + i_2 + \cdots + i_n, \quad d(I) := d_1 i_1 + d_2 i_2 + \cdots + d_n i_n. \tag{1.16}$$

Here $|I|$ is called an order of the differential operators X^I and Y^I, and $d(I)$ is their degree of homogeneity or the homogeneous degree. If we denote by Δ the set of all numbers $d(I)$ as I ranges over $\mathbb{N}^n$ then we have $\mathbb{N} \subset \Delta$ as $d_1 = 1$.

There are two useful facts. On the one hand, left translations are isometries on $L^2(\mathbb{G})$, and the operators X_k and Y_k are formally skew-adjoint. Therefore,

$$\int_{\mathbb{G}} (X^I f) g = (-1)^{|I|} \int_{\mathbb{G}} f(X^I g), \quad \int_{\mathbb{G}} f(Y^I g) = (-1)^{|I|} \int_{\mathbb{G}} (Y^I f) g,$$

for all smooth functions f and g for which the integrands decay suitably at infinity. On the other hand, the operators X^I and Y^I interact with convolutions by the formulae

$$X^I(f * g) = f * (X^I g), \quad Y^I(f * g) = (Y^I f) * g, \quad (X^I f) * g = f * (Y^I g).$$

Except for the last one these equalities are direct consequences of differentiating, and the third can be obtained by integration by parts:

$$\begin{aligned}(X^I f) * g(x) &= \int_{\mathbb{G}} X^I f(xy) g(y^{-1}) dy = (-1)^{|I|} \int_{\mathbb{G}} f(xy) X^I [g(y^{-1})] dy \\ &= \int_{\mathbb{G}} f(xy)(Y^I g)(y^{-1}) dy = f * (Y^I g)(x).\end{aligned}$$

Definition 1.2.16 (Polynomials on the homogeneous group $\mathbb{G}$). A function P on $\mathbb{G}$ will be called a *polynomial* if $P \circ \exp$ is a polynomial on $\mathfrak{g}$.

We can form a global coordinate system on $\mathbb{G}$ and generate the algebra of polynomials on $\mathbb{G}$ by setting

$$\eta_k = \xi_k \circ \exp^{-1},$$

where $\eta_1, \ldots, \eta_n$ are polynomials on $\mathbb{G}$, and $\xi_1, \ldots, \xi_n$ are the basis for the linear forms on $\mathfrak{g}$ dual to the basis $X_1, \ldots, X_n$ for $\mathfrak{g}$. Therefore, each polynomial on $\mathbb{G}$ can be defined uniquely in the form

$$P = \sum_I a_I \eta^I,$$

where

$$\eta^I = \eta_1^{i_1} \cdots \eta_n^{i_n},$$

$a_I \in \mathbb{C}$, and all but finitely many of the coefficients a_I vanish. Since η^I is homogeneous of degree $d(I)$, the set of possible homogeneous degrees for polynomials coincides with the set Δ. The *isotropic degree* of a polynomial P is

$$\max\{|I| : a_I \neq 0\}.$$

And the *homogeneous degree* of a polynomial P is

$$\max\{d(I) : a_I \neq 0\}.$$

By $\mathcal{P}_N^{\text{iso}}$, for $N \in \mathbb{N}$, we denote the space of polynomials of isotropic degree $\leq N$, and by $\mathcal{P}_a$, for $a \in \Delta$, we denote the space of polynomials of homogeneous degree $\leq a$. Since $1 \leq d_k \leq \bar{d}$ for $k = 1, \ldots, n$, we observe that

$$\mathcal{P}_N \subset \mathcal{P}_N^{\text{iso}} \subset \mathcal{P}_{\bar{d}N}$$

for $N \in \mathbb{N}$.

There is a more explicit description of the group law in terms of the coordinates η_k since the map

$$(x, y) \mapsto \eta_k(xy)$$

is a polynomial on $\mathbb{G} \times \mathbb{G}$. Thus, we have

$$\eta_k((rx)(ry)) = r^{d_k}\eta_k(xy),$$

that is, it is jointly homogeneous of degree d_k and, according to the Baker–Campbell–Hausdorff formula, we have $\eta_k(xy) = \eta_k(x) + \eta_k(y)$ modulo terms of isotropic degree ≥ 2. It follows that

$$\eta_k(xy) = \eta_k(x) + \eta_k(y) + \sum_{I \neq 0, J \neq 0, d(I)+d(J)=d_k} C_k^{IJ}\eta^I(x)\eta^J(y), \tag{1.17}$$

where C_k^{IJ} are constants. It is easy to see that the monomials η^I, η^J can only involve coordinates with homogeneous degree less than d_k, since the multi-indices I and J in (1.17) must satisfy $d(I) < d_k$ and $d(J) < d_k$. In particular, only the coordinates $\eta_1, \ldots, \eta_{j-1}$ can be involved, for instance:

$$\begin{aligned} d_k = 1: &\quad \eta_k(xy) = \eta_k(x) + \eta_k(y), \\ d_k = 2: &\quad \eta_k(xy) = \eta_k(x) + \eta_k(y) + \sum_{d_j = d_l = 1} C_k^{jl}\eta_j(x)\eta_l(y). \end{aligned}$$

Proposition 1.2.17 (Polynomials are translation invariant). *For any $a \in \Delta$, $\mathcal{P}_a$ is left translation invariant.*

Proof. According the formula (1.17), it is easy to see that $\eta_k(xy)$ is in $\mathcal{P}_{d_k}$ (as a function of x for each y, and also as a function of y for each x). On the other hand, the η_k's generate all polynomials, therefore $\mathcal{P}_a$ is left translation invariant for all $a \in \Delta$. □

Definition 1.2.18 (Coordinate functions on the group). For $x \in \mathbb{G}$ and $k = 1, \ldots, n$, we can think of

$$x_k := \eta_k(x)$$

as the *coordinates* of the variable x. Thus, each x_k becomes a polynomial of homogeneous degree k.

We now establish a link between left and right invariant differential operators and derivatives with respect to coordinate functions on the group.

Proposition 1.2.19 (Formulae for invariant derivatives). *We have*

$$X_k = \sum P_{kj}(\partial/\partial x_j), \quad Y_k = \sum Q_{kj}(\partial/\partial x_j), \tag{1.18}$$

where $P_{kk} = Q_{kk} = 1$, $P_{kj} = Q_{kj} = 0$ if $d_j < d_k$ or if $d_j = d_k$ and $j \neq k$, and P_{kj}, Q_{kj} are homogeneous polynomials of degree $d_k - d_j$ if $d_j > d_k$.

Proof. Let us define the operator $L_x : \mathbb{G} \to \mathbb{G}$ by $L_x(y) := xy$ for $x \in \mathbb{G}$. Then, using the fact that X_k agrees with $\partial/\partial x_k$ at 0, for each differentiable function f on $\mathbb{G}$ and $x \in \mathbb{G}$, we have

$$X_k f(x) = (X_k f) \circ L_x(0) = X_k(f \circ L_x)(0) = (\partial/\partial x_k)(f \circ L_x)(0).$$

Therefore, by the chain rule, we obtain

$$X_k f(x) = \sum_{k=1}^{n} \frac{\partial f}{\partial x_k}(x) \frac{\partial [x_k \circ L_x]}{\partial x_k}(0).$$

But by formula (1.17) it follows that

$$\frac{\partial [x_j \circ L_x]}{\partial x_k}(0) = \delta_{kj} + \sum_{d(I)=d_j-d_k} C_j^{I[k]} \eta^I(x),$$

where $[k]$ is the multi-index with 1 in the kth place and zeros elsewhere. The desired result for X_k follows from this, and for Y_k it can be proved in a similar way. □

There are also similar expressions for $\partial/\partial x_k$ in terms of X_j or Y_j:

$$\partial/\partial x_k = \sum P'_{kj} X_j = \sum Q'_{kj} Y_j,$$

where P'_{kj}, Q'_{kj} are of the same form as P_{kj}, Q_{kj} in (1.18). Above formulae can be directly obtained from (1.18) with $j = n$, that is, we have

$$\begin{aligned} X_n &= \partial/\partial x_n, \\ X_{n-1} &= \partial/\partial x_{n-1} + P_{(n-1)n}\partial/\partial x_n, \\ X_{n-2} &= \partial/\partial x_{n-2} + P_{(n-2)(n-1)}\partial/\partial x_{n-1} + \ P_{(n-2)n}\partial/\partial x_n, \end{aligned}$$

therefore, we obtain

$$\begin{aligned} \partial/\partial x_n &= X_n, \\ \partial/\partial x_{n-1} &= X_{n-1} - P_{(n-1)n}\partial/\partial x_n, \\ \partial/\partial x_{n-2} &= X_{n-2} - P_{(n-2)(n-1)}\partial/\partial x_{n-1} - \ P_{(n-2)n}\partial/\partial x_n, \end{aligned}$$

and so on. Similarly, one can obtain expressions for higher-order derivatives. For instance,

$$X^I = \sum_{|J|\le|I|, d(J)\ge d(I)} P_{IJ}(\partial/\partial x)^J, \tag{1.19}$$

where P_{IJ} is a homogeneous polynomial of degree $d(J) - d(I)$. Analogously, we obtain formulae for Y^I in terms of $(\partial/\partial x)^J$ and for $(\partial/\partial x)^I$ in terms of X^J or

Y^J, that is,

$$X^I = \sum_{|J|\leq|I|, d(J)\leq d(I)} P_{IJ}Y^J,$$
$$Y^I = \sum_{|J|\leq|I|, d(J)\leq d(I)} Q_{IJ}X^J,$$

where P_{IJ} and Q_{IJ} are homogeneous polynomials of degree $d(J) - d(I)$.

Proposition 1.2.20 (Determination of invariant differential operators). *Let $a \in \Delta$ and let $\mu := \dim P_a$. Then the following maps are linear isomorphisms from P_a to $\mathbb{C}^\mu$:*

(i) $P \to ((\partial/\partial x)^I P(0))_{d(I)\leq a}$,

(ii) $P \to (X^I P(0))_{d(I)\leq a}$,

(iii) $P \to (Y^I P(0))_{d(I)\leq a}$.

Proof. Note that Case (i) is a simple consequence of Taylor's theorem. Also, in view of (1.19), since P_{IJ} is a constant function when $d(I) = d(J)$ and $P_{IJ}(0) = 0$ when $d(J) > d(I)$, we have

$$X^I|_0 = \sum_{|J|\leq|I|, d(J)=d(I)} P_{IJ}(\partial/\partial x)^J|_0,$$

and similarly for the other formulae relating X^I, Y^I and $(\partial/\partial\eta)^I$. Cases (ii) and (iii) follow easily from this observation together with Case (i). □

The properties above motivate the following:

Definition 1.2.21 (Taylor polynomials). Let $x \in \mathbb{G}$, $a \in \Delta$, and let f be a function whose (distributional) derivatives $X^I f$ (resp. $Y^I f$) are continuous functions in a neighborhood of x for $d(I) \leq a$. The left (resp. right) *Taylor polynomial* of f at x of homogeneous degree a is the unique $P \in P_a$ such that $X^I P(0) = X^I f(x)$ (resp. $Y^I P(0) = Y^I f(x)$) for all I such that $d(I) \leq a$.

Now we provide simple proofs of an explicit expression of the Taylor formula and the Taylor inequality in the spirit of [Bon09].

Let $X \in \mathfrak{g}$ be given, and suppose $\gamma(t)$ is any integral curve of X, i.e.

$$\dot{\gamma} = X(\gamma(t))$$

for all $t \in \mathbb{R}$. If $m \in \mathbb{N} \cup \{0\}$ and $u \in C^{m+1}(\mathbb{G})$ is real-valued, then since

$$\frac{d^k}{dt^k}(u(\gamma(t))) = (X^k u)(\gamma(t)),$$

for each $k \in \mathbb{N} \cup \{0\}$, by applying the usual Taylor formula (with integral reminder) to $t \mapsto u(\gamma(t))$, we obtain

$$u(\gamma(t)) = \sum_{k=0}^{m} \frac{t^k}{k!}(X^k u)(\gamma(0)) + \frac{1}{m!}\int_0^t (t-s)^m (X^{m+1}u)(\gamma(s))ds. \tag{1.20}$$

Moreover, it is easy to see that the integral curve γ of $\log h$ starting at x is $s \mapsto \gamma(s) = x\exp(s\log h)$. With $\gamma(0) = x, \gamma(1) = xh$, we get

$$u(xh) = \sum_{k=0}^{m} \frac{1}{k!}((\log h)^k u)(x) + \frac{1}{m!}\int_0^1 (1-s)^m ((\log h)^{m+1}u)(x\exp(s\log h))ds. \tag{1.21}$$

On the other hand, there always exist (polynomial) functions $\mathbb{G} \ni h \mapsto \zeta_i(h) \in \mathbb{R}$ such that

$$\log h = \zeta_1(h)X_1 + \cdots + \zeta_n(h)X_n,$$

for all $h \in \mathbb{G}$, where $\{X_1, \ldots, X_n\}$ is a basis of the Lie algebra of $\mathbb{G}$. Thus, we have

$$(\log h)^k = \left(\sum_{i=1}^{n} \zeta_i(h)X_i\right)^k = \sum_{i_1,\ldots,i_k=1}^{n} \zeta_{i_1}(h)\cdots\zeta_{i_k}(h)X_{i_1}\cdots X_{i_k},$$

for every $k \in \mathbb{N}$.

Therefore, (1.21) implies that

$$\begin{aligned} u(xh) = u(x) &+ \sum_{k=1}^{m} \sum_{I=(i_1,\ldots,i_k),i_1,\ldots,i_k \le n} \frac{X_I u(x)}{k!}\zeta_{i_1}(h)\cdots\zeta_{i_k}(h) \\ &+ \sum_{I=(i_1,\ldots,i_{m+1}),i_1,\ldots,i_{n+1}\le n} \zeta_{i_1}(h)\cdots\zeta_{i_{m+1}}(h) \\ &\times \int_0^1 (X_I u)\left(x\exp\left(\sum_{i\le n} s\zeta_i(h)X_i\right)\right)\frac{(1-s)^m}{m!}ds, \end{aligned} \tag{1.22}$$

where $X_I = X_{i_1}\cdots X_{i_k}$ and $I = (i_1, \ldots, i_k)$ with $i_1, \ldots, i_k \in \{1, \ldots, n\}$.

For a multi-index α, we will be using the notations (1.16), i.e.,

$$|\alpha| = \alpha_1 + \cdots + \alpha_n, \quad d(\alpha) = d_1\alpha_1 + \cdots + d_n\alpha_n,$$

where d_j is the homogeneous degree of X_k; these are called the Euclidean length and the homogeneous length of α, respectively. One also sets

$$\mathfrak{G} := \{d(\alpha) : \alpha \in (\mathbb{N} \cup \{0\})^n\}.$$

As usual, $[\beta]$ below is the integer part of the real number β. Now we are in a position to state the Taylor formula on homogeneous groups.

Theorem 1.2.22 (Taylor formula). *Let $\mathbb{G}$ be a homogeneous group (identified with $\mathbb{R}^n$ as a topological space). Suppose $\{X_1, \ldots, X_n\}$ is the Jacobian basis for its Lie algebra, $m \in \mathfrak{G}$ and $u \in C^{[m]+1}(\mathbb{G})$. Let also $x_0 \in \mathbb{G}$ be fixed. Then, for every $x \in \mathbb{G}$ we have*

$$\begin{aligned} u(x) &= P_m(u, x_0)(x) + R_m(x, x_0) \\ &= u(x_0) + \sum_{k=1}^{[m]} \sum_{\substack{I=(i_1,\ldots,i_k),\\ i_1,\ldots,i_k \le n,\\ d(I)\le m}} \frac{X_I u(x_0)}{k!} \zeta_{i_1}(x_0^{-1}x) \cdots \zeta_{i_k}(x_0^{-1}x) + R_m(x, x_0), \end{aligned} \tag{1.23}$$

where the reminder term $R_m(x, x_0)$ is given by

$$\begin{aligned} R_m(x, x_0) = & \sum_{k=1}^{[m]} \sum_{\substack{I=(i_1,\ldots,i_k),\\ i_1,\ldots,i_k \le n,\\ d(I)> m}} \frac{X_I u(x_0)}{k!} \zeta_{i_1}(x_0^{-1}x) \cdots \zeta_{i_k}(x_0^{-1}x) \\ & + \sum_{\substack{I=(i_1,\ldots,i_{[n]+1}),\\ i_1,\ldots,i_{[m]+1} \le n}} \zeta_{i_1}(x_0^{-1}x) \cdots \zeta_{i_{[m]+1}}(x_0^{-1}x) \\ & \times \int_0^1 (X_I u)\left(x_0 \exp\left(\sum_{i\le n} s\zeta_i(x_0^{-1}x)X_i\right)\right) \frac{(1-s)^{[m]}}{[m]!} ds. \end{aligned}$$

Proof of Theorem 1.2.22. If $x_0 \in \mathbb{G}$ is any fixed element, by replacing x and h in the formula (1.22) by respectively x_0 and $x_0^{-1}x$, we obtain the following:

$$\begin{aligned} u(x) = u(x_0) + & \sum_{k=1}^{[m]} \sum_{\substack{I=(i_1,\ldots,i_k),\\ i_1,\ldots,i_k \le n,\\ d(I)> m}} \frac{X_I u(x_0)}{k!} \zeta_{i_1}(x_0^{-1}x) \cdots \zeta_{i_k}(x_0^{-1}x) \\ & + \sum_{\substack{I=(i_1,\ldots,i_{[m]+1}),\\ i_1,\ldots,i_{[m]+1} \le n}} \zeta_{i_1}(x_0^{-1}x) \cdots \zeta_{i_{[m]+1}}(x_0^{-1}x) \\ & \times \int_0^1 (X_I u)\left(x_0 \exp\left(\sum_{i\le n} s\zeta_i(x_0^{-1}x)X_i\right)\right) \frac{(1-s)^{[m]}}{[m]!} ds. \end{aligned} \tag{1.24}$$

Since a polynomial $\zeta_i(x)$ is homogeneous of degree σ_i, there exists $C_1 > 0$ such that

$$C_1^{-1}|x|^{\sigma_i} \le |\zeta_i(x)| \le C_1|x|^{\sigma_i}, \quad \forall x \in \mathbb{G}, \quad i = 1, \ldots, n.$$

As a consequence, for every $k \in \mathbb{N}$, $\zeta_{i_1} \cdots \zeta_{i_k}$ is a homogeneous polynomial of degree $d_{i_1} + \cdots + d_{i_k}$. Similarly, $\zeta_{i_1} \cdots \zeta_{i_{[m]+1}}$ is homogeneous of degree $\ge [m] + 1$ (since the d_i's are ≥ 1) and, there appear only derivatives $X_I u$ with $d(I) \geq [m]+1 > m$ in the integral summands.

We restate (1.24) emphasizing out the polynomial of degree $\leq m$ in the right-hand side:

$$
\begin{aligned}
u(x) &= u(x_0) + \sum_{k=1}^{[m]} \sum_{\substack{I=(i_1,\dots,i_k),\\ i_1,\dots,i_k \leq n,\\ d(I)\leq m}} \frac{X_I u(x_0)}{k!} \zeta_{i_1}(x_0^{-1}x) \cdots \zeta_{i_k}(x_0^{-1}x) \\
&\quad + \sum_{k=1}^{[m]} \sum_{\substack{I=(i_1,\dots,i_k),\\ i_1,\dots,i_k \leq n,\\ d(I)> m}} \frac{X_I u(x_0)}{k!} \zeta_{i_1}(x_0^{-1}x) \cdots \zeta_{i_k}(x_0^{-1}x) \\
&\quad + \sum_{\substack{I=(i_1,\dots,i_{[m]+1}),\\ i_1,\dots,i_{[m]+1} \leq n}} \zeta_{i_1}(x_0^{-1}x) \cdots \zeta_{i_{[m]+1}}(x_0^{-1}x) \\
&\quad \times \int_0^1 (X_I u) \left(x_0 \exp \left(\sum_{i \leq n} s \zeta_i (x_0^{-1}x) X_i \right) \right) \frac{(1-s)^{[m]}}{[m]!} ds \\
&=: P_m(u, x_0)(x) + R_m(x, x_0).
\end{aligned}
$$

By construction, $P_m(u, x_0)(x)$ is a polynomial of homogeneous degree $\leq m$. $\square$

Theorem 1.2.23 (Taylor inequality). *Assume the hypotheses of Theorem* 1.2.22. *Then for every fixed homogeneous norm* $|\cdot|$ *on* $\mathbb{G}$ *and every* $m \in \mathfrak{G}$, *there exists* $C > 0$ *(depending on* $\mathbb{G}$ *and* $|\cdot|$*) such that*

$$
|R_m(x, x_0)| \leq \sum_{k=1}^{[m]+1} \frac{C^k}{k!} \sum_{\substack{I=(i_1,\dots,i_k),\\ i_1,\dots,i_k \leq n,\\ d(I)>m}} |x_0^{-1}x|^{d(I)} \sup_{|y| \leq C|x_0^{-1}x|} |X_I u(x_0^{-1}y)|. \tag{1.25}
$$

Moreover, an explicit formula for the Taylor polynomial $P_m(u, x_0)$ *of degree* $m \in \mathfrak{G}$ *related to* u *about* x_0 *is*

$$
\begin{aligned}
P_m(u, x_0)(x) &= u(x_0) \\
&+ \sum_{k=1}^{[m]} \sum_{I=(i_1,\dots,i_k), i_1,\dots,i_k \leq n, d(I) \leq m} \frac{X_I u(x_0)}{k!} \zeta_{i_1}(x_0^{-1}x) \cdots \zeta_{i_k}(x_0^{-1}x).
\end{aligned} \tag{1.26}
$$

Proof of Theorem 1.2.23. Since

$$
\sum_{i=1}^{n} s\zeta_i(x) X_i = s \log(x)
$$

by Proposition 1.2.5 we obtain

$$\left|(X_I u)\left(\exp\left(\sum_{i\leq n} s\zeta_i(x)X_i\right)\right)\right| \leq \sup_{|y|\leq C_0|x|} |X_I u(y)|.$$

This implies that

$$\begin{aligned}|R_m(x_0^{-1}x)| &\leq \sum_{k=1}^{[m]+1} \sum_{\substack{I=(i_1,\ldots,i_k),\\ i_1,\ldots,i_k\leq n,\\ d(I)>m}} \sup_{|y|\leq C_0|x_0^{-1}x|} |X_I u(x_0^{-1}y)| \frac{\zeta_{i_1}(x_0^{-1}x)\cdots\zeta_{i_k}(x_0^{-1}x)}{k!} \\ &\leq \sum_{k=1}^{[m]+1} \frac{C_1^k}{k!} \sum_{\substack{I=(i_1,\ldots,i_k),\\ i_1,\ldots,i_k\leq n,\\ d(I)>m}} |x_0^{-1}x|^{\sigma(I)} \sup_{|y|\leq C_0|x_0^{-1}x|} |X_I u(x_0^{-1}y)|.\end{aligned}$$

Choosing $C := \max\{C_0, C_1\}$ we complete the proof of the Taylor inequality (1.25).

First, note that the above estimate of R_m gives $R_m(x) = \mathcal{O}(\rho^{m+\varepsilon}(x))$ as $x \to 0$, where

$$\varepsilon := \min_{k=1,\ldots,[m]+1} \{d(I) - m : I = (i_1, \ldots, i_k), i_1, \ldots, i_k \leq n, d(I) > m\}. \tag{1.27}$$

Thus, the Taylor formula (1.23) can be rewritten as $u(x) = P_m(x) + \mathcal{O}(|x|^{m+\varepsilon})$ as $x \to 0$, with $\varepsilon > 0$ as in (1.27).

Now let us see that there exists at most one polynomial function P on $\mathbb{G}$, with degree $\leq m$, such that, for some $\varepsilon > 0$ (depending on P and m) it holds

$$u(x) = P(x) + \mathcal{O}_{x\to x_0}(|x_0^{-1}x|^{m+\varepsilon}). \tag{1.28}$$

Indeed, suppose there are two such polynomials, A and B (with related $\varepsilon_1, \varepsilon_2 > 0$). Then setting $\varepsilon := \min\{\varepsilon_1, \varepsilon_2\}$ we have

$$Q(x) := B - A = \mathcal{O}_{x\to x_0}(|x_0^{-1}x|^{m+\varepsilon}).$$

Setting $\widetilde{Q}(z) := Q(x_0 z)$, this is equivalent to

$$\widetilde{Q}(z) = \mathcal{O}_{z\to 0}(|z|^{m+\varepsilon}). \tag{1.29}$$

The fact that Q is a polynomial of degree at most m and the ith component function of $x_0 z$ is a polynomial in z of degree at most d_i, it follows that $Q(x_0 z)$ is a polynomial in z of degree at most m.

Therefore, (1.29) is valid if and only if $\widetilde{Q} \equiv 0$, that is, $Q(x_0 z) = 0$ for all $z \in \mathbb{G}$. This is in turn equivalent to $Q \equiv 0$, i.e., $A \equiv B$. Note that the equivalence of all homogeneous norms (cf. Proposition 1.2.3) implies that a polynomial P as in (1.28) is independent of $|\cdot|$. Thus, P_m is the Taylor polynomial of degree m related to u, which has the explicit formula (1.26). □

1.3 Radial and Euler operators

An important tool for working on homogeneous groups will be an extensive use of radial and Euler operators. We now discuss them in some detail and establish a number of properties used throughout the book.

1.3.1 Radial derivative

First we introduce a radial derivative (acting on a differentiable function f) on a homogeneous group $\mathbb{G}$ by

$$\mathcal{R}f(x) := \frac{df(x)}{d|x|}, \tag{1.30}$$

where $|x|$ is a homogeneous quasi-norm of $\mathbb{G}$. Note that the homogeneous quasi-norm $|x|$ in the formula (1.30) can be arbitrary, that is, in general the radial operator $\mathcal{R}$ depends on a chosen homogeneous quasi-norm.

Let $\{X_1, \ldots, X_n\}$ be a basis of the Lie algebra $\mathfrak{g}$ of $\mathbb{G}$ such that we have

$$AX_k = \nu_k X_k \quad \text{for every } k = 1, \ldots, n.$$

Then the matrix A can be taken to be $A = \operatorname{diag}(\nu_1, \ldots, \nu_n)$ and each X_k is homogeneous of degree ν_k. By decomposing the vector $\exp_{\mathbb{G}}^{-1}(x)$ in $\mathfrak{g}$ with respect to the basis $\{X_1, \ldots, X_n\}$, we get the vector

$$e(x) = (e_1(x), \ldots, e_n(x))$$

given by the formula

$$\exp_{\mathbb{G}}^{-1}(x) = e(x) \cdot \nabla \equiv \sum_{j=1}^{n} e_j(x) X_j,$$

where

$$\nabla = (X_1, \ldots, X_n)$$

is the full gradient. It gives the equality

$$x = \exp_{\mathbb{G}}\left(e_1(x)X_1 + \cdots + e_n(x)X_n\right). \tag{1.31}$$

By homogeneity and denoting $x = ry$, with $y \in \wp$ being on the quasi-sphere (1.12), we get

$$e(x) = e(ry) = (r^{\nu_1} e_1(y), \ldots, r^{\nu_n} e_n(y)).$$

Indeed, since each X_k is homogeneous of degree ν_k, from (1.31) we get that

$$rx = \exp_{\mathbb{G}}\left(r^{\nu_1} e_1(x)X_1 + \cdots + r^{\nu_n} e_n(x)X_n\right),$$

and hence

$$e(rx) = (r^{\nu_1} e_1(x), \ldots, r^{\nu_n} e_n(x)).$$

Thus, since $r > 0$ is arbitrary, without loss of generality taking $|x| = 1$, we can write

$$\frac{d}{d|rx|} f(rx) = \frac{d}{dr} f(\exp_{\mathbb{G}}(r^{\nu_1} e_1(x) X_1 + \cdots + r^{\nu_n} e_n(x) X_n)). \tag{1.32}$$

So, summarizing, one obtains

$$\frac{d}{d|x|}(f(x)) = \frac{d}{dr}(f(ry)) = \frac{d}{dr}(f(\exp_{\mathbb{G}}(r^{\nu_1} e_1(y) X_1 + \cdots + r^{\nu_n} e_n(y) X_n))). \tag{1.33}$$

Throughout this book we will be often abbreviate the notation by writing

$$\mathcal{R} := \frac{d}{dr}, \tag{1.34}$$

meaning that the derivative is taken with respect to the radial direction with respect to the quasi-norm $|\cdot|$.

We can also observe that for any differentiable function f we have

$$\frac{d}{d|x|} f(x) = \frac{d}{d|x|} f\left(\frac{x}{|x|}|x|\right) = \frac{x}{|x|}\frac{d}{dx} f(x) = \frac{x \cdot \nabla_E}{|x|} f(x), \tag{1.35}$$

since for $x \in \mathbb{G}$, we have that $\frac{x}{|x|}$ does not depend on $|x|$, and where

$$\nabla_E = \left(\frac{\partial}{\partial x_1}, \ldots, \frac{\partial}{\partial x_n}\right)$$

is an anisotropic (Euclidean) gradient on $\mathbb{G}$ consisting of partial derivatives with respect to coordinate functions.

Although x_j and $\frac{\partial}{\partial x_j}$ may have degrees of homogeneity depending on j, the operator

$$\mathcal{R} = \frac{x \cdot \nabla_E}{|x|} = \frac{d}{d|x|} \tag{1.36}$$

is homogeneous of degree -1.

1.3.2 Euler operator

Given the radial derivative operator $\mathcal{R}$, we define the *Euler operator* on $\mathbb{G}$ by

$$\mathbb{E} := |x|\mathcal{R}. \tag{1.37}$$

Since $\mathcal{R}$ is homogeneous of degree -1, the operator $\mathbb{E}$ is homogeneous of degree 0.

We can note the following useful property shedding some more light on the link between the radial derivative and the Euler operators, also clarifying how to take derivatives with respect to points that are not on the quasi-sphere $\wp$. Thus, for $x \in \mathbb{G}$, we can write $x = ry$ with $y \in \wp$. Then, denoting $\rho := e^t r$ for $t \in \mathbb{R}$, we have

$$\frac{d}{dt}(f(e^t x)) = \frac{d}{dt}(f(e^t ry)) = \frac{d}{dt}(f(\rho y)) = \rho\frac{d}{d\rho}(f(\rho y)) = \mathbb{E}f(\rho y) = \mathbb{E}f(e^t x),$$

that is,

$$\frac{d}{dt}f(e^t x) = \mathbb{E}f(e^t x). \tag{1.38}$$

The Euler operator has the following useful properties, also justifying the name of Euler associated to this operator.

Proposition 1.3.1 (Properties of the Euler operator). *We have the following properties:*

(i) *Let $\nu \in \mathbb{R}$. If $f : \mathbb{G}\backslash\{0\} \to \mathbb{R}$ is differentiable, then*

$$\mathbb{E}(f) = \nu f \quad \textit{if and only if} \quad f(rx) = r^\nu f(x) \quad (\forall r > 0, x \neq 0).$$

(ii) *The formal adjoint operator of $\mathbb{E}$ has the form*

$$\mathbb{E}^* = -Q\mathbb{I} - \mathbb{E}, \tag{1.39}$$

where $\mathbb{I}$ is the identity operator.

(iii) *For all complex-valued functions $f \in C_0^\infty(\mathbb{G}\backslash\{0\})$ we have*

$$\|\mathbb{E}f\|_{L^2(\mathbb{G})} = \|\mathbb{E}^* f\|_{L^2(\mathbb{G})}\,. \tag{1.40}$$

Proof. Part (i). If a function f is positively homogeneous of order ν, that is, if

$$f(rx) = r^\nu f(x)$$

holds for all $r > 0$ and $x := \rho y \neq 0$, $y \in \wp$, then using (1.30) for such an f, it follows that

$$\mathbb{E}f = \nu f(x).$$

Conversely, let us fix $x \neq 0$ and define

$$g(r) := f(rx).$$

Using (1.30), the equality $\mathbb{E}f(rx) = \nu f(rx)$ means that

$$g'(r) = \frac{d}{dr}f(rx) = \frac{1}{r}\mathbb{E}f(rx) = \frac{\nu}{r}f(rx) = \frac{\nu}{r}g(r).$$

Consequently, $g(r) = g(1)r^\nu$, i.e., $f(rx) = r^\nu f(x)$ and thus f is positively homogeneous of order ν.

Part (ii). We can calculate the formal adjoint operator of $\mathbb{E}$ on $C_0^\infty(\mathbb{G}\backslash\{0\})$ as follows:

$$\begin{aligned}\int_{\mathbb{G}} \mathbb{E}f(x)\overline{g(x)}dx &= \int_0^\infty \int_{\wp} \frac{d}{dr} f(ry)\overline{g(ry)}r^Q d\sigma(y)dr \\ &= -\int_0^\infty \int_{\wp} f(ry)\left(Qr^{Q-1}\overline{g(ry)} + r^Q \frac{d}{dr}\overline{g(ry)}\right) d\sigma(y)dr \\ &= -\int_{\mathbb{G}} f(x)(Q+\mathbb{E})\overline{g(x)}dx,\end{aligned}$$

by the polar decomposition in Proposition 1.2.10 and the integration by parts using formula (1.30).

Part (iii). By using the representation of $\mathbb{E}^*$ in (1.39), we get

$$\begin{aligned}\|\mathbb{E}^* f\|^2_{L^2(\mathbb{G})} &= \|(-Q\mathbb{I} - \mathbb{E})f\|^2_{L^2(\mathbb{G})} \\ &= Q^2 \|f\|^2_{L^2(\mathbb{G})} + 2Q\mathrm{Re}\int_{\mathbb{G}} f(x)\overline{\mathbb{E}f(x)}dx + \|\mathbb{E}f\|^2_{L^2(\mathbb{G})}.\end{aligned} \tag{1.41}$$

Then we have

$$\begin{aligned}2Q\mathrm{Re}\int_{\mathbb{G}} f(x)\overline{\mathbb{E}f(x)}dx &= 2Q\mathrm{Re}\int_0^\infty \int_{\wp} f(ry)\frac{d}{dr}\overline{f(ry)}r^Q d\sigma(y)dr \\ &= Q\int_0^\infty r^Q \int_{\wp} \frac{d}{dr}(|f(ry)|^2)d\sigma(y)dr \\ &= -Q^2\int_0^\infty \int_{\wp} |f(ry)|^2 r^{Q-1}d\sigma(y)dr \\ &= -Q^2 \|f\|^2_{L^2(\mathbb{G})}.\end{aligned} \tag{1.42}$$

Combining this with (1.41) we obtain (1.40). □

Let us introduce the following operator that will be of importance in the sequel,

$$\mathbb{A} := \mathbb{E}\mathbb{E}^*.$$

It is easy to see that this operator is formally self-adjoint, that is,

$$\mathbb{A} = \mathbb{E}\mathbb{E}^* = \mathbb{E}^*\mathbb{E} = \mathbb{A}^*,$$

where we can use Proposition 1.3.1, Part (ii), to also write

$$\mathbb{E}\mathbb{E}^* = \mathbb{E}^*\mathbb{E} = -Q\mathbb{E} - \mathbb{E}^2.$$

Then by replacing f by $\mathbb{E}f$ in (1.40), we obtain the equality

$$\|\mathbb{A}f\|_{L^2(\mathbb{G})} = \left\|\mathbb{E}^2 f\right\|_{L^2(\mathbb{G})} \tag{1.43}$$

for all complex-valued functions $f \in C_0^\infty(\mathbb{G}\backslash\{0\})$. Moreover, the operator $\mathbb{A}$ is *Komatsu-non-negative* in $L^2(\mathbb{G})$, which means that $(-\infty, 0)$ is included in the resolvent set $\rho(\mathbb{A})$ of $\mathbb{A}$ and we have the property

$$\exists M > 0,\ \forall \lambda > 0, \quad \|(\lambda + \mathbb{A})^{-1}\|_{L^2(\mathbb{G})\to L^2(\mathbb{G})} \leq M\lambda^{-1}.$$

Indeed, and more precisely, we have the following:

Lemma 1.3.2 ($\mathbb{A} = \mathbb{E}\mathbb{E}^*$ is Komatsu-non-negative). *The operator* $\mathbb{A} = \mathbb{E}\mathbb{E}^*$ *is Komatsu-non-negative in* $L^2(\mathbb{G})$*:*

$$\|(\lambda + \mathbb{A})^{-1}\|_{L^2(\mathbb{G})\to L^2(\mathbb{G})} \leq \lambda^{-1} \quad \textit{for all } \lambda > 0. \tag{1.44}$$

Proof. We start with $f \in C_0^\infty(\mathbb{G}\backslash\{0\})$. Using Proposition 1.3.1, Part (ii), a direct calculation shows that we have the equality

$$\begin{aligned} \|(\lambda\mathbb{I} + \mathbb{A})f\|^2_{L^2(\mathbb{G})} &= \|(\lambda\mathbb{I} - \mathbb{E}(Q\mathbb{I} + \mathbb{E}))f\|^2_{L^2(\mathbb{G})} \\ &= \lambda^2 \|f\|^2_{L^2(\mathbb{G})} + \|\mathbb{E}(Q\mathbb{I} + \mathbb{E})f\|^2_{L^2(\mathbb{G})} - 2\lambda \mathrm{Re} \int_{\mathbb{G}} f(x)\overline{Q\mathbb{E}f + \mathbb{E}^2 f}dx. \end{aligned} \tag{1.45}$$

Since

$$\begin{aligned} \mathrm{Re} \int_{\mathbb{G}} f(x)\overline{\mathbb{E}^2 f}dx &= \mathrm{Re} \int_0^\infty \int_{\wp} f(ry)\overline{\frac{d}{dr}(\mathbb{E}f(ry))}r^Q d\sigma(y)dr \\ &= -\mathrm{Re} \int_0^\infty \int_{\wp} \overline{(\mathbb{E}f(ry))}\left(r^Q \frac{d}{dr}f(ry) + Qr^{Q-1}f(ry)\right) d\sigma(y)dr \\ &= -\|\mathbb{E}f\|^2_{L^2(\mathbb{G})} - Q\mathrm{Re}\int_{\mathbb{G}} \overline{\mathbb{E}f(x)}f(x)dx, \end{aligned}$$

we have

$$\begin{aligned} &-2\lambda\mathrm{Re}\int_{\mathbb{G}} f(x)\overline{Q\mathbb{E}f(x) + \mathbb{E}^2 f(x)}dx \\ &= -2\lambda Q\mathrm{Re}\int_{\mathbb{G}} f(x)\overline{\mathbb{E}f(x)}dx - 2\lambda\mathrm{Re}\int_{\mathbb{G}} f(x)\overline{\mathbb{E}^2 f(x)}dx = 2\lambda \|\mathbb{E}f\|^2_{L^2(\mathbb{G})}. \end{aligned} \tag{1.46}$$

Combining (1.45) with (1.46), we obtain the equality

$$\|(\lambda\mathbb{I} - \mathbb{E}(Q\mathbb{I} + \mathbb{E}))f\|^2_{L^2(\mathbb{G})} = \lambda^2 \|f\|^2_{L^2(\mathbb{G})} + 2\lambda \|\mathbb{E}f\|^2_{L^2(\mathbb{G})} + \|\mathbb{E}(Q\mathbb{I} + \mathbb{E})f\|^2_{L^2(\mathbb{G})}.$$

By dropping positive terms, it follows that

$$\|(\lambda\mathbb{I} - \mathbb{E}(Q\mathbb{I} + \mathbb{E}))f\|^2_{L^2(\mathbb{G})} \geq \lambda^2 \|f\|^2_{L^2(\mathbb{G})},$$

which implies (1.44). □

We can refer to [FR16, Section A.3] for more details on general further properties of Komatsu-non-negative operators and their use in the theory of fractional powers of operators.

1.3.3 From radial to non-radial inequalities

Now we show that the Euler operator and, consequently, also the radial derivative operator, have a very useful property that in order to prove certain inequalities on $\mathbb{G}$ is may be enough to prove them only for radially symmetric functions. We summarize it in the following proposition. Such ideas will be of use, for example, in the analysis of remainder estimates in Theorem 2.3.1 or in Theorem 3.2.6.

As usual, we will say that a function $f = f(x)$ on $\mathbb{G}$ is *radial*, or *radially symmetric*, if it depends only on $|x|$; clearly, this notion depends on the seminorm that we are using.

Proposition 1.3.3 (Radialisation of functions). *Let $\phi_1, \phi_2, \phi_3 \in L^1_{\mathrm{loc}}(\mathbb{G})$ be arbitrary radially symmetric functions. For $f \in L^p_{\mathrm{loc}}(\mathbb{G})$, define its radial average by*

$$\widetilde{f}(|x|) := \left(\frac{1}{|\wp|}\int_{\wp} |f(|x|y)|^p d\sigma(y)\right)^{1/p}. \tag{1.47}$$

Then for any $f \in L^p_{\mathrm{loc}}(\mathbb{G})$ and $1 < p < \infty$ we have the equality

$$\int_{\mathbb{G}} \phi_1(x)\left|\widetilde{f}(|x|)\right|^p dx = \int_{\mathbb{G}} \phi_1(x)\,|f(x)|^p\,dx. \tag{1.48}$$

Moreover, if $\phi_2, \phi_3 \geq 0$, we have the inequalities

$$\int_{\mathbb{G}} \phi_2(x)\left|\mathbb{E}^k \widetilde{f}(|x|)\right|^p dx \leq \int_{\mathbb{G}} \phi_2(x)\left|\mathbb{E}^k f(x)\right|^p dx \tag{1.49}$$

and

$$\int_{\mathbb{G}} \phi_3(x)\left|\mathcal{R}^k \widetilde{f}(|x|)\right|^p dx \leq \int_{\mathbb{G}} \phi_3(x)\left|\mathcal{R}^k f(x)\right|^p dx, \tag{1.50}$$

for all $1 < p < \infty$, any $k \in \mathbb{N}$ and all $f \in L^p_{\mathrm{loc}}(\mathbb{G})$ such that $\mathbb{E}^k f \in L^p_{\mathrm{loc}}(\mathbb{G})$ or $\mathcal{R}^k f \in L^p_{\mathrm{loc}}(\mathbb{G})$, respectively. The constants in these inequalities are sharp, and are attained when $f = \widetilde{f}$.

Proof. Using definition (1.47) and the polar decomposition formula in Proposition 1.2.10, we have

$$\begin{aligned}\int_{\mathbb{G}} |\widetilde{f}(|x|)|^p \phi_1(x) dx &= |\wp| \int_0^{\infty} |\widetilde{f}(r)|^p \phi_1(r) r^{Q-1} dr \\ &= |\wp| \int_0^{\infty} \frac{1}{|\wp|}\int_{\wp} |f(ry)|^p d\sigma(y) \phi_1(r) r^{Q-1} dr = \int_{\mathbb{G}} |f(x)|^p \phi_1(x) dx,\end{aligned} \tag{1.51}$$

which proves the identity (1.48).

To prove (1.49) let us show first that

$$|\mathbb{E}^k \widetilde{f}| \leq \left(\frac{1}{|\wp|}\int_{\wp} \left|\mathbb{E}^k f(ry)\right|^p d\sigma(y)\right)^{\frac{1}{p}},\ r = |x|, \tag{1.52}$$

holds for any $k \in \mathbb{N}$. We use the induction. For $k = 1$, by the Hölder inequality we obtain

$$
\begin{aligned}
|\mathbb{E}\widetilde{f}| &= r\left(\frac{1}{|\wp|}\int_{\wp}|f(ry)|^p d\sigma(y)\right)^{\frac{1}{p}-1}\frac{1}{|\wp|}\left|\int_{\wp}|f(ry)|^{p-2}f(ry)\overline{\frac{d}{dr}f(ry)}d\sigma(y)\right| \\
&\leq \left(\frac{1}{|\wp|}\int_{\wp}|f(ry)|^p d\sigma(y)\right)^{\frac{1-p}{p}}\frac{1}{|\wp|}\int_{\wp}|f(ry)|^{p-1}\left|r\frac{d}{dr}f(ry)\right|d\sigma(y) \\
\leq \left(\frac{1}{|\wp|}\int_{\wp}|f(ry)|^p d\sigma(y)\right)^{\frac{1-p}{p}} &\frac{1}{|\wp|}\left(\int_{\wp}\left|r\frac{d}{dr}f(ry)\right|^p d\sigma(y)\right)^{\frac{1}{p}}\left(\int_{\wp}|f(ry)|^p d\sigma(y)\right)^{\frac{p-1}{p}} \\
&= \left(\frac{1}{|\wp|}\int_{\wp}|\mathbb{E}f(ry)|^p\, d\sigma(y)\right)^{\frac{1}{p}}.
\end{aligned}
$$

For the induction step, we assume that for some $\ell \in \mathbb{N}$ we have

$$
|\mathbb{E}^{\ell}\widetilde{f}| \leq \left(\frac{1}{|\wp|}\int_{\wp}\left|\mathbb{E}^{\ell}f(ry)\right|^p d\sigma(y)\right)^{1/p}, \tag{1.53}
$$

and we want to prove that it then follows that

$$
|\mathbb{E}^{\ell+1}\widetilde{f}| \leq \left(\frac{1}{|\wp|}\int_{\wp}\left|\mathbb{E}^{\ell+1}f(ry)\right|^p d\sigma(y)\right)^{1/p}.
$$

So, using (1.53), similarly to the case $\ell = 1$ above, we calculate

$$
\begin{aligned}
|\mathbb{E}^{\ell+1}\widetilde{f}(r)| &\leq \left|\mathbb{E}\left(\left(\frac{1}{|\wp|}\int_{\wp}\left|\mathbb{E}^{\ell}f(ry)\right|^p d\sigma(y)\right)^{\frac{1}{p}}\right)\right| \\
&= r\left(\frac{1}{|\wp|}\int_{\wp}\left|\mathbb{E}^{\ell}f(ry)\right|^p d\sigma(y)\right)^{\frac{1}{p}-1}\left|\frac{1}{|\wp|}\int_{\wp}\left|\mathbb{E}^{\ell}f(ry)\right|^{p-2}\mathbb{E}f(ry)\overline{\frac{d}{dr}(\mathbb{E}^{\ell}f(ry))d\sigma(y)}\right| \\
&\leq \left(\frac{1}{|\wp|}\int_{\wp}\left|\mathbb{E}^{\ell}f(ry)\right|^p d\sigma(y)\right)^{\frac{1}{p}-1}\left(\frac{1}{|\wp|}\int_{\wp}\left|\mathbb{E}^{\ell}f(ry)\right|^{p-1}\left|\mathbb{E}^{\ell+1}f(ry)\right|d\sigma(y)\right) \\
&\leq \left(\frac{1}{|\wp|}\int_{\wp}|\mathbb{E}^{\ell}f(ry)|^p d\sigma(y)\right)^{\frac{1}{p}-1}\frac{1}{|\wp|}\left(\int_{\wp}\left|\mathbb{E}^{\ell+1}f(ry)\right|^p d\sigma(y)\right)^{\frac{1}{p}} \\
&\qquad\times\left(\int_{\wp}|\mathbb{E}^{\ell}f(ry)|^p d\sigma(y)\right)^{\frac{p-1}{p}} \\
&= \left(\frac{1}{|\wp|}\int_{\wp}\left|\mathbb{E}^{\ell+1}f(ry)\right|^p d\sigma(y)\right)^{\frac{1}{p}}.
\end{aligned}
$$

Here in the last line we have used Hölder's inequality. It proves (1.52).

Now (1.52) yields

$$\begin{aligned}\int_{\mathbb{G}} \left|\mathbb{E}^k \widetilde{f}(x)\right|^p \phi_2(x)dx &= |\wp| \int_0^\infty \left|\mathbb{E}^k \widetilde{f}(r)\right|^p \phi_2(r) r^{Q-1} dr \\ &\leq |\wp| \int_0^\infty \frac{1}{|\wp|} \int_{\wp} \left|\mathbb{E}^k f(ry)\right|^p \phi_2(r) r^{Q-1} d\sigma(y) dr \\ &= \int_{\mathbb{G}} \left|\mathbb{E}^k f(x)\right|^p \phi_2(x) dx.\end{aligned}$$

This completes the proof of (1.49). The proof of (1.50) is similar. □

1.3.4 Euler semigroup $e^{-t\mathbb{E}^*\mathbb{E}}$

Here we will describe the operator semigroup $\{e^{-t\mathbb{E}^*\mathbb{E}}\}_{t>0}$ associated with the Euler operator on homogeneous groups.

Theorem 1.3.4 (Euler semigroup). *Let $\mathbb{G}$ be a homogeneous group of homogeneous dimension Q. Let $x \in \mathbb{G}$, $x \neq 0$, and let $y := \frac{x}{|x|}$, and $t > 0$. Then the semigroup $e^{-t\mathbb{E}^*\mathbb{E}}$ is given by*

$$(e^{-t\mathbb{E}^*\mathbb{E}} f)(x) = \frac{e^{-tQ^2/4}}{\sqrt{4\pi t}} |x|^{-Q/2} \int_0^\infty e^{-\frac{(\ln|x| - \ln s)^2}{4t}} s^{-Q/2} f(sy) s^{Q-1} ds. \tag{1.54}$$

Before we prove formula (1.54), let us introduce some notation that will be useful in the sequel. Thus, let us define the map $F : L^2(\mathbb{G}) \to L^2(\mathbb{R} \times \wp)$ by the formula

$$(Ff)(s, y) := e^{sQ/2} f(e^s y), \tag{1.55}$$

for $y \in \wp$ and $s \in \mathbb{R}$. Its inverse map $F^{-1} : L^2(\mathbb{R} \times \wp) \to L^2(\mathbb{G})$ can be given by the formula

$$(F^{-1} g)(x) := r^{-Q/2} g(\ln r, y), \tag{1.56}$$

and one can readily check that F preserves the L^2 norm. The map F can be also described as

$$(Ff)(s, y) = (U(s)f)(y)$$

for all $y \in \wp$ and $s \in \mathbb{R}$, with the dilation mapping $U(t)$ defined by

$$U(t)f(x) := e^{tQ/2} f(e^t x). \tag{1.57}$$

We then immediately have

$$(F(U(t)f))(s, y) = (U(s)(U(t)f))(y) = (U(s + t)f)(y) = (Ff)(s + t, y). \tag{1.58}$$

The dilations $U(t)$ can be linked to the Euler operator through the relation

$$\frac{d}{dt} f(e^t x) = \mathbb{E} f(e^t x), \quad x \in \mathbb{G},$$

see (1.38). It then follows that these dilations can be also seen as a group of unitary operators $U(t) = e^{iAt}$ with the generator

$$Af = \frac{1}{i}\frac{d}{dt}U(t)f|_{t=0} = \frac{1}{i}\left(\mathbb{E} + \frac{Q}{2}\right)f = -i\mathbb{E}f - i\frac{Q}{2}f. \tag{1.59}$$

Since $\mathbb{E}^* = -Q\mathbb{I} - \mathbb{E}$ by Proposition 1.3.1, Part (ii), the formula (1.59) implies that

$$A = A^* = -i\mathbb{E} - i\frac{Q}{2}, \tag{1.60}$$

which yields the relation

$$\mathbb{A} = \mathbb{E}^*\mathbb{E} = \left(-iA - \frac{Q}{2}\right)\left(iA - \frac{Q}{2}\right) = A^2 + \frac{Q^2}{4}. \tag{1.61}$$

The family $U(t)$ is mapped to the multiplication by exponents $e^{it\tau}$ through the Mellin transformation $M : L^2(\mathbb{G}) \to L^2(\mathbb{R} \times \wp)$ defined by the formula $M = \mathcal{F} \circ F$, where $\mathcal{F}$ is the Fourier transform on $\mathbb{R}$, that is,

$$(Mf)(\tau, y) := \frac{1}{\sqrt{2\pi}}\int_{\mathbb{R}} e^{-is\tau}(Ff)(s, y)ds. \tag{1.62}$$

Indeed, using (1.58) and changing variables, we have

$$\begin{aligned} (MU(t)f)(\tau, y) &= \frac{1}{\sqrt{2\pi}}\int_{\mathbb{R}} e^{-is\tau}(Ff)(s+t, y)ds \\ &= \frac{e^{it\tau}}{\sqrt{2\pi}}\int_{\mathbb{R}} e^{-is\tau}(Ff)(s, y)ds = e^{it\tau}(Mf)(\tau, y). \end{aligned} \tag{1.63}$$

Before finally proving Theorem 1.3.4, let us point out that it implies the following representation of the semigroup e^{-tA^2}.

Corollary 1.3.5 (Semigroup e^{-tA^2}). *Let F and F^{-1} be mappings as in* (1.55) *and* (1.56), *respectively. Then we have*

$$Fe^{-tA^2}F^{-1}f(r, y) = \frac{1}{\sqrt{4\pi t}}\int_{\mathbb{R}} \exp\left(-\frac{(r-s)^2}{4t}\right)f(sy)ds. \tag{1.64}$$

Proof of Corollary 1.3.5. Setting $e^{-tA^2} = e^{tQ^2/4}e^{-t\mathbb{E}^*\mathbb{E}}$ as well as combining (1.55) and (1.56), and using (1.54) we get

$$\begin{aligned} &Fe^{-tA^2}F^{-1}f(r, y) \\ &= F\left(e^{tQ^2/4}\frac{e^{-tQ^2/4}r^{-Q/2}}{\sqrt{4\pi t}}\int_0^\infty e^{-\frac{(\ln r - \ln s)^2}{4t}}s^{Q/2-1}(s^{-Q/2}f(\ln s, y))ds\right) \\ &= e^{rQ/2}\frac{e^{-rQ/2}}{\sqrt{4\pi t}}\int_0^\infty e^{-\frac{(r-\ln s)^2}{4t}}\frac{f(\ln s, y)}{s}ds = \frac{1}{\sqrt{4\pi t}}\int_{-\infty}^\infty e^{-\frac{(r-s_1)^2}{4t}}f(s_1 y)ds_1, \end{aligned}$$

which is (1.64), where we have used the new variable $s = e^{s_1}$ in the last line. □

Let us finally prove Theorem 1.3.4.

Proof of Theorem 1.3.4. Noting that from the definition we have $iAe^{itA} = \partial_t U(t)$, we can calculate

$$(MiAe^{iAt}f)(\tau, y) = (M\partial_t U(t)f)(\tau, y) = \partial_t(MU(t)f)(\tau, y).$$

Now using (1.63) we get from above that

$$(MiAe^{iAt}f)(\tau, y) = \partial_t e^{it\tau}(Mf)(\tau, y) = i\tau e^{it\tau}(Mf)(\tau, y),$$

which (after setting $t = 0$) implies

$$(MAf)(\tau, y) = \tau(Mf)(\tau, y) \tag{1.65}$$

for f in the domain $\mathcal{D}(A)$. It follows that the condition $f \in \mathcal{D}(A)$ can be described by the property that the function $(\tau, y) \mapsto \tau(Mf)(\tau, y) \in L^2(\mathbb{R} \times \wp)$.

So, first we prove that

$$(Me^{-tA^2}f)(\tau, y) = e^{-t\tau^2}(Mf)(\tau, y). \tag{1.66}$$

We have

$$(Me^{-tA^2}f)(\tau, y) = \sum_{k=0}^{\infty} \frac{(-t)^k}{k!}(MA^{2k}f)(\tau, y). \tag{1.67}$$

Moreover, by iterating (1.65), it follows that

$$(MA^{2k}f)(\tau, y) = \tau^{2k}(Mf)(\tau, y), \quad k = 0, 1, 2, \ldots.$$

Combining this with (1.67), we get

$$(Me^{-tA^2}f)(\tau, y) = \sum_{k=0}^{\infty} \frac{(-t)^k}{k!}\tau^{2k}(Mf)(\tau, y) = e^{-t\tau^2}(Mf)(\tau, y).$$

That is, we have showed that (1.66) holds. Thus, it follows that

$$e^{-tA^2} = M^{-1}e^{-t\tau^2}M.$$

Here by using that $M = \mathcal{F} \circ F$, we have

$$e^{-tA^2} = F^{-1} \circ \mathcal{F}^{-1}(e^{-t\tau^2}\mathcal{F} \circ F). \tag{1.68}$$

Furthermore, we calculate

$$\begin{aligned}
\mathcal{F}^{-1}(e^{-t\tau^2}Mf)(\lambda, y) &= \mathcal{F}^{-1}(e^{-t\tau^2}\mathcal{F} \circ F)(\lambda, y) \\
&= \frac{1}{2\pi}\int_{\mathbb{R}}\int_{\mathbb{R}} e^{i\lambda\tau}e^{-t\tau^2}e^{-is\tau}(Ff)(s, y)\,ds\,d\tau \\
&= \frac{1}{2\pi}\int_{\mathbb{R}}\left(\int_{\mathbb{R}} e^{-t\tau^2+i(\lambda-s)\tau}d\tau\right)(Ff)(s, y)\,ds \\
&= \frac{1}{\sqrt{4\pi t}}\int_{\mathbb{R}} e^{-\frac{(\lambda-s)^2}{4t}}(Ff)(s, y)\,ds =: \varphi_t(\lambda, y),
\end{aligned}$$

since

$$\int_{\mathbb{R}} e^{-t\tau^2+i(\lambda-s)\tau}d\tau = \sqrt{\frac{\pi}{t}}e^{-\frac{(\lambda-s)^2}{4t}}.$$

From this and (1.68), using (1.55), (1.56) and $M = \mathcal{F} \circ F$ with $x = |x|y$, we compute

$$\begin{aligned}(e^{-tA^2}f)(|x|y) &= (F^{-1}\varphi_t)(|x|y) = r^{-Q/2}\varphi_t(\ln|x|, y) \\ &= \frac{1}{\sqrt{4\pi t}}|x|^{-Q/2}\int_{\mathbb{R}} e^{-\frac{(\ln|x|-s)^2}{4t}}(Ff)(s,y)ds \\ &= \frac{1}{\sqrt{4\pi t}}|x|^{-Q/2}\int_0^{\infty} e^{-\frac{(\ln|x|-\ln z)^2}{4t}}z^{\frac{Q}{2}-1}f(zy)dz,\end{aligned}$$

where we have used the change of variables $z = e^s$ in the last line.

Since we have $e^{-t\mathbb{E}^*\mathbb{E}} = e^{-tQ^2/4}e^{-tA^2}$ by (1.61), we arrive at

$$\begin{aligned}(e^{-t\mathbb{E}^*\mathbb{E}}f)(|x|y) &= e^{-tQ^2/4}(e^{-tA^2}f)(|x|y) \\ &= \frac{1}{\sqrt{4\pi t}}|x|^{-Q/2}e^{-tQ^2/4}\int_0^{\infty} e^{-\frac{(\ln|x|-\ln z)^2}{4t}}z^{\frac{Q}{2}-1}f(zy)dz \\ &= \frac{1}{\sqrt{4\pi t}}|x|^{-Q/2}e^{-tQ^2/4}\int_0^{\infty} e^{-\frac{(\ln|x|-\ln z)^2}{4t}}z^{-\frac{Q}{2}}f(zy)z^{Q-1}dz,\end{aligned}$$

completing the proof of (1.54). □

Remark 1.3.6.

1. The representation of the Euler semigroup in Theorem 1.3.4 becomes instrumental in deriving several forms of the Hardy–Sobolev and Gagliardo–Nirenberg type inequalities for the Euler operator, as we will show in the sequel, see, e.g., Section 10.4.

2. In the Euclidean case $\mathbb{R}^n$, the results of this section have been obtained in [BEHL08]. For general homogeneous groups, our presentation followed the results obtained in [RSY18a].

1.4 Stratified groups

An important special case of homogeneous groups is that of stratified groups introduced in Definition 1.1.5. Because this is an important class that will be analysed in Chapter 6 from the point of view of Hardy and other inequalities, here we will provide more details on it and fix the corresponding notation.

1.4.1 Stratified Lie groups

We recall the definition of stratified groups.

Definition 1.4.1 (Stratified groups). A Lie group $\mathbb{G} = (\mathbb{R}^n, \circ)$ is called a *stratified group* (or a *homogeneous Carnot group*) if it satisfies the following conditions:

(a) For some natural numbers $N + N_2 + \cdots + N_r = n$, that is $N = N_1$, the decomposition $\mathbb{R}^n = \mathbb{R}^N \times \cdots \times \mathbb{R}^{N_r}$ is valid, and for every $\lambda > 0$ the dilation $\delta_\lambda : \mathbb{R}^n \to \mathbb{R}^n$ given by

$$\delta_\lambda(x) \equiv \delta_\lambda(x', x^{(2)}, \ldots, x^{(r)}) := (\lambda x', \lambda^2 x^{(2)}, \ldots, \lambda^r x^{(r)})$$

is an automorphism of the group $\mathbb{G}$. Here $x' \equiv x^{(1)} \in \mathbb{R}^N$ and $x^{(k)} \in \mathbb{R}^{N_k}$ for $k = 2, \ldots, r$.

(b) Let N be as in (a) and let $X_1, \ldots, X_N$ be the left invariant vector fields on $\mathbb{G}$ such that $X_k(0) = \frac{\partial}{\partial x_k}|_0$ for $k = 1, \ldots, N$. Then

$$\operatorname{rank}(\operatorname{Lie}\{X_1, \ldots, X_N\}) = n,$$

for every $x \in \mathbb{R}^n$, i.e., the iterated commutators of $X_1, \ldots, X_N$ span the Lie algebra of $\mathbb{G}$.

The number r is called the *step of* $\mathbb{G}$ and the left invariant vector fields $X_1, \ldots, X_N$ are called the (Jacobian) generators of $\mathbb{G}$. The homogeneous dimension of a stratified Lie group $\mathbb{G}$ is given by

$$Q = \sum_{k=1}^{r} k N_k, \quad N_1 = N.$$

The second-order differential operator

$$\mathcal{L} = \sum_{k=1}^{N} X_k^2 \tag{1.69}$$

is called the (*canonical*) *sub-Laplacian* on $\mathbb{G}$. The sub-Laplacian $\mathcal{L}$ is a left invariant homogeneous hypoelliptic differential operator and it is elliptic if and only if the step of $\mathbb{G}$ is equal to 1.

The *hypoellipticity* of $\mathcal{L}$ means that for a distribution $f \in \mathcal{D}'(\Omega)$ in any open set Ω, if $\mathcal{L}f \in C^\infty(\Omega)$ then $f \in C^\infty(\Omega)$. It is a special case of Hörmander's sum of squares theorem [Hör67].

The left invariant vector field X_k has an explicit form given in Proposition 1.2.19, namely,

$$X_k = \frac{\partial}{\partial x'_k} + \sum_{l=2}^{r} \sum_{m=1}^{N_l} a_{k,m}^{(l)}(x', \ldots, x^{(l-1)}) \frac{\partial}{\partial x_m^{(l)}}, \tag{1.70}$$

where $a_{k,m}^{(l)}$ is a homogeneous (with respect to δ_λ) polynomial function of degree $l-1$. We will also use the following notation for the *horizontal gradient*

$$\nabla_H := (X_1, \ldots, X_N),$$

for the *horizontal divergence*

$$\operatorname{div}_H v := \nabla_H \cdot v,$$

and for the *horizontal p-Laplacian (or p-sub-Laplacian)*

$$\mathcal{L}_p f := \operatorname{div}_H(|\nabla_H f|^{p-2} \nabla_H f), \quad 1 < p < \infty. \tag{1.71}$$

Denoting the Euclidean distance by

$$|x'| = \sqrt{x_1'^2 + \cdots + x_N'^2}$$

for the Euclidean norm on $\mathbb{R}^N$, the representation (1.70) for derivatives leads to the identities

$$|\nabla_H |x'|^\gamma| = \gamma |x'|^{\gamma-1}, \tag{1.72}$$

and

$$\operatorname{div}_H \left(\frac{x'}{|x'|^\gamma} \right) = \frac{\sum_{j=1}^N |x'|^\gamma X_j x'_j - \sum_{j=1}^N x'_j \gamma |x'|^{\gamma-1} X_j |x'|}{|x'|^{2\gamma}} = \frac{N-\gamma}{|x'|^\gamma} \tag{1.73}$$

for all $\gamma \in \mathbb{R}$, $|x'| \neq 0$.

It was shown by Folland [Fol75] that the sub-Laplacian $\mathcal{L}$ in (1.69) on a general stratified group $\mathbb{G}$ has a unique *fundamental solution* ε, that is,

$$\mathcal{L}\varepsilon = \delta, \tag{1.74}$$

where δ is the delta-distribution at the unit element of $\mathbb{G}$. Moreover, the function ε is homogeneous of degree $2-Q$.

The function

$$d(x) := \begin{cases} \varepsilon(x)^{\frac{1}{2-Q}}, & \text{for } x \neq 0, \\ 0, & \text{for } x = 0, \end{cases} \tag{1.75}$$

is called the *$\mathcal{L}$-gauge* on $\mathbb{G}$. It is a *homogeneous quasi-norm* on $\mathbb{G}$, that is, it is a continuous function $d : \mathbb{G} \to [0, \infty)$, smooth away from the origin, which satisfies the conditions

$$d(\lambda x) = \lambda d(x), \; d(x^{-1}) = d(x) \text{ and } d(x) = 0 \text{ if only if } x = 0. \tag{1.76}$$

We refer to the original paper [Fol75] by Folland as well as to a recent presentation in [FR16, Section 3.2.7] for further details and properties of these fundamental solutions.

For future use, we record the action of $\mathcal{L}$ on d and its powers. Since $\mathcal{L}d^{2-Q} = 0$ in $\mathbb{G}\backslash\{0\}$, a straightforward calculation shows that for $Q \geq 3$ we have

$$\mathcal{L}d = (Q-1)\frac{|\nabla_H d|^2}{d} \text{ in } \mathbb{G}\backslash\{0\}, \tag{1.77}$$

as well as, consequently, for all $\alpha \in \mathbb{R}$,

$$\mathcal{L}d^{\alpha} = \alpha(\alpha + Q - 2)d^{\alpha-2}|\nabla_H d|^2 \text{ in } \mathbb{G}\backslash\{0\}. \tag{1.78}$$

1.4.2 Extended sub-Laplacians

In general, most of the results described in this book in the setting of stratified groups can be extended to any second-order hypoelliptic differential operators which are "equivalent" to the sub-Laplacian $\mathcal{L}$. Let us very briefly discuss this matter in the sprit of [BLU07].

Let $A = (a_{k,j})_{1\leq k,j\leq N_1}$ be a positive-definite symmetric matrix. Consider the following second-order hypoelliptic differential operator based on the matrix A and the vector fields $\{X_1, \ldots, X_{N_1}\}$ from the first stratum, given by

$$\mathcal{L}_A = \sum_{k,j=1}^{N_1} a_{k,j} X_k X_j. \tag{1.79}$$

For instance, in the Euclidean case, that is, for $\mathbb{G} = (\mathbb{R}^N, +)$ and $N_1 = N$, the constant coefficients second-order elliptic operator

$$\Delta_A = \sum_{k,j=1}^{N} a_{k,j} \frac{\partial^2}{\partial x_k \partial x_j}$$

is transformed into the Laplacian

$$\Delta = \sum_{k=1}^{N} \frac{\partial^2}{\partial x_k^2}$$

under a linear change of coordinates in $\mathbb{R}^N$. Thus, the operator Δ_A is "equivalent" to the operator Δ by a linear change of the coordinate system.

In general, to apply the above argument to transform $\mathcal{L}_A$ to the sub-Laplacian $\mathcal{L}$ it is not enough to change the basis by a linear transformation. However, it is enough in the setting of free stratified groups. We say that a stratified group $\mathbb{G}$ is a *free stratified group* if its Lie algebra is (isomorphic to) a free Lie algebra. For instance, the Heisenberg group $\mathbb{H}^1$ is a free stratified group. In this case we have the following result.

Theorem 1.4.2 ([BLU07]). *Let $\mathbb{G}$ be a free stratified group and let A be a given positive-definite symmetric matrix. Let $X = \{X_1, \ldots, X_{N_1}\}$ be left invariant vector fields in the first stratum of the Lie algebra of $\mathbb{G}$. Let*

$$Y_k := \sum_{j=1}^{N_1} \left(A^{\frac{1}{2}}\right)_{k,j} X_j, \quad k = 1, \ldots, N_1.$$

Consider the related second-order differential operator

$$\mathcal{L}_A = \sum_{k=1}^{N_1} Y_k^2 = \sum_{k,j=1}^{N_1} a_{k,j} X_k X_j.$$

Then there exists a Lie group automorphism T_A of $\mathbb{G}$ such that

$$Y_k(u \circ T_A) = (X_k u) \circ T_A, \quad k = 1, \ldots, N_1,$$

$$\mathcal{L}_A(u \circ T_A) = (\mathcal{L}u) \circ T_A,$$

for every smooth function $u : \mathbb{G} \to \mathbb{R}$. Moreover, T_A has polynomial component functions and commutes with the dilations of $\mathbb{G}$.

Remark 1.4.3. The automorphism T_A may not exist when $\mathbb{G}$ is not a free stratified group. However, for any stratified group $\mathbb{G}$ one can find a different stratified group $\mathbb{G}_* = (\mathbb{R}^N, *, \delta_\lambda)$, that is, the stratified group with the same underlying manifold $\mathbb{R}^N$ and the same group of dilations δ_λ as $\mathbb{G}$, and a Lie-group isomorphism from $\mathbb{G}$ to $\mathbb{G}_*$ turning the extended sub-Laplacian $\mathcal{L}_A$ on $\mathbb{G}$ into the sub-Laplacian $\mathcal{L}$ on $\mathbb{G}_*$, see [BLU07, Chapter 16.3].

1.4.3 Divergence theorem

Here we discuss the divergence theorem on stratified Lie groups that will be useful for our analysis at different places of the book.

Let $d\nu$ denote the volume element on $\mathbb{G}$ corresponding to the first stratum on $\mathbb{G}$:

$$d\nu := d\nu(x) = \bigwedge_{j=1}^{N} dx_j. \tag{1.80}$$

However, for simplicity of the exposition, we will mainly use the notation $dx := d\nu(x)$. Regarding it as a differential form, let $\langle X_k, d\nu \rangle$ denote the natural pairing between vector fields and differential forms. As it will follow from the proof of Theorem 1.4.5, using formula (1.70) for the left invariant operators X_k expressing them in terms of the Euclidean derivatives, the pairing $\langle X_k, d\nu\rangle$ can be also expressed in terms of the differential forms corresponding to the Euclidean coordinates in the form

$$\langle X_k, d\nu(x)\rangle = \bigwedge_{j=1, j\neq k}^{N_1} dx_j^{(1)} \bigwedge_{l=2}^{r} \bigwedge_{m=1}^{N_l} \theta_{l,m}, \tag{1.81}$$

with

$$\theta_{l,m} = -\sum_{k=1}^{N_1} a_{k,m}^{(l)}(x^{(1)}, \ldots, x^{(l-1)})dx_k^{(1)} + dx_m^{(l)}, \tag{1.82}$$

for $l = 2, \ldots, r$ and $m = 1, \ldots, N_l$, where $a_{k,m}^{(l)}$ is a homogeneous polynomial of degree $l-1$ from (1.70).

Definition 1.4.4 (Admissible domains). A bounded open set $\Omega \subset \mathbb{G}$ will be called an *admissible domain* if its boundary $\partial\Omega$ is piecewise smooth and simple, that is, it has no self-intersections. The condition for the boundary to be simple amounts to $\partial\Omega$ being orientable.

The following divergence theorem can be regarded as a consequence of the abstract Stokes formula. However, we give a detailed local proof which will also lead to the explicit representation formula (1.82) that will be of use in the sequel.

Theorem 1.4.5 (Divergence formula). *Let $\Omega \subset \mathbb{G}$ be an admissible domain. Let $f_k \in C^1(\Omega)\bigcap C(\overline{\Omega})$, $k = 1, \ldots, N_1$. Then for each $k = 1, \ldots, N_1$, we have*

$$\int_\Omega X_k f_k d\nu = \int_{\partial\Omega} f_k\langle X_k, d\nu\rangle. \tag{1.83}$$

Consequently, we also have

$$\int_\Omega \sum_{k=1}^{N_1} X_k f_k d\nu = \int_{\partial\Omega} \sum_{k=1}^{N_1} f_k\langle X_k, d\nu\rangle. \tag{1.84}$$

Proof of Theorem 1.4.5. Using (1.70), for any function f we obtain the following differentiation formula

$$\begin{aligned}
df &= \sum_{k=1}^{N_1} \frac{\partial f}{\partial x_k^{(1)}} dx_k^{(1)} + \sum_{l=2}^{r}\sum_{m=1}^{N_l} \frac{\partial f}{\partial x_m^{(l)}} dx_m^{(l)} \\
&= \sum_{k=1}^{N_1} X_k f dx_k^{(1)} - \sum_{k=1}^{N_1}\sum_{l=2}^{r}\sum_{m=1}^{N_l} a_{k,m}^{(l)}(x^{(1)}, \ldots, x^{(l-1)}) \frac{\partial f}{\partial x_m^{(l)}} dx_k^{(l)} \sum_{l=2}^{r}\sum_{m=1}^{N_l} \frac{\partial f}{\partial x_m^{(l)}} dx_m^{(l)} \\
&= \sum_{k=1}^{N_1} X_k f dx_k^{(1)} + \sum_{l=2}^{r}\sum_{m=1}^{N_l} \frac{\partial f}{\partial x_m^{(l)}} \left(-\sum_{k=1}^{N_1} a_{k,m}^{(l)}(x^{(1)}, \ldots, x^{(l-1)})dx_k^{(1)} + dx_m^{(l)}\right) \\
&= \sum_{k=1}^{N_1} X_k f dx_k^{(1)} + \sum_{l=2}^{r}\sum_{m=1}^{N_l} \frac{\partial f}{\partial x_m^{(l)}} \theta_{l,m},
\end{aligned}$$

where for each $l = 2, \ldots, r$ and $m = 1, \ldots, N_l$,

$$\theta_{l,m} = -\sum_{k=1}^{N_1} a_{k,m}^{(l)}(x^{(1)}, \ldots, x^{(l-1)})dx_k^{(1)} + dx_m^{(l)}.$$

That is, we can write

$$df = \sum_{k=1}^{N_1} X_k f dx_k^{(1)} + \sum_{l=2}^{r}\sum_{m=1}^{N_l} \frac{\partial f}{\partial x_m^{(l)}}\theta_{l,m}. \tag{1.85}$$

It is simple to see that

$$\langle X_s, dx_j^{(1)}\rangle = \frac{\partial}{\partial x_s^{(1)}} dx_j^{(1)} = \delta_{sj},$$

where δ_{sj} is the Kronecker delta. Moreover, we have

$$\begin{aligned}
&\langle X_s, \theta_{l,m}\rangle \\
&\quad= \left(\frac{\partial}{\partial x_s^{(1)}} + \sum_{h=2}^{r}\sum_{g=1}^{N_h} a_{s,g}^{(h)}(x^{(1)}, \ldots, x^{(h-1)})\frac{\partial}{\partial x_g^{(h)}}\right) \\
&\qquad \times \left(-\sum_{k=1}^{N_1} a_{k,m}^{(l)}(x^{(1)}, \ldots, x^{(l-1)})dx_k^{(1)} + dx_m^{(l)}\right) \\
&\quad= -\sum_{k=1}^{N_1}\left(\frac{\partial}{\partial x_s^{(1)}} a_{k,m}^{(l)}(x^{(1)}, \ldots, x^{(l-1)})\right) dx_k^{(1)} \\
&\qquad -\sum_{k=1}^{N_1} a_{k,m}^{(l)}(x^{(1)}, \ldots, x^{(l-1)})\frac{\partial}{\partial x_s^{(1)}} dx_k^{(1)} + \frac{\partial}{\partial x_s^{(1)}} dx_m^{(l)} \\
&\qquad -\sum_{k=1}^{N_1}\sum_{h=2}^{r}\sum_{g=1}^{N_h} a_{s,g}^{(h)}(x^{(1)}, \ldots, x^{(h-1)})\left(\frac{\partial}{\partial x_g^{(h)}} a_{k,m}^{(l)}(x^{(1)}, \ldots, x^{(l-1)})\right) dx_k^{(1)} \\
&\qquad -\sum_{k=1}^{N_1}\sum_{h=2}^{r}\sum_{g=1}^{N_h} a_{s,g}^{(h)}(x^{(1)}, \ldots, x^{(h-1)}) a_{k,m}^{(l)}(x^{(1)}, \ldots, x^{(l-1)})\frac{\partial}{\partial x_g^{(h)}} dx_k^{(1)} \\
&\qquad +\sum_{h=2}^{r}\sum_{g=1}^{N_h} a_{s,g}^{(h)}(x^{(1)}, \ldots, x^{(h-1)})\frac{\partial}{\partial x_g^{(h)}} dx_m^{(l)} \\
&\quad= -\sum_{k=1}^{N_1}\left(\frac{\partial}{\partial x_s^{(1)}} a_{k,m}^{(l)}(x^{(1)}, \ldots, x^{(l-1)})\right) dx_k^{(1)} - \sum_{k=1}^{N_1} a_{k,m}^{(l)}(x^{(1)}, \ldots, x^{(l-1)})\delta_{sk} \\
&\qquad -\sum_{k=1}^{N_1}\sum_{h=2}^{r}\sum_{g=1}^{N_h} a_{s,g}^{(h)}(x^{(1)}, \ldots, x^{(h-1)})\left(\frac{\partial}{\partial x_g^{(h)}} a_{k,m}^{(l)}(x^{(1)}, \ldots, x^{(l-1)})\right) dx_k^{(1)}
\end{aligned}$$

$$
\begin{aligned}
&+\sum_{h=2}^{r}\sum_{g=1}^{N_h} a_{s,g}^{(h)}(x^{(1)},\ldots,x^{(h-1)})\delta_{gm}\delta_{hl}\\
&= -\sum_{k=1}^{N_1}\sum_{h=2}^{r}\sum_{g=1}^{N_h} a_{s,g}^{(h)}(x^{(1)},\ldots,x^{(h-1)})\left(\frac{\partial}{\partial x_g^{(h)}}a_{k,m}^{(l)}(x^{(1)},\ldots,x^{(l-1)})\right)dx_k^{(1)}\\
&\quad -\sum_{k=1}^{N_1}\left(\frac{\partial}{\partial x_s^{(1)}}a_{k,m}^{(l)}(x^{(1)},\ldots,x^{(l-1)})\right)dx_k^{(1)}\\
&= -\sum_{k=1}^{N_1}\Bigg[\sum_{h=2}^{r}\sum_{g=1}^{N_l} a_{s,g}^{(h)}(x^{(1)},\ldots,x^{(h-1)})\left(\frac{\partial}{\partial x_g^{(h)}}a_{k,m}^{(l)}(x^{(1)},\ldots,x^{(l-1)})\right)\\
&\quad +\frac{\partial}{\partial x_s^{(1)}}a_{k,m}^{(l)}(x^{(1)},\ldots,x^{(l-1)})\Bigg]dx_k^{(1)}.
\end{aligned}
$$

That is, we have

$$\langle X_s, dx_j^{(1)}\rangle = \delta_{sj},$$

for $s, j = 1, \ldots, N_1$, and

$$\langle X_s, \theta_{l,m}\rangle = \sum_{k=1}^{N_1}\mathcal{C}_k dx_k^{(1)},$$

for $s = 1, \ldots, N_1,\ l = 2, \ldots, r,\ m = 1, \ldots, N_l$. Here we used the notation

$$
\begin{aligned}
\mathcal{C}_k = &-\sum_{h=2}^{r}\sum_{g=1}^{N_l} a_{s,g}^{(h)}(x^{(1)},\ldots,x^{(h-1)})\frac{\partial}{\partial x_g^{(h)}}a_{k,m}^{(l)}(x^{(1)},\ldots,x^{(l-1)})\\
&-\frac{\partial}{\partial x_s^{(1)}}a_{k,m}^{(l)}(x^{(1)},\ldots,x^{(l-1)}).
\end{aligned}
$$

By (1.80) we have

$$d\nu = d\nu(x) = \bigwedge_{j=1}^{N} dx_j = \bigwedge_{j=1}^{N_1} dx_j^{(1)}\bigwedge_{l=2}^{r}\bigwedge_{m=1}^{N_l} dx_m^{(l)} = \bigwedge_{j=1}^{N_1} dx_j^{(1)}\bigwedge_{l=2}^{r}\bigwedge_{m=1}^{N_l}\theta_{l,m},$$

so that we obtain

$$\langle X_k, d\nu(x)\rangle = \bigwedge_{j=1, j\neq k}^{N_1} dx_j^{(1)}\bigwedge_{l=2}^{r}\bigwedge_{m=1}^{N_l}\theta_{l,m}.$$

Therefore, using (1.85) we get

$$
\begin{aligned}
d(f_s\langle X_s, d\nu(x)\rangle) &= df_s\wedge\langle X_s, d\nu(x)\rangle\\
&= \sum_{k=1}^{N_1} X_k f_s dx_k^{(1)}\wedge\langle X_s, d\nu(x)\rangle + \sum_{l=2}^{r}\sum_{m=1}^{N_l}\frac{\partial f_s}{\partial x_m^{(l)}}\theta_{l,m}\wedge\langle X_s, d\nu(x)\rangle\\
&= X_s f_s d\nu(x),
\end{aligned}
$$

that is, we have

$$d(\langle f_k X_k, d\nu(x)\rangle) = X_k f_k d\nu(x), \quad k = 1, \ldots, N_1.$$

Now using the classical Stokes theorem (see, e.g., [DFN84, Theorem 26.3.1]) we obtain (1.83). Taking a sum over k we also obtain (1.84). □

1.4.4 Green's identities for sub-Laplacians

In this section we prove Green's first and second formulae for the sub-Laplacian on stratified groups. These formulae will be useful throughout the book when we will be dealing with inequalities and with the potential theory on stratified groups. We will formulate them in admissible domains in the sense of Definition 1.4.4.

Theorem 1.4.6 (Green's first and second identities). *Let $\mathbb{G}$ be a stratified group and let $\Omega \subset \mathbb{G}$ be an admissible domain.*

(1) *Green's first identity: Let $v \in C^1(\Omega)\bigcap C(\overline{\Omega})$ and $u \in C^2(\Omega)\bigcap C^1(\overline{\Omega})$. Then*

$$\int_\Omega \left((\widetilde{\nabla} v)u + v\mathcal{L}u\right) d\nu = \int_{\partial\Omega} v\langle\widetilde{\nabla} u, d\nu\rangle, \tag{1.86}$$

where $\mathcal{L}$ is the sub-Laplacian on $\mathbb{G}$ and where the vector field $\widetilde{\nabla} u$ is defined by

$$\widetilde{\nabla} u := \sum_{k=1}^{N_1} (X_k u)\, X_k. \tag{1.87}$$

(2) *Green's second identity: Let $u, v \in C^2(\Omega)\bigcap C^1(\overline{\Omega})$. Then*

$$\int_\Omega (u\mathcal{L}v - v\mathcal{L}u) d\nu = \int_{\partial\Omega} (u\langle\widetilde{\nabla} v, d\nu\rangle - v\langle\widetilde{\nabla} u, d\nu\rangle). \tag{1.88}$$

Remark 1.4.7.

1. The definition (1.87) means that $\widetilde{\nabla} u$ is a vector field. Consequently, the expression $(\widetilde{\nabla} v)u$ is a scalar, given by

$$\left(\widetilde{\nabla} v\right) u = \widetilde{\nabla} v u = \sum_{k=1}^{N_1} (X_k v)\,(X_k u) = \sum_{k=1}^{N_1} X_k v X_k u.$$

At the same time the expression $\widetilde{\nabla}(vu)$ is a vector field, also understood as an operator.

2. Although we formulate Green's identities in bounded domains, they are still applicable in unbounded domains for functions with necessary decay rates at infinity. It can be readily shown by the standard argument using quasi-balls with radii $R \to \infty$.

3. The version (1.86) of Green's first identity was proved for the ball in [Gav77] and for any smooth domain of the complex Heisenberg group in [Rom91]. Other analogues have been also obtained in [BLU07] and [CGN08] but using different terminologies. Also, the group structure is not needed for it, see Proposition 12.2.1, with other versions also known, see, e.g., [CGL93]. The version given in Theorem 1.4.6 was obtained in [RS17c].

Proof of Theorem 1.4.6. Part (1). Let $f_k := vX_k u$, so that

$$\sum_{k=1}^{N_1} X_k f_k = (\widetilde{\nabla} v)u + v\mathcal{L}u.$$

By using the divergence formula in Theorem 1.4.5 we obtain

$$\begin{aligned}\int_\Omega \left(\widetilde{\nabla} vu + v\mathcal{L}u\right) d\nu &= \int_\Omega \sum_{k=1}^{N_1} X_k f_k d\nu = \int_{\partial\Omega} \sum_{k=1}^{N_1} \langle f_k X_k, d\nu\rangle \\ &= \int_{\partial\Omega} \sum_{k=1}^{N_1} \langle vX_k u X_k, d\nu\rangle = \int_{\partial\Omega} v\langle \widetilde{\nabla} u, d\nu\rangle,\end{aligned}$$

yielding (1.86).

Part (2). Rewriting (1.86) we have

$$\int_\Omega \left((\widetilde{\nabla} u)v + u\mathcal{L}v\right) d\nu = \int_{\partial\Omega} u\langle \widetilde{\nabla} v, d\nu\rangle,$$

$$\int_\Omega \left((\widetilde{\nabla} v)u + v\mathcal{L}u\right) d\nu = \int_{\partial\Omega} v\langle \widetilde{\nabla} u, d\nu\rangle.$$

By subtracting the second identity from the first one and using

$$(\widetilde{\nabla} u)v = (\widetilde{\nabla} v)u,$$

we obtain (1.88). □

Taking $v = 1$ in Theorem 1.4.6 we obtain the following analogue of Gauss' mean value formula for harmonic functions:

Corollary 1.4.8 (Gauss' mean value formula). *If $\mathcal{L}u = 0$ in an admissible domain $\Omega \subset \mathbb{G}$, then*

$$\int_{\partial\Omega} \langle \widetilde{\nabla} u, d\nu\rangle = 0.$$

As in the classical theory, we can approximate functions with (weak) singularities such as smooth functions because the Green formulae are still valid for them. In this sense, without further justification and using these Green formulae, in particular, we apply them to the fundamental solution ε of the sub-Laplacian $\mathcal{L}$ as in (1.74). We define the function

$$\varepsilon(x,y) := \varepsilon(x^{-1}y). \tag{1.89}$$

The properties of the $\mathcal{L}$-gauge imply that $\varepsilon(x,y) = \varepsilon(y,x)$.

Thus, for $x \in \Omega$, taking $v = 1$ and $u(y) = \varepsilon(x,y)$ we can record the following consequence of Theorem 1.4.6, (1):

Corollary 1.4.9. *If $\Omega \subset \mathbb{G}$ is an admissible domain, and $x \in \Omega$, then*

$$\int_{\partial\Omega} \langle \widetilde{\nabla}\varepsilon(x,y), d\nu(y)\rangle = 1,$$

where ε is the fundamental solution of the sub-Laplacian $\mathcal{L}$.

Putting the fundamental solution ε instead of v in (1.88) we obtain the following representation formulae.

Corollary 1.4.10 (Representation formulae for functions on stratified groups)**.** *Let $\mathbb{G}$ be a stratified group and let $\Omega \subset \mathbb{G}$ be an admissible domain.*

(1) *Let $u \in C^2(\Omega)\bigcap C^1(\overline{\Omega})$. Then for $x \in \Omega$ we have*

$$\begin{aligned} u(x) = & \int_{\Omega} \varepsilon(x,y)\mathcal{L}u(y)d\nu(y) \\ & + \int_{\partial\Omega} u(y)\langle \widetilde{\nabla}\varepsilon(x,y), d\nu(y)\rangle - \int_{\partial\Omega} \varepsilon(x,y)\langle \widetilde{\nabla}u(y), d\nu(y)\rangle. \end{aligned}$$

(2) *Let $u \in C^2(\Omega)\bigcap C^1(\overline{\Omega})$ and $\mathcal{L}u = 0$ on Ω, then for $x \in \Omega$ we have*

$$u(x) = \int_{\partial\Omega} u(y)\langle \widetilde{\nabla}\varepsilon(x,y), d\nu(y)\rangle - \int_{\partial\Omega} \varepsilon(x,y)\langle \widetilde{\nabla}u(y), d\nu(y)\rangle.$$

(3) *Let $u \in C^2(\Omega)\bigcap C^1(\overline{\Omega})$ and $u(x) = 0,\ x \in \partial\Omega$, then*

$$u(x) = \int_{\Omega} \varepsilon(x,y)\mathcal{L}u(y)d\nu(y) - \int_{\partial\Omega} \varepsilon(x,y)\langle \widetilde{\nabla}u(y), d\nu(y)\rangle.$$

(4) *Let $u \in C^2(\Omega)\bigcap C^1(\overline{\Omega})$ and $\sum_{j=1}^{N_1} X_j u\langle X_j, d\nu\rangle = 0$ on $\partial\Omega$, then*

$$u(x) = \int_{\Omega} \varepsilon(x,y)\mathcal{L}u(y)d\nu(y) + \int_{\partial\Omega} u(y)\langle \widetilde{\nabla}\varepsilon(x,y), d\nu(y)\rangle.$$

1.4.5 Green's identities for p-sub-Laplacians

In this section we show how Green's first and second formulae for the sub-Laplacian from Section 1.4.4 on stratified groups can be extended to the p-sub-Laplacian for all $1 < p < \infty$. As before, we will formulate them in admissible domains in the sense of Definition 1.4.4. We recall the definition of the p-sub-Laplacian from (1.71) as

$$\mathcal{L}_p f := \operatorname{div}_H(|\nabla_H f|^{p-2}\nabla_H f), \quad 1<p<\infty.$$

Theorem 1.4.11 (Green's first and second identities for p-sub-Laplacian). *Let $\mathbb{G}$ be a stratified group and let $\Omega \subset \mathbb{G}$ be an admissible domain. Let $1<p<\infty$.*

(1) *Green's first identity: Let $v \in C^1(\Omega)\bigcap C(\overline{\Omega})$ and $u \in C^2(\Omega)\bigcap C^1(\overline{\Omega})$. Then*

$$\int_\Omega \left((|\nabla_{\mathbb{G}} u|^{p-2}\widetilde{\nabla} v)u + v\mathcal{L}_p u\right) d\nu = \int_{\partial\Omega} |\nabla_{\mathbb{G}} u|^{p-2} v\langle \widetilde{\nabla} u, d\nu\rangle, \tag{1.90}$$

where

$$\widetilde{\nabla} u = \sum_{k=1}^{N_1} (X_k u)\, X_k.$$

(2) *Green's second identity: Let $u, v \in C^2(\Omega)\bigcap C^1(\overline{\Omega})$. Then*

$$\begin{aligned} &\int_\Omega \left(u\mathcal{L}_p v - v\mathcal{L}_p u + (|\nabla_{\mathbb{G}} v|^{p-2} - |\nabla_{\mathbb{G}} u|^{p-2})(\widetilde{\nabla} v)u\right) d\nu \\ &\quad = \int_{\partial\Omega} (|\nabla_{\mathbb{G}} v|^{p-2} u\langle \widetilde{\nabla} v, d\nu\rangle - |\nabla_{\mathbb{G}} u|^{p-2} v\langle \widetilde{\nabla} u, d\nu\rangle). \end{aligned} \tag{1.91}$$

Proof of Theorem 1.4.11. Part (1). Let $f_k := v|\nabla_{\mathbb{G}} u|^{p-2} X_k u$, then

$$\sum_{k=1}^{N_1} X_k f_k = (|\nabla_{\mathbb{G}} u|^{p-2}\widetilde{\nabla} v)u + v\mathcal{L}_p u.$$

By integrating both sides of this equality over Ω and using Proposition 1.4.5 we obtain

$$\begin{aligned} \int_\Omega \left((|\nabla_{\mathbb{G}} u|^{p-2}\widetilde{\nabla} v)u + v\mathcal{L}_p u\right) d\nu &= \int_\Omega \sum_{k=1}^{N_1} X_k f_k d\nu = \int_{\partial\Omega} \sum_{k=1}^{N_1} \langle f_k X_k, d\nu\rangle \\ &= \int_{\partial\Omega} \sum_{k=1}^{N_1} \langle v|\nabla_{\mathbb{G}} u|^{p-2} X_k u X_k, d\nu\rangle = \int_{\partial\Omega} |\nabla_{\mathbb{G}} u|^{p-2} v\langle \widetilde{\nabla} u, d\nu\rangle, \end{aligned}$$

showing (1.90).

Part (2). Using (1.90) we have

$$\int_{\Omega}\Big((|\nabla_{\mathbb{G}}v|^{p-2}\widetilde{\nabla}u)v+u\mathcal{L}_p v\Big)\,d\nu=\int_{\partial\Omega}|\nabla_{\mathbb{G}}v|^{p-2}u\langle\widetilde{\nabla}v,d\nu\rangle,$$
$$\int_{\Omega}\Big((|\nabla_{\mathbb{G}}u|^{p-2}\widetilde{\nabla}v)u+v\mathcal{L}_p u\Big)\,d\nu=\int_{\partial\Omega}|\nabla_{\mathbb{G}}u|^{p-2}v\langle\widetilde{\nabla}u,d\nu\rangle.$$

By subtracting the second identity from the first one, the equality

$$(\widetilde{\nabla}u)v=(\widetilde{\nabla}v)u$$

implies (1.91). □

Taking $v=1$ in Theorem 1.4.11 we get the following analogue of Gauss' mean value formula for p-harmonic functions:

Corollary 1.4.12 (Gauss' mean value formula for p-harmonic functions). *If* $1<p<\infty$ *and* $\mathcal{L}_p u=0$ *in an admissible domain* $\Omega\subset\mathbb{G}$*, then*

$$\int_{\partial\Omega}|\nabla_{\mathbb{G}}u|^{p-2}\langle\widetilde{\nabla}u,d\nu\rangle=0.$$

1.4.6 Sub-Laplacians with drift

In this section we briefly describe the so-called sub-Laplacians with drift. While such operators can be analysed on more general groups, we restrict our presentation to stratified groups $\mathbb{G}$ only since this will be the setting where we will be using these operators.

Definition 1.4.13 (Sub-Laplacian with drift). The (extended) positive sub-Laplacian with drift is defined on $C_0^\infty(\mathbb{G})$ as the operator

$$\mathcal{L}_X:=-\sum_{i,j=1}^{N}a_{i,j}X_iX_j-\gamma X,\tag{1.92}$$

where $\gamma\in\mathbb{R}$, the matrix $(a_{i,j})_{i,j=1}^N$ is real, symmetric, positive definite, and $X\in\mathfrak{g}$ is a left invariant vector field on $\mathbb{G}$.

Similar to Section 1.4.2 the operator (1.92) can be transformed to the *(positive) sub-Laplacian with drift* of the form

$$\mathcal{L}_X=-\sum_{j=1}^{N}X_j^2-\gamma X:=\mathcal{L}_0-\gamma X,\tag{1.93}$$

where $\mathcal{L}_0$ is the positive sub-Laplacian on $\mathbb{G}$ defined by

$$\mathcal{L}_0=-\sum_{j=1}^{N}X_j^2.\tag{1.94}$$

The details of such a transformation can be found in [HMM05] or [MOV17].

If $X = \sum_{j=1}^{N} a_j X_j$, then we denote $\|X\| := \left(\sum_{j=1}^{N} a_j^2\right)^{1/2}$ and

$$b_X := \frac{\|X\|}{2}, \tag{1.95}$$

where $a_j \in \mathbb{R}$ for $j = 1, \ldots, N$.

Let us collect the following spectral properties of (positive) sub-Laplacians with drift (1.93) which are true on more general groups than the stratified ones. In the case $\gamma = 1$ this was shown in [HMM05, Proposition 3.1] while here we follow [RY18b] for general $\gamma \in \mathbb{R}$:

Proposition 1.4.14 (Spectral properties of sub-Laplacians with drift). *Let $\mathbb{G}$ be a connected Lie group with unit e, $X_1, \ldots, X_N$ an algebraic basis of $\mathfrak{g}$ and let $X \in \mathfrak{g}\backslash\{0\}$. Let $\gamma \in \mathbb{R}$. Then we have for the operator $\mathcal{L}_X$, with domain $C_0^\infty(\mathbb{G})$, the following properties:*

(i) *the operator $\mathcal{L}_X$ is symmetric on $L^2(\mathbb{G}, \mu)$ for some positive measure μ on $\mathbb{G}$ if and only if there exists a positive character χ of $\mathbb{G}$ and a constant C such that $\mu = C\mu_X$ and $\nabla_H \chi|_e = \gamma X|_e$, where μ_X is the measure absolutely continuous with respect to the Haar measure μ with density χ;*

(ii) *assume that $\nabla_H \chi|_e = \gamma X|_e$ for some positive character χ of $\mathbb{G}$. Then the operator $\mathcal{L}_X$ is essentially self-adjoint on $L^2(\mathbb{G}, \mu_X)$ and its spectrum is contained in the interval $[\gamma^2 b_X^2, \infty)$.*

Proof of Proposition 1.4.14. Let μ be a positive measure on $\mathbb{G}$. Then for all test functions $\phi, \psi \in C_0^\infty(\mathbb{G})$ we can calculate

$$\begin{aligned}
\int_{\mathbb{G}} (\mathcal{L}_X \phi)\psi d\mu &= -\sum_{j=1}^{N} \left(\int_{\mathbb{G}} (X_j^2 \phi)\psi d\mu \right) - \gamma \int_{\mathbb{G}} \psi X \phi d\mu \\
&= \sum_{j=1}^{N} \left(\int_{\mathbb{G}} X_j \phi X_j \psi d\mu + \int_{\mathbb{G}} \psi X_j \phi X_j \mu \right) + \gamma \int_{\mathbb{G}} \phi X \psi d\mu + \gamma \int_{\mathbb{G}} \phi \psi X \mu \\
&= \sum_{j=1}^{N} \left(- \int_{\mathbb{G}} \phi X_j^2 \psi d\mu - 2 \int_{\mathbb{G}} \phi X_j \psi X_j \mu - \int_{\mathbb{G}} \phi \psi X_j^2 \mu \right) \\
&\quad + \gamma \int_{\mathbb{G}} \phi X \psi d\mu + \gamma \int_{\mathbb{G}} \phi \psi X \mu \\
&= \int_{\mathbb{G}} \phi (\mathcal{L}_X \psi) d\mu + 2\gamma \int_{\mathbb{G}} \phi X \psi d\mu - 2 \int_{\mathbb{G}} \phi \nabla_H \psi \nabla_H \mu + \int_{\mathbb{G}} \phi \psi (\mathcal{L}_0 + \gamma X)\mu \\
&= \int_{\mathbb{G}} \phi (\mathcal{L}_X \psi) d\mu + \langle \phi, 2\gamma (X\psi)\mu - 2\nabla_H \psi \nabla_H \mu + \psi (\mathcal{L}_0 + \gamma X)\mu \rangle \\
&=: \int_{\mathbb{G}} \phi (\mathcal{L}_X \psi) d\mu + \mathfrak{I}(\phi, \psi, \mu),
\end{aligned} \tag{1.96}$$

where $\mathcal{L}_0$ is defined in (1.94), $\nabla_H = (X_1, \dots, X_N)$ and $\langle \cdot, \cdot \rangle$ is the pairing between distributions and test functions on $\mathbb{G}$. From this we see that $\mathcal{L}_X$ is symmetric on $L^2(\mathbb{G}, \mu)$ if and only if $\mathfrak{I}(\phi, \psi, \mu) = 0$ for all functions ϕ and ψ, that is,

$$(\mathcal{L}_0 + \gamma X)\mu = 0, \;\; \gamma(X\psi)\mu - \nabla_H\psi\nabla_H\mu = 0, \;\; \forall \psi \in C_0^\infty(\mathbb{G}). \tag{1.97}$$

The vector fields $X_1, \dots, X_N$ satisfy Hörmander's condition, so that $\mathcal{L}_0 + \gamma X$ is hypoelliptic, which implies with the condition $(\mathcal{L}_0 + \gamma X)\mu = 0$ that μ has a smooth density ω with respect to the Haar measure. Then, as in [HMM05, Proof of Proposition 3.1], we show that

$$X = \sum_{j=1}^{N} a_j X_j, \tag{1.98}$$

for some coefficients $a_1, \dots, a_N$. Using the fact that $X_1, \dots, X_N$ are linearly independent and the second equation of (1.97), we obtain that

$$X_k\omega = \gamma a_k \omega, \tag{1.99}$$

where $k = 1, \dots, N$. The solution of (1.99) is given by

$$\omega(x) = \omega(e) \exp\left(\gamma \int_0^1 \sum_{k=1}^{N} a_k \vartheta_k(t) dt \right),$$

which is a positive and uniquely determined by its value at the identity, where $\vartheta_k(t)$ is the piecewise C^1 path. By normalizing ω, we get that $\omega(e) = 1$, and that it is a character of $\mathbb{G}$. Then, we see that the function $x \mapsto \omega(xy)/\omega(y)$ is a solution of (1.99) for any y in $\mathbb{G}$. Since the value of this function at the identity is 1, we have $\omega(xy) = \omega(x)\omega(y)$ for any $x, y \in \mathbb{G}$, and ω is a character of $\mathbb{G}$. From (1.98) and (1.99), we get $\nabla_H \chi|_e = \gamma X|_e$ with $\chi = \omega$. This proves Part (i) of Proposition 1.4.14.

As in the case $\gamma = 1$ (see [HMM05, Proposition 3.1]), by considering the isometry $\mathcal{U}_2 f = \chi^{-1/2} f$ of $L^2(\mathbb{G}, \mu)$ onto $L^2(\mathbb{G}, \mu_X)$, we have

$$\chi^{\frac{1}{2}} \mathcal{L}_X (\chi^{-\frac{1}{2}} f) = (\mathcal{L}_0 + \gamma^2 b_X^2) f, \tag{1.100}$$

which is an essentially self-adjoint operator on $L^2(\mathbb{G}, \mu)$, where b_X is defined in (1.95). Since the spectrum of this operator is contained in $[\gamma^2 b_X^2, \infty)$, we obtain that $\mathcal{L}_X$ is essentially self-adjoint on $L^2(\mathbb{G}, \mu_X)$ and its spectrum is contained in $[\gamma^2 b_X^2, \infty)$.

This completes the proof of Proposition 1.4.14. □

As a corollary of Proposition 1.4.14 let us collect the properties that will be important for us in the sequel.

Corollary 1.4.15 (Transformation of sub-Laplacian with drift). *Let $\mathbb{G}$ be a stratified group and assume conditions of Proposition* 1.4.14. *Assume that there exists a positive character χ of $\mathbb{G}$ such that*

$$\nabla_H \chi|_e = \gamma X|_e.$$

Then the operator $\mathcal{L}_X$ is formally self-adjoint with respect to the positive measure $\mu_X = \chi\mu$, where μ is the Haar measure of $\mathbb{G}$. The operator $\mathcal{L}_X$ is self-adjoint on $L^2(\mathbb{G}, \mu_X)$ and the mapping

$$L^2(\mathbb{G}, \mu) \ni f \mapsto \chi^{-1/2} f \in L^2(\mathbb{G}, \mu_X) \tag{1.101}$$

is an isometric isomorphism.

For a detailed discussion about more properties of the sub-Laplacians with drift we refer to [HMM04], [HMM05] and [MOV17].

1.4.7 Polarizable Carnot groups

In (1.74) we recalled the result of Folland that the sub-Laplacian $\mathcal{L}$ on general stratified groups always has a unique fundamental solution ε. The explicit formula (1.75) relating the fundamental solution to the $\mathcal{L}$-gauge turns out to be useful in many explicit calculations.

In applications to nonlinear partial differential equations, a natural question arises to express the fundamental solution of the p-sub-Laplacian (1.71) in terms of the fundamental solution of the sub-Laplacian or, equivalently, in terms of the $\mathcal{L}$-gauge. One of the largest classes of stratified Lie groups, for which the fundamental solution of the p-sub-Laplacian is known to be expressed explicitly in terms of the $\mathcal{L}$-gauge are the so-called polarizable Carnot groups which we now briefly discuss.

A Lie group $\mathbb{G}$ is called a *polarizable Carnot group* if the $\mathcal{L}$-gauge d satisfies the following ∞-sub-Laplacian equality

$$\mathcal{L}_\infty d := \frac{1}{2}\langle \nabla_H(|\nabla_H d|^2), \nabla_H d\rangle = 0 \quad \text{in } \mathbb{G}\backslash\{0\}. \tag{1.102}$$

It is known that the Euclidean space, the Heisenberg group $\mathbb{H}^n$ and Kaplan's H-type groups are polarizable Carnot groups.

It was shown by Balogh and Tyson in [BT02b] that if $\mathbb{G}$ is a polarizable Carnot group, then the fundamental solutions of the p-sub-Laplacian (1.71) are given by the explicit formulae

$$\varepsilon_p := \begin{cases} c_p d^{\frac{p-Q}{p-1}}, & \text{if } p \neq Q, \\ -c_Q \log d, & \text{if } p = Q. \end{cases} \tag{1.103}$$

This class of groups also admits an advantageous version of the polar coordinates decomposition. In particular, it can be shown (see [BT02b, Proposition

2.20]) that (1.103) implies a useful identity

$$\mathcal{L}_\infty u = \frac{Q-1}{Q-2}\frac{|\nabla_H u|^4}{u},$$

which can be also written as

$$\sum_{j=1}^{N} \frac{uX_j uX_j |\nabla_H u|}{|\nabla_H u|^3} = \frac{Q-1}{Q-2}. \tag{1.104}$$

It can be shown that the $\mathcal{L}$-gauge d on polarizable Carnot groups satisfies a number of further useful relations. For example, the following formula established in [BT02b] will be useful for some calculations in the sequel:

$$\nabla_H \left(\frac{d}{|\nabla_H d|^2} \nabla_H d \right) = Q \text{ in } \mathbb{G}\backslash \mathcal{Z}, \tag{1.105}$$

where the set

$$\mathcal{Z} := \{0\} \bigcup \{x \in \mathbb{G}\backslash\{0\} : \nabla_H d = 0\}$$

has Haar measure zero, and we have $\nabla_H d \neq 0$ for a.e. $x \in \mathbb{G}$.

As usual, the Green identities are still valid for functions with (weak) singularities provided we can approximate them by smooth functions. Thus, for example, for $x \in \Omega$ in a polarizable Carnot group, taking $v = 1$ and $u(y) = \varepsilon_p(x, y)$ we have the following corollary of Theorem 1.4.11 as an extension of Corollary 1.4.9:

Corollary 1.4.16. *Let Ω be an admissible domain in a polarizable Carnot group $\mathbb{G}$ and let $x \in \Omega$. Then we have*

$$\int_{\partial\Omega} |\nabla_{\mathbb{G}} \varepsilon_p|^{p-2} \langle \widetilde{\nabla} \varepsilon_p(x, y), d\nu(y) \rangle = 1.$$

Note that there are stratified Lie groups other than polarizable Carnot groups where the fundamental solution of the sub-Laplacian can be expressed explicitly (see, e.g., [BT02b, Section 6]).

In particular, since on the polarizable Carnot groups we have the fundamental solution ε_p, putting it instead of v in (1.91) we get the following representation type formulae extending those for $p = 2$ from Corollary 1.4.10:

Corollary 1.4.17 (Representation formulae for functions on polarizable Carnot groups)**.** *Let Ω be an admissible domain in a polarizable Carnot group $\mathbb{G}$.*

1. *Let $u \in C^2(\Omega) \bigcap C^1(\overline{\Omega})$. Then for $x \in \Omega$ we have*

$$u(x) = \int_\Omega \varepsilon_p \mathcal{L}_p u - (|\nabla_{\mathbb{G}} \varepsilon_p|^{p-2} - |\nabla_{\mathbb{G}} u|^{p-2})(\widetilde{\nabla} \varepsilon_p) u d\nu$$
$$+ \int_{\partial\Omega} (|\nabla_{\mathbb{G}} \varepsilon_p|^{p-2} u \langle \widetilde{\nabla} \varepsilon_p, d\nu \rangle - |\nabla_{\mathbb{G}} u|^{p-2} \varepsilon_p \langle \widetilde{\nabla} u, d\nu \rangle).$$

2. *Let $u \in C^2(\Omega)\bigcap C^1(\overline{\Omega})$ and $\mathcal{L}_p u = 0$ on Ω, then for $x \in \Omega$ we have*

$$u(x) = \int_\Omega (|\nabla_{\mathbb{G}} u|^{p-2} - |\nabla_{\mathbb{G}} \varepsilon_p|^{p-2})(\widetilde{\nabla}\varepsilon_p) u d\nu$$
$$+ \int_{\partial\Omega} (|\nabla_{\mathbb{G}}\varepsilon_p|^{p-2} u \langle \widetilde{\nabla}\varepsilon_p, d\nu\rangle - |\nabla_{\mathbb{G}} u|^{p-2}\varepsilon_p \langle \widetilde{\nabla} u, d\nu\rangle).$$

3. *Let $u \in C^2(\Omega)\bigcap C^1(\overline{\Omega})$ and*

$$u(x) = 0,\ x \in \partial\Omega,$$

then

$$u(x) = \int_\Omega \varepsilon_p \mathcal{L}_p u - (|\nabla_{\mathbb{G}}\varepsilon_p|^{p-2} - |\nabla_{\mathbb{G}} u|^{p-2})(\widetilde{\nabla}\varepsilon_p) u d\nu$$
$$- \int_{\partial\Omega} |\nabla_{\mathbb{G}} u|^{p-2}\varepsilon_p \langle \widetilde{\nabla} u, d\nu\rangle.$$

4. *Let $u \in C^2(\Omega)\bigcap C^1(\overline{\Omega})$ and $\sum_{j=1}^{N_1} X_j u \langle X_j, d\nu\rangle = 0$ on $\partial\Omega$, then*

$$u(x) = \int_\Omega \varepsilon_p \mathcal{L}_p u - (|\nabla_{\mathbb{G}}\varepsilon_p|^{p-2} - |\nabla_{\mathbb{G}} u|^{p-2})(\widetilde{\nabla}\varepsilon_p) u d\nu$$
$$+ \int_{\partial\Omega} |\nabla_{\mathbb{G}}\varepsilon_p|^{p-2} u \langle \widetilde{\nabla}\varepsilon_p, d\nu\rangle.$$

1.4.8 Heisenberg group

One of the important examples of the stratified groups is the Heisenberg group that was introduced in Example 1.1.8. Here we collect several of its basic properties that will be of use later in the book. We will give both real and complex descriptions of the Heisenberg group as both will be of use to us in the sequel.

Real description of the Heisenberg group. The Heisenberg group $\mathbb{H}^n$ is the manifold $\mathbb{R}^{2n+1}$ but with the group law given by

$$(x^{(1)}, y^{(1)}, t^{(1)})(x^{(2)}, y^{(2)}, t^{(2)})$$
$$:= \left(x^{(1)} + x^{(2)}, y^{(1)} + y^{(2)}, t^{(1)} + t^{(2)} + \frac{1}{2}(x^{(1)} \cdot y^{(2)} - x^{(2)} \cdot y^{(1)})\right), \tag{1.106}$$

for $(x^{(1)}, y^{(1)}, t^{(1)}), (x^{(2)}, y^{(2)}, t^{(2)}) \in \mathbb{R}^n \times \mathbb{R}^n \times \mathbb{R} \sim \mathbb{H}^n$, where $x^{(1)} \cdot y^{(2)}$ and $x^{(2)} \cdot y^{(1)}$ are the usual scalar products on $\mathbb{R}^n$. The canonical basis of the Lie algebra $\mathfrak{h}^n$ of the Heisenberg group $\mathbb{H}^n$ is given by the left invariant vector fields

$$X_j = \partial_{x_j} - \frac{y_j}{2}\partial_t,\ Y_j = \partial_{y_j} + \frac{x_j}{2}\partial_t,\ j = 1, \dots, n,\ T = \partial_t. \tag{1.107}$$

It follows that the basis elements X_j, Y_j, T, $j = 1, \ldots, n$, have the following commutator relations,

$$[X_j, Y_j] = T, \quad j = 1, \ldots, n,$$

with all the other commutators being zero. The Heisenberg (Lie) algebra $\mathfrak{h}^n$ is stratified via the decomposition

$$\mathfrak{h}^n = V_1 \oplus V_2,$$

where V_1 is linearly spanned by the X_j's and Y_j's, and $V_2 = \mathbb{R}T$. Therefore, the natural dilations on $\mathfrak{h}^n$ are given by

$$\delta_r(X_j) = rX_j, \quad \delta_r(Y_j) = rY_j, \quad \delta_r(T) = r^2 T.$$

On the level of the Heisenberg group $\mathbb{H}^n$ this can be expressed as

$$\delta_r(x, y, t) = r(x, y, t) = (rx, ry, r^2 t), \quad (x, y, t) \in \mathbb{H}^n, \ r > 0.$$

Consequently, $Q = 2n + 2$ is the homogeneous dimension of the Heisenberg group $\mathbb{H}^n$. The (negative) sub-Laplacian on $\mathbb{H}^n$ is given by

$$\mathcal{L} := \sum_{j=1}^{n} (X_j^2 + Y_j^2) = \sum_{j=1}^{n} \left(\partial_{x_j} - \frac{y_j}{2}\partial_t\right)^2 + \left(\partial_{y_j} + \frac{x_j}{2}\partial_t\right)^2,$$

corresponding to the horizontal gradient

$$\nabla_H := (X_1, \ldots, X_n, Y_1, \ldots, Y_n).$$

We can also write

$$\mathcal{L} = \Delta_{x,y} + \frac{|x|^2 + |y|^2}{4}\partial_t^2 + Z\partial_t, \qquad \text{with } Z = \sum_{j=1}^{n} (x_j \partial_{y_j} - y_j \partial_{x_j}),$$

where $\Delta_{x,y}$ is the Euclidean Laplacian with respect to x, y, and Z is the tangential derivative in the (x, y)-variables.

Complex description of the Heisenberg group. There is an alternative description of the Heisenberg group using complex rather than real variables. It is easy to see that both descriptions are equivalent.

The Heisenberg group $\mathbb{H}^n$ is the space $\mathbb{C}^n \times \mathbb{R}$ with the group operation given by

$$(\zeta, t) \circ (\eta, \tau) = (\zeta + \eta, t + \tau + 2 \operatorname{Im} \zeta\eta), \tag{1.108}$$

for $(\zeta, t), (\eta, \tau) \in \mathbb{C}^n \times \mathbb{R}$.

Comparing (1.106) with (1.108) we can note the change of the constant from $\frac{1}{2}$ to 2. As a result, we are getting different constants in the group law and in the formulae for the left invariant vector fields. We chose to give two descriptions with different constants since the adaptation of such constants in real and complex descriptions of the Heisenberg group seems to be happening in most of the literature. As a result, it will make it more convenient to refer to the relevant literature when needed. This should lead to no confusion since we will never be using these two descriptions at the same time. We note that in general, one can put any constant instead of $\frac{1}{2}$ or 2, the appearing objects are all isomorphic. We can refer the reader to [FR16, Section 6.1.1] for a detailed discussion on the choice of constants in the descriptions of the Heisenberg group.

Writing $\zeta = x + iy$ with x_j, y_j, $j = 1, \ldots, n$, the real coordinates on $\mathbb{H}^n$, the left invariant vector fields

$$\begin{aligned}\tilde{X}_j &= \frac{\partial}{\partial x_j} + 2y_j \frac{\partial}{\partial t}, \quad j = 1, \ldots, n,\\ \tilde{Y}_j &= \frac{\partial}{\partial y_j} - 2x_j \frac{\partial}{\partial t}, \quad j = 1, \ldots, n,\\ T &= \frac{\partial}{\partial t},\end{aligned}$$

form a basis for the Lie algebra $\mathfrak{h}^n$ of $\mathbb{H}^n$; again, this can be compared to (1.107).

At the same time, $\mathbb{H}^n$ can be seen as the boundary of the Siegel upper half-space in $\mathbb{C}^{n+1}$,

$$\mathbb{H}^n = \{(\zeta, z_{n+1}) \in \mathbb{C}^{n+1} : \operatorname{Im} z_{n+1} = |\zeta|^2, \zeta = (z_1, \ldots, z_n)\}.$$

Again, we can refer the reader, e.g., to [FR16, Section 6.1.1] for more details on different descriptions of the Heisenberg group.

Parametrising $\mathbb{H}_n$ by $z = (\zeta, t)$ where $t = \operatorname{Re} z_{n+1}$, a basis for the complex tangent space of $\mathbb{H}_n$ at the point z is given by the left invariant vector fields

$$X_j = \frac{\partial}{\partial z_j} + i\overline{z}\frac{\partial}{\partial t}, \quad j = 1, \ldots, n.$$

We denote their conjugates by

$$X_{\overline{j}} \equiv \overline{X}_j = \frac{\partial}{\partial \overline{z}_j} - iz\frac{\partial}{\partial t}, \quad j = 1, \ldots, n.$$

The operator

$$\mathcal{L}_{a,b} = \sum_{j=1}^{n} (aX_j X_{\overline{j}} + bX_{\overline{j}} X_j), \quad a + b = n, \tag{1.109}$$

is a left invariant, rotation invariant differential operator that is homogeneous of degree two. We can refer to the book of Folland and Kohn [FK72] for further properties of such operators. However, we can note that this operator is a slight generalization of the standard sub-Laplacian or Kohn-Laplacian $\mathcal{L}_b$ on the

Heisenberg group $\mathbb{H}^n$ which, when acting on the coefficients of a $(0,q)$-form can be written as

$$\mathcal{L}_b = -\frac{1}{n}\sum_{j=1}^{n}\big((n-q)X_jX_{\bar{j}} + qX_{\bar{j}}X_j\big).$$

Folland and Stein [FS74] obtained the fundamental solution of the operator $\mathcal{L}_{a,b}$ as a constant multiple of the function

$$\varepsilon_{a,b}(z) = \varepsilon(z) = \varepsilon(\zeta,t) = \frac{1}{(t+i|\zeta|^2)^a(t-i|\zeta|^2)^b}. \tag{1.110}$$

More precisely, the distribution $\frac{1}{c_{a,b}}\varepsilon$ is the fundamental solution of $\mathcal{L}_{a,b}$ since ε from (1.110) satisfies the equation

$$\mathcal{L}_{a,b}\varepsilon = c_{a,b}\delta. \tag{1.111}$$

The constant $c_{a,b}$ is zero if a and $b = -1,-2,\ldots,n,n+1,\ldots$, and $c_{a,b} \neq 0$ if a or $b \neq -1,-2,\ldots,n,n+1,\ldots$. In fact, then we can take

$$c_{a,b} = \frac{2(a^2+b^2)\mathrm{Vol}(B_1)}{(2i)^{n+1}}\frac{n!}{a(a-1)\cdots(a-n)}(1-\exp(-2ia\pi)) \tag{1.112}$$

for $a \notin \mathbb{Z}$, see Romero [Rom91, Proof of Theorem 1.6]. We will use the above description of the Heisenberg group and of the (rescaled) fundamental solution (1.110) to $\mathcal{L}_{a,b}$ in Section 11.3.3.

1.4.9 Quaternionic Heisenberg group

In this section we describe the basics of the quaternionic version of the Heisenberg group. We start by recalling the notion of quaternions and summarizing their main properties. As the space of quaternions is usually denoted by $\mathbb{H}$, we keep this notation here as well. There should be no notational confusion with the Heisenberg group since the quaternionic notation will be mostly localized to this section only.

Let $\mathbb{H}$ be the set of quaternions

$$x := x_0 + x_1 i_1 + x_2 i_2 + x_3 i_3,$$

where $(x_0,x_1,x_2,x_3) \in \mathbb{R}^4$, and $1, i_1, i_2, i_3$ are the basis elements of $\mathbb{H}$ with the following rules of multiplication:

$$i_1^2 = i_2^2 = i_3^2 = i_1 i_2 i_3 = -1,$$
$$i_1 i_2 = -i_2 i_1 = i_3,\ i_2 i_3 = -i_3 i_2 = i_1,\ i_3 i_1 = -i_1 i_3 = i_2.$$

The usual convention is that the real part of $x \in \mathbb{H}$ is the real number x_0 and its imaginary part is the point $(x_1,x_2,x_3) \in \mathbb{R}^3$. And so, the real and imaginary parts of x can be denoted by $\Re x$ and $\Im x$, respectively. In addition, we use more precise notations for the imaginary parts as

$$\Im_1 x := x_1, \quad \Im_2 x := x_2, \quad \Im_3 x := x_3.$$

The conjugate of x is denoted by

$$\overline{x} := x_0 - x_1 i_1 - x_2 i_2 - x_3 i_3,$$

and the modulus $|x|$ is defined by

$$|x|^2 := x\overline{x} = \sum_{j=0}^{3} x_j^2.$$

The Grassmanian product (or the quaternion product) of x and y is defined by

$$xy := (x_0 y_0 - \Im x \cdot \Im y) + (x_0 \Im y + y_0 \Im x + \Im x \times \Im y),$$

where

$$\Im x \times \Im y := \det \begin{pmatrix} i_1 & i_2 & i_3 \\ x_1 & x_2 & x_3 \\ y_1 & y_2 & y_3 \end{pmatrix}.$$

Let us denote $\mathbb{H}_q := \mathbb{H} \times \mathbb{R}^3$, it is called the *quaternion Heisenberg group.* Then $\mathbb{H}_q$ becomes a non-commutative group with the group law

$$\begin{aligned}&(x, t_1, t_2, t_3) \circ (y, \tau_1, \tau_2, \tau_3) \\ &\quad := (x + y, t_1 + \tau_1 - 2\Im_1(\overline{y}x), t_2 + y_2 - 2\Im_2(\overline{y}x), t_3 + \tau_3 - 2\Im_3(\overline{y}x)),\end{aligned}$$

for all $(x, t), (y, \tau) \in \mathbb{H}_q$. We note that $e = (0, 0, 0, 0)$ is the identity element of $\mathbb{H}_q$ and the inverse of every element $(x, t_1, t_2, t_3) \in \mathbb{H}_q$ is $(-x, -t_1, -t_2, -t_3)$.

The Haar measure on $\mathbb{H}_q$ coincides with the Lebesgue measure on $\mathbb{H} \times \mathbb{R}^3$ which is denoted by $d\nu = dxdt$. Let $\mathfrak{h}_q$ be the Lie algebra of left invariant vector fields on $\mathbb{H}_q$. A basis of $\mathfrak{h}_q$ is given by $\{X_0, X_1, X_2, X_3\}$ and $\{T_1, T_2, T_3\}$, where

$$\begin{aligned} X_0 &= \frac{\partial}{\partial x_0} - 2x_1 \frac{\partial}{\partial t_1} - 2x_2 \frac{\partial}{\partial t_2} - 2x_3 \frac{\partial}{\partial t_3}, \\ X_1 &= \frac{\partial}{\partial x_1} + 2x_0 \frac{\partial}{\partial t_1} - 2x_3 \frac{\partial}{\partial t_2} + 2x_2 \frac{\partial}{\partial t_3}, \\ X_2 &= \frac{\partial}{\partial x_2} + 2x_3 \frac{\partial}{\partial t_1} + 2x_0 \frac{\partial}{\partial t_2} - 2x_1 \frac{\partial}{\partial t_3}, \\ X_3 &= \frac{\partial}{\partial x_3} - 2x_2 \frac{\partial}{\partial t_1} + 2x_1 \frac{\partial}{\partial t_2} + 2x_0 \frac{\partial}{\partial t_3}, \end{aligned}$$

and

$$T_k = \frac{\partial}{\partial t_k}, \quad k = 1, 2, 3.$$

The Lie brackets of these vector fields are given by

$$[X_0, X_1] = [X_3, X_2] = 4T_1,$$
$$[X_0, X_2] = [X_1, X_3] = 4T_2,$$
$$[X_0, X_3] = [X_2, X_1] = 4T_3.$$

Thus, the sub-Laplacian on $\mathbb{H}_q$ is given by

$$\mathcal{L} = \sum_{j=0}^{3} X_j^2 = \Delta_x - 4|x|^2\Delta_t - 4\sum_{k=1}^{3}(i_k x \cdot \nabla_x)\frac{\partial}{\partial t_k}, \tag{1.113}$$

where

$$\Delta_x = \sum_{k=0}^{3}\frac{\partial^2}{\partial x_k^2}, \quad \text{and} \quad \Delta_t = \sum_{k=1}^{3}\frac{\partial^2}{\partial t_k^2}.$$

Note that the fundamental solution of the sub-Laplacian $\mathcal{L}$ on $\mathbb{H}_q$ was found by Tie and Wong in [TW09]. We restate their results in the following theorem.

Theorem 1.4.18 (Fundamental solutions for sub-Laplacian on quaternion Heisenberg groups). *The fundamental solution $\Gamma(\xi)$ of the sub-Laplacian $\mathcal{L}$ on the quaternion Heisenberg group $\mathbb{H}_q$ is given by*

$$\Gamma(\xi) := \Gamma(|x|, t) = \frac{2}{(2\pi)^{7/2}|x|^2}\int_{\mathbb{S}^2}\frac{1}{(|x|^2 - i(t\cdot n))^3}d\sigma, \tag{1.114}$$

where $\xi = (x, t) \in \mathbb{H}_q$, $n = (n_1, n_2, n_3)$ is a point on the unit sphere $\mathbb{S}^2$ in $\mathbb{R}^3$ with centre at the origin, and $d\sigma$ is the surface measure on $\mathbb{S}^2$. That is,

$$\mathcal{L}\Gamma_\zeta = -\delta_\zeta, \tag{1.115}$$

where $\Gamma_\zeta(\xi) = \Gamma(\zeta^{-1}\circ\xi)$ and δ_ζ is the Dirac distribution at $\zeta \equiv (y, \tau) \in \mathbb{H}_q$.

The quaternion Heisenberg group is a special case of the model step two nilpotent Lie group. It is a homogeneous group with respect to the dilation

$$\delta_\lambda : \mathbb{R}^7 \to \mathbb{R}^7, \quad \delta_\lambda = (\lambda x, \lambda^2 t).$$

Thus,

$$d(\xi) = \frac{1}{\Gamma^{1/8}(\xi)}, \quad \xi = (x, t) \in \mathbb{H}_q, \tag{1.116}$$

is a homogeneous quasi-norm on $\mathbb{H}_q$ with respect to the dilation δ_λ (see, e.g., [Cyg81]).

1.4.10 *H*-type groups

The H-type groups are a special family of stratified groups with a similar structure to that of the Heisenberg group; one of their important features is that the fundamental solutions to the sub-Laplacian are known explicitly.

We briefly recall the main notions related to this family of groups adopting the notation from [BLU07]; we refer to it for further details.

Definition 1.4.19 (Prototype H-type groups). The space $\mathbb{R}^{m+n}$ equipped with the group law

$$(x,t)\circ(y,\tau)=\begin{pmatrix} x_k+y_k, \quad k=1,\ldots,m \\ t_k+\tau_k+\frac{1}{2}\langle A^{(k)}x,y\rangle, \quad k=1,\ldots,n \end{pmatrix} \tag{1.117}$$

and with the dilations

$$\delta_\lambda(x,t)=(\lambda x,\lambda^2 t)$$

is called a *prototype H-type group*. Here $A^{(k)}$ is an $m\times m$ skew-symmetric orthogonal matrix, such that,

$$A^{(k)}A^{(l)}+A^{(l)}A^{(k)}=0$$

for all $k,l\in\{1,\ldots,n\}$ with $k\neq l$.

Clearly, the Euclidean (Abelian) group and the Heisenberg group are examples of prototype H-type groups.

We leave aside the general H-type groups since it can be shown that any (abstract) H-type group is naturally isomorphic to a prototype H-group (see [BLU07, Theorem 18.2.1]).

It can be directly checked that prototype H-groups are two step nilpotent Lie groups in which the identity of the group is the origin $(0,0)$ and the inverse of (x,t) is

$$(x,t)^{-1}=(-x,-t).$$

It can be also verified that the vector field in the Lie algebra $\mathfrak{g}$ of $\mathbb{G}$ that agrees at the origin with $\frac{\partial}{\partial x_j}$, $j=1,\ldots,m$, is given by

$$X_j=\frac{\partial}{\partial x_j}+\frac{1}{2}\sum_{k=1}^{n}\left(\sum_{i=1}^{m}a_{j,i}^{k}x_i\right)\frac{\partial}{\partial t_k}, \tag{1.118}$$

where $a_{j,i}^k$ is the (j,i)th element of the matrix $A^{(k)}$.

The prototype H-type groups are stratified with a basis of the first stratum given by these vector fields $X_1,\ldots,X_m$. Thus, the (negative) sub-Laplacian on a prototype H-type group $\mathbb{G}$ is given by

$$\mathcal{L}=\sum_{j=1}^{m}X_j^2=\Delta_x+\frac{1}{4}|x|^2\Delta_t+\sum_{k=1}^{n}\langle A^{(k)}x,\nabla_x\rangle\frac{\partial}{\partial t_k}, \tag{1.119}$$

where Δ and ∇ are the Euclidean Laplacian and the Euclidean gradient, respectively. There is no restriction to suppose that, if ϱ is the centre of the Lie algebra $\mathfrak{g}$ of $\mathbb{G}$, $\varrho^\perp$ is the orthogonal complement of ϱ and

$$m = \dim(\varrho^\perp), \quad n = \dim(\varrho).$$

So, the $\mathbb{R}^n$-component of the prototype H-type group $\mathbb{G} \simeq \mathbb{R}^{m+n}$ can be thought of as of its centre.

We have that the homogeneous dimension of the group is

$$Q = m + 2n.$$

We note that since for H-type groups we have $m \geq 2$ and $n \geq 1$, we actually always have $Q \geq 4$.

Now using a generic coordinate $\xi \equiv (x, t)$, $x \in \mathbb{R}^m$, $t \in \mathbb{R}^n$, let us introduce the following functions on $\mathbb{G}$:

$$v : \mathbb{G} \to \varrho^\perp, \quad v(\xi) := \sum_{j=1}^{m} \langle \exp_{\mathbb{G}}^{-1}(\xi), X_j \rangle X_j,$$

where $\{X_1, \ldots, X_m\}$ is an orthogonal basis of $\varrho^\perp$,

$$z : \mathbb{G} \to \varrho, \quad z(\xi) := \sum_{j=1}^{n} \langle \exp_{\mathbb{G}}^{-1}(\xi), Z_j \rangle Z_j,$$

where $\{Z_1, \ldots, Z_n\}$ is an orthogonal basis of ϱ. Thus, by the definition of v and z, for any $\xi \in \mathbb{G}$, one has

$$\xi = \exp(v(\xi) + z(\xi)), \quad v(\xi) \in \varrho^\perp, \quad z(\xi) \in \varrho,$$

and by a direct calculation we have (see, e.g., [BLU07, Proof of Remark 18.3.3]) that

$$|v(\xi)| = |x|, \quad |z(\xi)| = |t|.$$

The fundamental solutions for the sub-Laplacian on abstract H-type groups were found by A. Kaplan in [Kap80]. Such results boil down to the following statement.

Theorem 1.4.20 (Fundamental solutions for sub-Laplacian on H-type groups). *There exists a positive constant c such that*

$$\Gamma(\xi) := c \left(|x|^4 + 16|t|^2 \right)^{(2-Q)/4}$$

is the fundamental solution of the sub-Laplacian $\mathcal{L}$, that is,

$$\mathcal{L}\Gamma_\zeta = -\delta_\zeta, \tag{1.120}$$

where $\Gamma_\zeta(\xi) = \Gamma(\zeta^{-1} \circ \xi)$ and δ_ζ is the Dirac distribution at $\zeta \equiv (y, \tau) \in \mathbb{G}$.

For future use in Section 11.5, we will prefer to have the appearing function Γ positive, which leads to the appearance of the minus sign in (1.120).

For further details and analysis on H-type and related groups we may refer the reader to Kohn–Nirenberg [KN65], Folland [Fol75], Kaplan [Kap80], as well as to a more detailed exposition in [BLU07].

2

Homogeneous Groups: Hardy Inequalities

This chapter is devoted to Hardy inequalities and the analysis of their remainders in different forms. Moreover, we discuss several related inequalities such as Rellich inequalities and uncertainty principles.

In this chapter we will use all the notations given in Chapter 1 concerning homogeneous groups and the operators defined on it. In particular, $\mathbb{G}$ *is always a homogeneous group of homogeneous dimension* $Q \geq 1$. Some statements will hold for $Q \geq 2$ or for $Q \geq 3$ but we will be specifying this explicitly in formulations when needed.

2.1 Hardy inequalities and sharp remainders

In this section we analyse the anisotropic version of the classical L^p-Hardy inequality

$$\left\| \frac{f}{|x|_E} \right\|_{L^p(\mathbb{R}^n)} \leq \frac{p}{n-p} \|\nabla f\|_{L^p(\mathbb{R}^n)}, \quad n \geq 2, \quad 1 \leq p < n, \tag{2.1}$$

where ∇ is the standard gradient in $\mathbb{R}^n$, $|x|_E = \sqrt{x_1^2 + \cdots + x_n^2}$ is the Euclidean norm, $f \in C_0^\infty(\mathbb{R}^n)$, and the constant $\frac{p}{n-p}$ is known to be sharp. We also discuss in detail its critical cases and remainder estimates. As consequences, we derive Rellich type inequalities and the corresponding uncertainty principles.

2.1.1 Hardy inequality and uncertainty principle

First we establish the L^p-Hardy inequality and derive a formula for the remainder on a homogeneous group $\mathbb{G}$ of homogeneous dimension $Q \geq 2$. The radial operator $\mathcal{R}$ from (1.30) is entering the appearing expressions.

Theorem 2.1.1 (Hardy inequalities on homogeneous groups). *Let* $|\cdot|$ *be any homogeneous quasi-norm on* $\mathbb{G}$.

(i) *Let $f \in C_0^\infty(\mathbb{G}\backslash\{0\})$ be a complex-valued function. Then we have*

$$\left\| \frac{f}{|x|} \right\|_{L^p(\mathbb{G})} \leq \frac{p}{Q-p} \|\mathcal{R}f\|_{L^p(\mathbb{G})}, \quad 1 < p < Q, \tag{2.2}$$

where the constant $\frac{p}{Q-p}$ is sharp. Moreover, the equality in (2.2) is attained if and only if $f = 0$.

(ii) *For a real-valued function $f \in C_0^\infty(\mathbb{G}\backslash\{0\})$ and with the notations*

$$u := u(x) = -\frac{p}{Q-p}\mathcal{R}f(x),$$
$$v := v(x) = \frac{f(x)}{|x|},$$

we have

$$\|u\|^p_{L^p(\mathbb{G})} - \|v\|^p_{L^p(\mathbb{G})} = p\int_{\mathbb{G}} I_p(v,u)|v-u|^2 dx, \tag{2.3}$$

where

$$I_p(h,g) = (p-1)\int_0^1 |\xi h + (1-\xi)g|^{p-2}\xi d\xi. \tag{2.4}$$

(iii) *For $Q \geq 3$, for a complex-valued function $f \in C_0^\infty(\mathbb{G}\backslash\{0\})$ we have*

$$\|\mathcal{R}f\|^2_{L^2(\mathbb{G})} = \left(\frac{Q-2}{2}\right)^2 \left\| \frac{f}{|x|} \right\|^2_{L^2(\mathbb{G})} + \left\| \mathcal{R}f + \frac{Q-2}{2}\frac{f}{|x|} \right\|^2_{L^2(\mathbb{G})}, \tag{2.5}$$

that is, when $p = 2$, (2.3) holds for complex-valued functions as well.

Remark 2.1.2.

1. In the case of $\mathbb{G} = \mathbb{R}^n$ and $|x| = |x|_E = \sqrt{x_1^2 + \cdots + x_n^2}$ the Euclidean norm, we have $Q = n$ and $\mathcal{R} = \partial_r$ is the usual radial derivative, and (2.2) implies the classical Hardy inequality (2.1). Indeed, in this case for $1 < p < n$ inequality (2.2) yields

$$\begin{aligned} \left\| \frac{f}{|x|_E} \right\|_{L^p(\mathbb{R}^n)} &\leq \frac{p}{n-p}\|\mathcal{R}f\|_{L^p(\mathbb{R}^n)} = \frac{p}{n-p}\|\partial_r f\|_{L^p(\mathbb{R}^n)} \\ &= \frac{p}{n-p}\left\| \frac{x}{|x|_E}\cdot \nabla f \right\|_{L^p(\mathbb{R}^n)} \leq \frac{p}{n-p}\|\nabla f\|_{L^p(\mathbb{R}^n)}, \end{aligned} \tag{2.6}$$

   in view of the Cauchy–Schwarz inequality for the Euclidean norm.

   An interesting feature of the Hardy inequality in Part (i) is that the constant in (2.2) is sharp for any homogeneous quasi-norm $|\cdot|$.

2. In the setting of Part 1 above the remainder formula (2.3) for the Euclidean norm $|\cdot|_E$ in $\mathbb{R}^n$ was analysed by Ioku, Ishiwata and Ozawa [IIO17].

3. In Theorem 2.1.1, Part (ii) implies Part (i). To show it one can notice that the right-hand side of (2.3) is non-negative, which implies that

$$\left\| \frac{f}{|x|} \right\|_{L^p(\mathbb{G})} \leq \frac{p}{Q-p} \|\mathcal{R}f\|_{L^p(\mathbb{G})}, \quad 1 < p < Q, \tag{2.7}$$

for any real-valued $f \in C_0^\infty(\mathbb{G}\backslash\{0\})$. Moreover, by using the following identity we obtain the same inequality for all complex-valued functions: for all $z \in \mathbb{C}$ we have

$$|z|^p = \left(\int_{-\pi}^{\pi} |\cos\theta|^p d\theta \right)^{-1} \int_{-\pi}^{\pi} |\mathrm{Re}(z)\cos\theta + \mathrm{Im}(z)\sin\theta|^p \, d\theta, \tag{2.8}$$

which is a consequence of the decomposition of a complex number $z = r(\cos\phi + i\sin\phi)$.

That is, we obtain inequality (2.2), and also that the constant $\frac{p}{Q-p}$ is sharp, in view of the remainder formula. Now let us show that this constant is attained only for $f = 0$. Identity (2.8) says that it is sufficient to look only for real-valued functions f. If the right-hand side of (2.3) vanishes, then we must have $u = v$, that is,

$$-\frac{p}{Q-p}\mathcal{R}f(x) = \frac{f(x)}{|x|}.$$

This also means that $\mathbb{E}f = -\frac{Q-p}{p}f$. Lemma 1.3.1 implies that f is positively homogeneous of order $-\frac{Q-p}{p}$, i.e., there exists a function $h : \wp \to \mathbb{C}$ such that

$$f(x) = |x|^{-\frac{Q-p}{p}} h\left(\frac{x}{|x|} \right), \tag{2.9}$$

where $\wp$ is the unit sphere for the quasi-norm $|\cdot|$. It confirms that f cannot be compactly supported unless it is identically zero.

4. The identity (2.8) has been often used in similar estimates for passing from real-valued to complex-valued functions, see, e.g., Davies [Dav80, p. 176].
5. Let us denote by $H^1_{\mathcal{R}}(\mathbb{G})$ the functional space of the functions $f \in L^2(\mathbb{G})$ with $\mathcal{R}f \in L^2(\mathbb{G})$. Then Theorem 2.1.1 can be extended for functions in $H^1_{\mathcal{R}}(\mathbb{G})$, that is, the proof of (2.2) given above works in this case. As for the sharpness and the equality in (2.2), having (2.9) also implies that $\frac{f(x)}{|x|} = |x|^{-\frac{Q}{p}} h\left(\frac{x}{|x|}\right)$ is not in $L^p(\mathbb{G})$ unless $h = 0$ and $f = 0$.

Remark 2.1.2, Part 2, shows that (2.3) implies Part (i) of Theorem 2.1.1, that is, we only need to prove Parts (ii) and (iii). However, we now give an independent proof of (2.2) for complex-valued functions without relying on the formula (2.8). We see that this argument will be also useful in the proof of Part (ii).

Proof of Theorem 2.1.1. Proof of Part (i). Using the polar decomposition from Proposition 1.2.10, a direct calculation shows that

$$\begin{aligned}\int_{\mathbb{G}} \frac{|f(x)|^p}{|x|^p} dx &= \int_0^\infty \int_{\wp} \frac{|f(ry)|^p}{r^p} r^{Q-1} d\sigma(y) dr \\ &= -\frac{p}{Q-p} \int_0^\infty r^{Q-p} \operatorname{Re} \int_{\wp} |f(ry)|^{p-2} f(ry) \overline{\frac{df(ry)}{dr}} d\sigma(y) dr \\ &= -\frac{p}{Q-p} \operatorname{Re} \int_{\mathbb{G}} \frac{|f(x)|^{p-2} f(x)}{|x|^{p-1}} \overline{\frac{df(x)}{d|x|}} dx. \end{aligned} \tag{2.10}$$

Now by the Hölder inequality with $\frac{1}{p} + \frac{1}{q} = 1$ we obtain

$$\begin{aligned}\int_{\mathbb{G}} \frac{|f(x)|^p}{|x|^p} dx &= -\frac{p}{Q-p} \operatorname{Re} \int_{\mathbb{G}} \frac{|f(x)|^{p-2} f(x)}{|x|^{p-1}} \overline{\frac{df(x)}{d|x|}} dx \\ &\leq \frac{p}{Q-p} \left(\int_{\mathbb{G}} \left| \frac{|f(x)|^{p-2} f(x)}{|x|^{p-1}} \right|^q dx \right)^{\frac{1}{q}} \left(\int_{\mathbb{G}} \left| \frac{df(x)}{d|x|} \right|^p dx \right)^{\frac{1}{p}} \\ &= \frac{p}{Q-p} \left(\int_{\mathbb{G}} \frac{|f(x)|^p}{|x|^p} dx \right)^{1-\frac{1}{p}} \left\| \frac{df(x)}{d|x|} \right\|_{L^p(\mathbb{G})}. \end{aligned}$$

This proves inequality (2.2) in Part (i).

Proof of Part (ii). Since

$$u := u(x) = -\frac{p}{Q-p} \mathcal{R} f, \quad \text{and} \quad v := v(x) = \frac{f(x)}{|x|},$$

the formula (2.10) can be restated as

$$\|v\|^p_{L^p(\mathbb{G})} = \operatorname{Re} \int_{\mathbb{G}} |v|^{p-2} v \bar{u} dx. \tag{2.11}$$

For a real-valued f the formula (2.10) becomes

$$\int_{\mathbb{G}} \frac{|f(x)|^p}{|x|^p} dx = -\frac{p}{Q-p} \int_{\mathbb{G}} \frac{|f(x)|^{p-2} f(x)}{|x|^{p-1}} \frac{df(x)}{d|x|} dx$$

and (2.11) becomes

$$\|v\|^p_{L^p(\mathbb{G})} = \int_{\mathbb{G}} |v|^{p-2} v u dx. \tag{2.12}$$

Moreover, for any L^p-integrable real-valued functions u and v, we have

$$\begin{aligned} &\|u\|^p_{L^p(\mathbb{G})} - \|v\|^p_{L^p(\mathbb{G})} + p \int_{\mathbb{G}} (|v|^p - |v|^{p-2} v u) dx \\ &= \int_{\mathbb{G}} (|u|^p + (p-1)|v|^p - p|v|^{p-2} v u) dx = p \int_{\mathbb{G}} I_p(v,u) |v-u|^2 dx, \end{aligned} \tag{2.13}$$

where

$$I_p(v,u) = (p-1)\int_0^1 |\xi v + (1-\xi)u|^{p-2}\xi d\xi.$$

To show the last equality in (2.13), we observe the identity, for real numbers $u \neq v$,

$$\begin{aligned}
&\frac{1}{p}|u|^p + \left(1-\frac{1}{p}\right)|v|^p - |v|^{p-2}vu \\
&= \left(1-\frac{1}{p}\right)(|v|^p - |u|^p) - u(|v|^{p-2}v - |u|^{p-2}u) \\
&= (p-1)\int_0^1 |\xi v + (1-\xi)u|^{p-2}(\xi v + (1-\xi)u)d\xi\,(v-u) \\
&\quad - (p-1)\int_0^1 |\xi v + (1-\xi)u|^{p-2}d\xi\, u(v-u) \\
&= (p-1)\int_0^1 |\xi v + (1-\xi)u|^{p-2}\xi d\xi\,(v-u)^2,
\end{aligned}$$

using the integral expression for the remainder in the Taylor expansion formula.

Combining (2.13) with (2.12) we arrive at

$$\|u\|^p_{L^p(\mathbb{G})} - \|v\|^p_{L^p(\mathbb{G})} = p\int_{\mathbb{G}} I_p(v,u)|v-u|^2 dx.$$

It completes the proof of Part (ii).

Proof of Part (iii). When $p=2$, the equality (2.11) for complex-valued functions reduces to

$$\|v\|^2_{L^2(\mathbb{G})} = \mathrm{Re}\int_{\mathbb{G}} v\overline{u}dx.$$

Then we have

$$\begin{aligned}
\|u\|^2_{L^2(\mathbb{G})} - \|v\|^2_{L^2(\mathbb{G})} &= \|u\|^2_{L^2(\mathbb{G})} - \|v\|^2_{L^2(\mathbb{G})} + 2\int_{\mathbb{G}}(|v|^2 - \mathrm{Re}\, v\overline{u})dx \\
&= \int_{\mathbb{G}}(|u|^2 + |v|^2 - 2\mathrm{Re}\, v\overline{u})dx = \int_{\mathbb{G}} |u-v|^2 dx,
\end{aligned}$$

that is, (2.5) is proved. □

As a direct consequence of the inequality (2.2) we obtain the corresponding uncertainty principle:

Corollary 2.1.3 (Uncertainty principle on homogeneous groups). *For every complex-valued function $f \in C_0^\infty(\mathbb{G}\backslash\{0\})$ we have*

$$\left(\int_{\mathbb{G}}\left|\frac{df(x)}{d|x|}\right|^p dx\right)^{1/p}\left(\int_{\mathbb{G}}|x|^q|f|^q dx\right)^{1/q} \geq \frac{Q-p}{p}\int_{\mathbb{G}}|f|^2 dx. \tag{2.14}$$

Here $Q \geq 2$, $|\cdot|$ is an arbitrary homogeneous quasi-norm on $\mathbb{G}$, $1 < p < Q$ and $\frac{1}{p}+\frac{1}{q}=1$.

Proof. The inequality (2.7) and the Hölder inequality imply that

$$\left(\int_{\mathbb{G}}\left|\frac{df(x)}{d|x|}\right|^{p}dx\right)^{\frac{1}{p}}\left(\int_{\mathbb{G}}|x|^{q}|f|^{q}dx\right)^{\frac{1}{q}}$$
$$\geq\frac{Q-p}{p}\left(\int_{\mathbb{G}}\frac{|f|^{p}}{|x|^{p}}dx\right)^{\frac{1}{p}}\left(\int_{\mathbb{G}}|x|^{q}|f|^{q}dx\right)^{\frac{1}{q}}\geq\frac{Q-p}{p}\int_{\mathbb{G}}|f|^{2}dx,$$

This shows (2.14). □

Remark 2.1.4. In the Abelian case $\mathbb{G}=(\mathbb{R}^n,+)$ with the standard Euclidean distance $|x|_E$, we have $Q=n$, so that (2.14) with $p=q=2$ and $n\geq 3$ implies the uncertainty principle

$$\int_{\mathbb{R}^n}\left|\frac{x}{|x|_E}\cdot\nabla u(x)\right|^2dx\int_{\mathbb{R}^n}|x|_E^2|u(x)|^2dx\geq\left(\frac{n-2}{2}\right)^2\left(\int_{\mathbb{R}^n}|u(x)|^2dx\right)^2, \tag{2.15}$$

which in turn implies the classical uncertainty principle for $\mathbb{G}\equiv\mathbb{R}^n$:

$$\int_{\mathbb{R}^n}|\nabla u(x)|^2dx\int_{\mathbb{R}^n}|x|_E^2|u(x)|^2dx\geq\left(\frac{n-2}{2}\right)^2\left(\int_{\mathbb{R}^n}|u(x)|^2dx\right)^2,\quad n\geq 3.$$

2.1.2 Weighted Hardy inequalities

In this section $\mathbb{G}$ is a homogeneous group of homogeneous dimension $Q\geq 3$. Let $|\cdot|$ be an arbitrary homogeneous quasi-norm on $\mathbb{G}$. Here, we are going to discuss weighted Hardy inequalities on $\mathbb{G}$ which are the consequences of exact equalities.

Theorem 2.1.5 (Weighted Hardy identity in $L^2(\mathbb{G})$). *Let $\mathbb{G}$ be a homogeneous group of homogeneous dimension $Q\geq 3$ and let $|\cdot|$ be a homogeneous quasi-norm on $\mathbb{G}$. Then for every complex-valued function $f\in C_0^\infty(\mathbb{G}\backslash\{0\})$ and for any $\alpha\in\mathbb{R}$ we have the equality*

$$\left\|\frac{1}{|x|^\alpha}\mathcal{R}f\right\|^2_{L^2(\mathbb{G})}=\left(\frac{Q-2}{2}-\alpha\right)^2\left\|\frac{f}{|x|^{\alpha+1}}\right\|^2_{L^2(\mathbb{G})}+\left\|\frac{1}{|x|^\alpha}\mathcal{R}f+\frac{Q-2-2\alpha}{2|x|^{\alpha+1}}f\right\|^2_{L^2(\mathbb{G})}. \tag{2.16}$$

The equality (2.16) implies many different inequalities. For instance, by taking $\alpha=1$ and simplifying its coefficient, for any $Q\geq 3$ we obtain the identity

$$\left\|\frac{1}{|x|}\mathcal{R}f\right\|^2_{L^2(\mathbb{G})}=\left(\frac{Q-4}{2}\right)^2\left\|\frac{f}{|x|^{2}}\right\|^2_{L^2(\mathbb{G})}+\left\|\frac{1}{|x|}\mathcal{R}f+\frac{Q-4}{2|x|^{2}}f\right\|^2_{L^2(\mathbb{G})}. \tag{2.17}$$

By dropping the last term in (2.16) which is non-negative we obtain:

Corollary 2.1.6 (Weighted Hardy inequality in $L^2(\mathbb{G})$). *Let $Q \geq 3$ and let $\alpha \in \mathbb{R}$ be such that $Q - 2 - 2\alpha \neq 0$. Then for all complex-valued functions $f \in C_0^\infty(\mathbb{G}\backslash\{0\})$ we have*

$$\left\| \frac{f}{|x|^{\alpha+1}} \right\|_{L^2(\mathbb{G})} \leq \frac{2}{|Q-2-2\alpha|} \left\| \frac{1}{|x|^\alpha} \mathcal{R} f \right\|_{L^2(\mathbb{G})}. \tag{2.18}$$

Here the constant $\frac{2}{|Q-2-2\alpha|}$ is sharp and it is attained if and only if $f = 0$.

Remark 2.1.7.

1. It is interesting to note that the constant $\frac{2}{|Q-2-2\alpha|}$ in (2.18) is sharp for any homogeneous quasi-norm $|\cdot|$ on $\mathbb{G}$.
2. When $\alpha = 1$, (2.17) or (2.18) also imply that

$$\left\| \frac{f}{|x|^2} \right\|_{L^2(\mathbb{G})} \leq \frac{2}{Q-4} \left\| \frac{1}{|x|} \mathcal{R} f \right\|_{L^2(\mathbb{G})}, \quad Q \geq 5, \tag{2.19}$$

   again with $\frac{2}{Q-4}$ being the sharp constant.
3. If $\alpha = 0$, the identity (2.16) recovers Part (iii) of Theorem 2.1.1. However, we will use Part (iii) of Theorem 2.1.1 in the proof of Theorem 2.1.5.
4. In the Abelian case $\mathbb{G} = (\mathbb{R}^n, +)$, $n \geq 3$, we have $Q = n$, so for any homogeneous quasi-norm $|\cdot|$ on $\mathbb{R}^n$ identity (2.16) implies the following inequality with the optimal constant:

$$\frac{|n-2-2\alpha|}{2} \left\| \frac{f}{|x|^{\alpha+1}} \right\|_{L^2(\mathbb{R}^n)} \leq \left\| \frac{1}{|x|^\alpha} \frac{x}{|x|} \cdot \nabla f \right\|_{L^2(\mathbb{R}^n)} \quad \text{for all } \alpha \in \mathbb{R}.$$

   In the case of the Euclidean distance $|x|_E = \sqrt{x_1^2 + \cdots + x_n^2}$, by the Cauchy–Schwarz inequality we obtain the following estimate:

$$\frac{|n-2-2\alpha|}{2} \left\| \frac{f}{|x|_E^{\alpha+1}} \right\|_{L^2(\mathbb{R}^n)} \leq \left\| \frac{1}{|x|_E^\alpha} \nabla f \right\|_{L^2(\mathbb{R}^n)} \tag{2.20}$$

   for all $\alpha \in \mathbb{R}$ and for any $f \in C_0^\infty(\mathbb{R}^n\backslash\{0\})$. The sharpness of the constant $\frac{|n-2-2\alpha|}{2}$ in the Euclidean case of $\mathbb{R}^n$ with the Euclidean norm, (2.20) was shown in [CW01, Theorem 1.1. (ii)].
5. Hardy inequalities with homogeneous weights have been also considered by Hoffmann-Ostenhof and Laptev [HOL15]. There are also further many-particle versions of such inequalities, see [HOHOLT08] and many further references therein. We will discuss some of such inequalities in Section 6.11 and Section 6.12.
6. Theorem 2.1.5 was established in [RS17b]. Its extension from L^2 to L^p spaces presented in Theorem 2.1.8 was made in [Ngu17].

Proof of Theorem 2.1.5. We first observe the equality

$$\frac{1}{|x|^{\alpha}}\mathcal{R}f=\mathcal{R}\frac{f}{|x|^{\alpha}}+\alpha\frac{f}{|x|^{\alpha+1}} \tag{2.21}$$

for any $\alpha\in\mathbb{R}$, which follows from

$$\mathcal{R}\frac{f}{|x|^{\alpha}}=\frac{1}{|x|^{\alpha}}\mathcal{R}f+f\mathcal{R}\frac{1}{|x|^{\alpha}}$$

and hence, by using (1.30), we have

$$\mathcal{R}\frac{1}{|x|^{\alpha}}=\frac{d}{dr}\frac{1}{r^{\alpha}}=-\alpha\frac{1}{r^{\alpha+1}}=-\alpha\frac{1}{|x|^{\alpha+1}},\quad r=|x|.$$

Then using (2.21) we can write

$$\begin{aligned}\left\|\frac{1}{|x|^{\alpha}}\mathcal{R}f\right\|^2_{L^2(\mathbb{G})}&=\left\|\mathcal{R}\frac{f}{|x|^{\alpha}}+\frac{\alpha f}{|x|^{\alpha+1}}\right\|^2_{L^2(\mathbb{G})}\\&=\left\|\mathcal{R}\frac{f}{|x|^{\alpha}}\right\|^2_{L^2(\mathbb{G})}+2\alpha\mathrm{Re}\int_{\mathbb{G}}\mathcal{R}\left(\frac{f}{|x|^{\alpha}}\right)\frac{\overline{f}}{|x|^{\alpha+1}}dx+\left\|\frac{\alpha f}{|x|^{\alpha+1}}\right\|^2_{L^2(\mathbb{G})}.\end{aligned}$$

By applying (2.5) to the function $\frac{f}{|x|^{\alpha}}$ and using (2.21) we have that

$$\left\|\mathcal{R}\frac{f}{|x|^{\alpha}}\right\|^2_{L^2(\mathbb{G})}=\left(\frac{Q-2}{2}\right)^2\left\|\frac{f}{|x|^{1+\alpha}}\right\|^2_{L^2(\mathbb{G})}+\left\|\frac{1}{|x|^{\alpha}}\mathcal{R}f+\frac{Q-2-2\alpha}{2|x|^{\alpha+1}}f\right\|^2_{L^2(\mathbb{G})}.$$

In addition, a direct calculation using the polar decomposition in Proposition 1.2.10 shows that

$$\begin{aligned}2\alpha\mathrm{Re}\int_{\mathbb{G}}\mathcal{R}\left(\frac{f}{|x|^{\alpha}}\right)\frac{\overline{f}}{|x|^{\alpha+1}}dx&=2\alpha\mathrm{Re}\int_0^{\infty}r^{Q-2}\int_{\wp}\frac{d}{dr}\left(\frac{f(ry)}{r^{\alpha}}\right)\frac{\overline{f(ry)}}{r^{\alpha}}d\sigma(y)dr\\&=\alpha\int_0^{\infty}r^{Q-2}\int_{\wp}\frac{d}{dr}\left(\frac{|f(ry)|^2}{r^{2\alpha}}\right)d\sigma(y)dr=-\alpha(Q-2)\left\|\frac{f}{|x|^{\alpha+1}}\right\|^2_{L^2(\mathbb{G})}.\end{aligned}$$

In conclusion, combining these identities we arrive at

$$\left\|\frac{1}{|x|^{\alpha}}\mathcal{R}f\right\|^2_{L^2(\mathbb{G})}=\left(\frac{Q-2}{2}-\alpha\right)^2\left\|\frac{f}{|x|^{\alpha+1}}\right\|^2_{L^2(\mathbb{G})}+\left\|\frac{1}{|x|^{\alpha}}\mathcal{R}f+\frac{Q-2-2\alpha}{2|x|^{\alpha+1}}f\right\|^2_{L^2(\mathbb{G})},$$

yielding (2.16). □

To present a weighted L^p-Hardy inequality on $\mathbb{G}$ we will use the following function R_p in analogy to I_p in (2.4). For $\xi,\eta\in\mathbb{C}$ we denote

$$R_p(\xi,\eta):=\frac{1}{p}|\eta|^p+\frac{p-1}{p}|\xi|^p-\mathrm{Re}(|\xi|^{p-2}\xi\overline{\eta}). \tag{2.22}$$

By the convexity of the function $z \mapsto |z|^p$ we see that $R_p(\xi, \eta) \geq 0$ is non-negative and $R_p(\xi, \eta) = 0$ and if and only if $\xi = \eta$. If $\xi, \eta \in \mathbb{R}$, we then have

$$R_p(\xi, \eta) = (p-1) \int_0^1 |t\xi + (1-t)\eta|^{p-2} t dt \, |\xi - \eta|^2,$$

analogous to I_p in (2.4).

Theorem 2.1.8 (Weighted Hardy identity in $L^p(\mathbb{G})$). *Let $\mathbb{G}$ be a homogeneous group of homogeneous dimension Q. Let $1 < p < Q$ and $\alpha \in \mathbb{R}$. Then for any homogeneous quasi-norm $|\cdot|$ on $\mathbb{G}$ and for all complex-valued functions $f \in C_0^\infty(\mathbb{G}\backslash\{0\})$ we have*

$$\begin{aligned} \left\| \frac{\mathcal{R}f}{|x|^\alpha} \right\|^p_{L^p(\mathbb{G})} &= \left| \frac{Q - p(1+\alpha)}{p} \right|^p \int_{\mathbb{G}} \frac{|f|^p}{|x|^{p(1+\alpha)}} dx \\ &+ p \int_{\mathbb{G}} \frac{1}{|x|^{p\alpha}} R_p \left(-\frac{Q-p(1+\alpha)}{p} \frac{f}{|x|}, \mathcal{R}f \right) dx. \end{aligned} \tag{2.23}$$

By dropping the last term in (2.23) which is non-negative we obtain:

Corollary 2.1.9 (Weighted Hardy inequality in $L^p(\mathbb{G})$). *Let $1 < p < Q$ and let $\alpha \in \mathbb{R}$. Then for all complex-valued functions $f \in C_0^\infty(\mathbb{G}\backslash\{0\})$ we have*

$$\frac{|Q-p(1+\alpha)|}{p} \left\| \frac{f}{|x|^{1+\alpha}} \right\|_{L^p(\mathbb{G})} \leq \left\| \frac{\mathcal{R}f}{|x|^\alpha} \right\|_{L^p(\mathbb{G})}. \tag{2.24}$$

If $Q - p(1+\alpha) \neq 0$, then the constant $\frac{|Q-p(1+\alpha)|}{p}$ in (2.24) is sharp and it is attained if and only if $f = 0$.

Proof of Theorem 2.1.8. We can assume that $Q - p(1+\alpha) \neq 0$, otherwise there is nothing to prove. A direct calculation gives

$$\begin{aligned} &\int_{\mathbb{G}} \frac{|f(x)|^p}{|x|^{p(1+\alpha)}} dx = \int_0^\infty r^{Q-p(1+\alpha)-1} \int_{\wp} |f(ry)|^p d\sigma(y) dr \\ &= \frac{1}{Q-p(1+\alpha)} \int_0^\infty (r^{Q-p(1+\alpha)})' \int_{\wp} |f(ry)|^p d\sigma(y) dr \\ &= -\frac{1}{Q-p(1+\alpha)} \mathrm{Re} \int_0^\infty r^{Q-p(1+\alpha)} \int_{\wp} |f(ry)|^{p-2} f(ry) \overline{\mathcal{R}f(ry)} d\sigma(y) dr \\ &= -\frac{1}{Q-p(1+\alpha)} \mathrm{Re} \int_{\mathbb{G}} \frac{|f(x)|^{p-2} f(x)}{|x|^{(p-1)(1+\alpha)}} \frac{\overline{\mathcal{R}f(x)}}{|x|^\alpha} dx \\ &= \frac{p-1}{p} \int_{\mathbb{G}} \frac{|f|^p}{|x|^{p(1+\alpha)}} dx + \frac{1}{p} \left(\frac{p}{|Q-p(1+\alpha)|} \right)^p \int_{\mathbb{G}} \frac{|\mathcal{R}f|^p}{|x|^{p\alpha}} dx \\ &\quad - \int_{\mathbb{G}} R_p \left(\frac{f}{|x|^{1+\alpha}}, -\frac{p}{Q-p(1+\alpha)} \frac{\mathcal{R}f}{|x|^\alpha} \right) dx, \end{aligned}$$

which implies the identity (2.23). ◻

Proof of Corollary 2.1.9. The equality (2.23) implies inequality (2.24) since the last term in (2.23) is non-negative. Let us now show the sharpness of the constant. For this we approximate the function $r^{-(Q-p(1+\alpha))/p}$ by smooth compactly supported functions, for details of such an argument see also the proof of Theorem 3.1.4. Using (2.23) it follows that the equality in (2.24) holds if and only if

$$R_p\left(-\frac{Q-p(1+\alpha)}{p}\frac{f}{|x|}, \mathcal{R}f\right) = 0,$$

or, equivalently,

$$\mathcal{R}f = -\frac{Q-p(1+\alpha)}{p}\frac{f}{|x|}.$$

In turn, this is equivalent to

$$\mathbb{E}f = -\frac{Q-p(1+\alpha)}{p}f.$$

By Proposition 1.3.1 it follows that f is positively homogeneous of order $-(Q-p(1+\alpha))/p$. Since $|f|/|x|^{1+\alpha}$ is in $L^p(\mathbb{G})$, it follows that $f=0$. □

2.1.3 Hardy inequalities with super weights

In this section we discuss sharp L^p-Hardy type inequalities with *super weights*, i.e., with weights of the form

$$\frac{(a+b|x|^\alpha)^{\frac{\beta}{p}}}{|x|^m}. \tag{2.25}$$

Such weights are sometimes called the super weights because of the arbitrariness of the choice of any homogeneous quasi-norm as well as a wide range of parameters. However, all the inequalities can be obtained with best constants.

Theorem 2.1.10 (Hardy inequalities with super weights). *Let $\mathbb{G}$ be a homogeneous group of homogeneous dimension $Q \geq 1$. Let $a, b > 0$ and $1 < p < \infty$. Then we have the following inequalities:*

(i) *If $\alpha\beta > 0$ and $pm \leq Q-p$, then for all $f \in C_0^\infty(\mathbb{G}\backslash\{0\})$ we have*

$$\frac{Q-pm-p}{p}\left\|\frac{(a+b|x|^\alpha)^{\frac{\beta}{p}}}{|x|^{m+1}}f\right\|_{L^p(\mathbb{G})} \leq \left\|\frac{(a+b|x|^\alpha)^{\frac{\beta}{p}}}{|x|^m}\mathcal{R}f\right\|_{L^p(\mathbb{G})}. \tag{2.26}$$

If $Q \neq pm+p$ then the constant $\frac{Q-pm-p}{p}$ is sharp.

(ii) *If $\alpha\beta < 0$ and $pm - \alpha\beta \leq Q-p$, then for all $f \in C_0^\infty(\mathbb{G}\backslash\{0\})$ we have*

$$\frac{Q-pm+\alpha\beta-p}{p}\left\|\frac{(a+b|x|^\alpha)^{\frac{\beta}{p}}}{|x|^{m+1}}f\right\|_{L^p(\mathbb{G})} \leq \left\|\frac{(a+b|x|^\alpha)^{\frac{\beta}{p}}}{|x|^m}\mathcal{R}f\right\|_{L^p(\mathbb{G})}. \tag{2.27}$$

If $Q \neq pm+p-\alpha\beta$ then the constant $\frac{Q-pm+\alpha\beta-p}{p}$ is sharp.

Proof of Theorem 2.1.10. Proof of Part (i). We can assume $Q \neq pm+p$ since in the case $Q = pm+p$ there is nothing to prove. As usual with $(r,y) = (|x|, \frac{x}{|x|}) \in (0,\infty) \times \wp$ on $\mathbb{G}$, where

$$\wp := \{x \in \mathbb{G} : |x| = 1\},$$

using the polar decomposition in Proposition 1.2.10 and integrating by parts, we obtain

$$\int_{\mathbb{G}} \frac{(a+b|x|^{\alpha})^{\beta}}{|x|^{pm+p}} |f(x)|^p dx = \int_0^{\infty} \int_{\wp} \frac{(a+br^{\alpha})^{\beta}}{r^{pm+p}} |f(ry)|^p r^{Q-1} d\sigma(y) dr. \qquad (2.28)$$

Since $a, b > 0$, $\alpha\beta > 0$ and $m < \frac{Q-p}{p}$ we obtain

$$\begin{aligned}
&\int_{\mathbb{G}} \frac{(a+b|x|^{\alpha})^{\beta}}{|x|^{pm+p}} |f(x)|^p dx \\
&\leq \int_0^{\infty} \int_{\wp} (a+br^{\alpha})^{\beta} r^{Q-1-pm-p} \left(\frac{\alpha\beta b r^{\alpha}}{(a+br^{\alpha})(Q-pm-p)} + 1 \right) |f(ry)|^p d\sigma(y) dr \\
&= \int_0^{\infty} \int_{\wp} \frac{d}{dr} \left(\frac{(a+br^{\alpha})^{\beta} r^{Q-pm-p}}{Q-pm-p} \right) |f(ry)|^p d\sigma(y) dr \\
&= -\frac{p}{Q-pm-p} \int_0^{\infty} (a+br^{\alpha})^{\beta} r^{Q-pm-p} \operatorname{Re} \int_{\wp} |f(ry)|^{p-2} f(ry) \overline{\frac{df(ry)}{dr}} d\sigma(y) dr \\
&\leq \left| \frac{p}{Q-pm-p} \right| \int_{\mathbb{G}} \frac{(a+b|x|^{\alpha})^{\beta} |\mathcal{R}f(x)| |f(x)|^{p-1}}{|x|^{pm+p-1}} dx \\
&= \frac{p}{Q-pm-p} \int_{\mathbb{G}} \frac{(a+b|x|^{\alpha})^{\frac{\beta(p-1)}{p}} |f(x)|^{p-1}}{|x|^{(m+1)(p-1)}} \frac{(a+b|x|^{\alpha})^{\frac{\beta}{p}}}{|x|^m} |\mathcal{R}f(x)| dx.
\end{aligned}$$

Now by using Hölder's inequality we arrive at the inequality

$$\begin{aligned}
&\int_{\mathbb{G}} \frac{(a+b|x|^{\alpha})^{\beta}}{|x|^{pm+p}} |f(x)|^p dx \\
&\leq \frac{p}{Q-pm-p} \left(\int_{\mathbb{G}} \frac{(a+b|x|^{\alpha})^{\beta}}{|x|^{pm+p}} |f(x)|^p dx \right)^{\frac{p-1}{p}} \left(\int_{\mathbb{G}} \frac{(a+b|x|^{\alpha})^{\beta}}{|x|^{pm}} |\mathcal{R}f(x)|^p dx \right)^{\frac{1}{p}},
\end{aligned}$$

which gives (2.26).

We need to check the equality condition in the above Hölder inequality in order to show the sharpness of the constant. Setting

$$g(x) = |x|^C,$$

where $C \in \mathbb{R}, C \neq 0$ and $Q \neq pm+p$ a direct calculation shows

$$\left| \frac{1}{C} \right|^p \left(\frac{(a+b|x|^{\alpha})^{\frac{\beta}{p}} |\mathcal{R}g(x)|}{|x|^m} \right)^p = \left(\frac{(a+b|x|^{\alpha})^{\frac{\beta(p-1)}{p}} |g(x)|^{p-1}}{|x|^{(m+1)(p-1)}} \right)^{\frac{p}{p-1}},$$

which satisfies the equality condition of the Hölder inequality. This gives the sharpness of the constant $\frac{Q-pm-p}{p}$ in the inequality (2.26).

Proof of Part (ii). Here we can assume that $Q \neq pm + p - \alpha\beta$ since for $Q = pm+p-\alpha\beta$ there is nothing to prove. Using the polar decomposition in Proposition 1.2.10, as before we have the equality (2.28). Since $\alpha\beta < 0$ and $pm - \alpha\beta < Q - p$ we obtain

$$\begin{aligned}
&\int_{\mathbb{G}} \frac{(a+b|x|^{\alpha})^{\beta}}{|x|^{pm+p}}|f(x)|^p dx \\
&\leq \int_0^{\infty} \int_{\wp} (a+br^{\alpha})^{\beta} r^{Q-1-pm-p} \\
&\quad \times \left(\frac{br^{\alpha}}{a+br^{\alpha}} + \frac{a}{a+br^{\alpha}} \cdot \frac{Q-pm-p}{Q-pm-p+\alpha\beta} \right) |f(ry)|^p d\sigma(y) dr \\
&= \int_0^{\infty} \int_{\wp} \frac{(a+br^{\alpha})^{\beta} r^{Q-1-pm-p}}{Q-pm-p+\alpha\beta} \left(\frac{\alpha\beta b r^{\alpha}}{a+br^{\alpha}} + Q - pm - p \right) |f(ry)|^p d\sigma(y) dr \\
&= \int_0^{\infty} \int_{\wp} \frac{d}{dr} \left(\frac{(a+br^{\alpha})^{\beta} r^{Q-pm-p}}{Q-pm-p+\alpha\beta} \right) |f(ry)|^p d\sigma(y) dr \\
&= -\frac{p}{Q-pm-p+\alpha\beta} \\
&\quad \times \int_0^{\infty} (a+br^{\alpha})^{\beta} r^{Q-pm-p} \operatorname{Re} \int_{\wp} |f(ry)|^{p-2} f(ry) \overline{\frac{df(ry)}{dr}} d\sigma(y) dr \\
&\leq \left| \frac{p}{Q-pm-p+\alpha\beta} \right| \int_{\mathbb{G}} \frac{(a+b|x|^{\alpha})^{\beta} |\mathcal{R}f(x)| |f(x)|^{p-1}}{|x|^{pm+p-1}} dx \\
&= \frac{p}{Q-pm-p+\alpha\beta} \int_{\mathbb{G}} \frac{(a+b|x|^{\alpha})^{\frac{\beta(p-1)}{p}} |f(x)|^{p-1}}{|x|^{(m+1)(p-1)}} \frac{(a+b|x|^{\alpha})^{\frac{\beta}{p}}}{|x|^m} |\mathcal{R}f(x)| dx.
\end{aligned}$$

By Hölder's inequality, it follows that

$$\begin{aligned}
\int_{\mathbb{G}} \frac{(a+b|x|^{\alpha})^{\beta}}{|x|^{pm+p}} |f(x)|^p dx \leq{}& \frac{p}{Q-pm-p+\alpha\beta} \left(\int_{\mathbb{G}} \frac{(a+b|x|^{\alpha})^{\beta}}{|x|^{pm+p}} |f(x)|^p dx \right)^{(p-1)/p} \\
&\times \left(\int_{\mathbb{G}} \frac{(a+b|x|^{\alpha})^{\beta}}{|x|^{pm}} |\mathcal{R}f(x)|^p dx \right)^{1/p},
\end{aligned}$$

which gives (2.27).

To show the sharpness of the constant we will check the equality condition in the above Hölder inequality. Thus, by taking

$$h(x) = |x|^C,$$

where $C \in \mathbb{R}, C \neq 0$ and $Q \neq pm + p - \alpha\beta$, we get

$$\left| \frac{1}{C} \right|^p \left(\frac{(a+b|x|^{\alpha})^{\frac{\beta}{p}} |\mathcal{R}h(x)|}{|x|^m} \right)^p = \left(\frac{(a+b|x|^{\alpha})^{\frac{\beta(p-1)}{p}} |h(x)|^{p-1}}{|x|^{(m+1)(p-1)}} \right)^{p/(p-1)},$$

which satisfies the equality condition in Hölder's inequality. This gives the sharpness of the constant $\frac{Q-pm-p+\alpha\beta}{p}$ in (2.27). □

2.1.4 Hardy inequalities of higher order with super weights

The iteration process gives the following higher-order L^p-Hardy type inequalities with super weights. Here as before, $\mathbb{G}$ is a homogeneous group of homogeneous dimension $Q \geq 1$ and $|\cdot|$ is a homogeneous quasi-norm on $\mathbb{G}$.

Theorem 2.1.11 (Higher-order Hardy inequalities with super weights). *Let $a, b > 0$ and $1 < p < \infty$, $Q \geq 1$, $k \in \mathbb{N}$. Then we have the following inequalities.*

(i) *If $\alpha\beta > 0$ and $pm \leq Q - p$, then for all $f \in C_0^\infty(\mathbb{G}\backslash\{0\})$ we have*

$$\left[\prod_{j=0}^{k-1}\left(\frac{Q-p}{p}-(m+j)\right)\right]\left\|\frac{(a+b|x|^\alpha)^{\frac{\beta}{p}}}{|x|^{m+k}}f\right\|_{L^p(\mathbb{G})} \leq \left\|\frac{(a+b|x|^\alpha)^{\frac{\beta}{p}}}{|x|^m}\mathcal{R}^k f\right\|_{L^p(\mathbb{G})}. \tag{2.29}$$

(ii) *If $\alpha\beta < 0$ and $pm - \alpha\beta \leq Q - p$, then for all $f \in C_0^\infty(\mathbb{G}\backslash\{0\})$ we have*

$$\left[\prod_{j=0}^{k-1}\left(\frac{Q-p+\alpha\beta}{p}-(m+j)\right)\right]\left\|\frac{(a+b|x|^\alpha)^{\frac{\beta}{p}}}{|x|^{m+k}}f\right\|_{L^p(\mathbb{G})} \leq \left\|\frac{(a+b|x|^\alpha)^{\frac{\beta}{p}}}{|x|^m}\mathcal{R}^k f\right\|_{L^p(\mathbb{G})}. \tag{2.30}$$

Remark 2.1.12.

1. In the case of $k = 1$, (2.29) gives inequality (2.26), and (2.30) gives inequality (2.27).
2. In the Euclidean case $\mathbb{G} = \mathbb{R}^n$ and $|\cdot| = |\cdot|_E$ the Euclidean norm, the super weights in the form (2.25) have appeared in [GM08], together with some applications to problems for differential equations. The case of homogeneous groups, as well as the iterative higher order estimates as in Theorem 2.1.11 were analysed in [RSY17b, RSY18b].

Proof of Theorem 2.1.11. We can iterate (2.26). That is, we start with

$$\frac{Q-pm-p}{p}\left\|\frac{(a+b|x|^\alpha)^{\frac{\beta}{p}}}{|x|^{m+1}}f\right\|_{L^p(\mathbb{G})} \leq \left\|\frac{(a+b|x|^\alpha)^{\frac{\beta}{p}}}{|x|^m}\mathcal{R}f\right\|_{L^p(\mathbb{G})}. \tag{2.31}$$

In (2.31) replacing f by $\mathcal{R}f$ we obtain

$$\frac{Q-pm-p}{p}\left\|\frac{(a+b|x|^\alpha)^{\frac{\beta}{p}}}{|x|^{m+1}}\mathcal{R}f\right\|_{L^p(\mathbb{G})} \leq \left\|\frac{(a+b|x|^\alpha)^{\frac{\beta}{p}}}{|x|^m}\mathcal{R}^2 f\right\|_{L^p(\mathbb{G})}. \tag{2.32}$$

On the other hand, replacing m by $m+1$, (2.31) gives

$$\frac{Q-p(m+1)-p}{p}\left\|\frac{(a+b|x|^{\alpha})^{\frac{\beta}{p}}}{|x|^{m+2}}f\right\|_{L^p(\mathbb{G})}\leq\left\|\frac{(a+b|x|^{\alpha})^{\frac{\beta}{p}}}{|x|^{m+1}}\mathcal{R}f\right\|_{L^p(\mathbb{G})}.$$

Combining this with (2.32) we obtain

$$\left(\frac{Q-pm-p}{p}\right)\left(\frac{Q-p(m+1)-p}{p}\right)\left\|\frac{(a+b|x|^{\alpha})^{\frac{\beta}{p}}}{|x|^{m+2}}f\right\|_{L^p(\mathbb{G})}\leq\left\|\frac{(a+b|x|^{\alpha})^{\frac{\beta}{p}}}{|x|^{m}}\mathcal{R}^2f\right\|_{L^p(\mathbb{G})}.$$

This iteration process gives

$$\prod_{j=0}^{k-1}\left(\frac{Q-p}{p}-(m+j)\right)\left\|\frac{(a+b|x|^{\alpha})^{\frac{\beta}{p}}}{|x|^{m+k}}f\right\|_{L^p(\mathbb{G})}\leq\left\|\frac{(a+b|x|^{\alpha})^{\frac{\beta}{p}}}{|x|^{m}}\mathcal{R}^kf\right\|_{L^p(\mathbb{G})}.$$

Similarly, we have for $\alpha\beta<0$, $pm-\alpha\beta\leq Q-2$ and $f\in C_0^{\infty}(\mathbb{G}\backslash\{0\})$ that

$$\prod_{j=0}^{k-1}\left(\frac{Q-p+\alpha\beta}{p}-(m+j)\right)\left\|\frac{(a+b|x|^{\alpha})^{\frac{\beta}{p}}}{|x|^{m+k}}f\right\|_{L^p(\mathbb{G})}\leq\left\|\frac{(a+b|x|^{\alpha})^{\frac{\beta}{p}}}{|x|^{m}}\mathcal{R}^kf\right\|_{L^p(\mathbb{G})},$$

completing the proof. □

2.1.5 Two-weight Hardy inequalities

In this section, using the method of factorization of differential expressions, we obtain Hardy type inequalities with two general weights $\phi(x)$ and $\psi(x)$. The idea of the factorization method can be best illustrated by the following example of an estimate due to Gesztesy and Littlejohn [GL17].

Example 2.1.13 (Gesztesy and Littlejohn two-parameter inequality)**.** Let $\alpha,\beta\in\mathbb{R}$, $x\in\mathbb{R}^n\backslash\{0\}$ and $n\geq 2$. Let us define the operator

$$T_{\alpha,\beta}:=-\Delta+\alpha|x|_E^{-2}x\cdot\nabla+\beta|x|_E^{-2}.$$

One readily checks that its formal adjoint is given by

$$T^{+}_{\alpha,\beta}:=-\Delta-\alpha|x|_E^{-2}x\cdot\nabla+(\beta-\alpha(n-2))|x|_E^{-2}.$$

Using the non-negativity of the operator $T^{+}_{\alpha,\beta}T_{\alpha,\beta}$ on $C_0^\infty(\mathbb{R}^n\backslash\{0\})$, for all $f \in C_0^\infty(\mathbb{R}^n\backslash\{0\})$ we can deduce that

$$\begin{aligned}\int_{\mathbb{R}^n} |(\Delta f)(x)|^2 dx \geq &((n-4)\alpha - 2\beta)\int_{\mathbb{R}^n} |x|_E^{-2}|(\nabla f)(x)|^2 dx \\ &-\alpha(\alpha-4)\int_{\mathbb{R}^n} |x|_E^{-4}|x\cdot(\nabla f)(x)|^2 dx \\ &+\beta((n-4)(\alpha-2)-\beta)\int_{\mathbb{R}^n} |x|_E^{-4}|f(x)|^2 dx.\end{aligned} \tag{2.33}$$

By choosing particular values of α and β it can be checked that this inequality yields classical Rellich and Hardy–Rellich type inequalities as special cases, see [GL17] for details.

On the other hand, using the non-negativity of the operator $T_{\alpha,\beta}T^{+}_{\alpha,\beta}$, it was shown in [RY17] that for $\alpha, \beta \in \mathbb{R}$ and $n \geq 2$, and for all $f \in C_0^\infty(\mathbb{R}^n\backslash\{0\})$ we can deduce another two-parameter inequality

$$\begin{aligned}&\int_{\mathbb{R}^n} |(\Delta f)(x)|^2 dx \\ &\geq (n\alpha - 2\beta)\int_{\mathbb{R}^n} |x|_E^{-2}|(\nabla f)(x)|^2 dx - \alpha(\alpha+4)\int_{\mathbb{R}^n} |x|_E^{-4}|x\cdot(\nabla f)(x)|^2 dx \\ &\quad + (2(n-4)(\alpha(n-2)-\beta) - 2\alpha^2(n-2) + \alpha\beta n - \beta^2)\int_{\mathbb{R}^n} |x|_E^{-4}|f(x)|^2 dx.\end{aligned} \tag{2.34}$$

The following result is a two-weight inequality on general homogeneous groups with general weights.

Theorem 2.1.14 (Two-weight Hardy inequality). *Let $\mathbb{G}$ be a homogeneous group of homogeneous dimension $Q \geq 3$ and let $|\cdot|$ be a homogeneous quasi-norm on $\mathbb{G}$. Let $\phi, \psi \in L^2_{\text{loc}}(\mathbb{G}\backslash\{0\})$ be any real-valued functions such that $\mathcal{R}\phi, \mathcal{R}\psi \in L^2_{\text{loc}}(\mathbb{G}\backslash\{0\})$. Let $\alpha \in \mathbb{R}$. Then for all complex-valued functions $f \in C_0^\infty(\mathbb{G}\backslash\{0\})$ we have inequalities*

$$\begin{aligned}&\int_{\mathbb{G}} (\phi(x))^2|\mathcal{R}f(x)|^2 dx \\ &\quad\geq \alpha\int_{\mathbb{G}} (\phi(x)\mathcal{R}\psi(x) + \psi(x)\mathcal{R}\phi(x))\,|f(x)|^2 dx \\ &\quad\quad + \alpha(Q-1)\int_{\mathbb{G}} \frac{\phi(x)\psi(x)}{|x|}|f(x)|^2 dx - \alpha^2\int_{\mathbb{G}} (\psi(x))^2|f(x)|^2 dx\end{aligned} \tag{2.35}$$

and

$$\begin{aligned}&\int_{\mathbb{G}} (\phi(x))^2|\mathcal{R}f(x)|^2 dx \\ &\quad\geq \alpha\int_{\mathbb{G}} (\psi(x)\mathcal{R}\phi(x) - \phi(x)\mathcal{R}\psi(x))\,|f(x)|^2 dx\end{aligned}$$

$$
\begin{aligned}
&+\alpha(Q-1)\int_{\mathbb{G}}\frac{\phi(x)\psi(x)}{|x|}|f(x)|^2dx-\alpha^2\int_{\mathbb{G}}(\psi(x))^2|f(x)|^2dx\\
&-(Q-1)\int_{\mathbb{G}}\frac{(\phi(x))^2}{|x|^2}|f(x)|^2dx+(Q-1)\int_{\mathbb{G}}\frac{\phi(x)\mathcal{R}\phi(x)}{|x|}|f(x)|^2dx\\
&+\int_{\mathbb{G}}\phi(x)\mathcal{R}^2\phi(x)|f(x)|^2dx. \qquad (2.36)
\end{aligned}
$$

From Theorem 2.1.14 one can get different weighted Hardy inequalities. Let us point out several examples.

Remark 2.1.15.

1. If we take $\phi(x)\equiv 1$ in Theorem 2.1.14, we obtain for all $\alpha\in\mathbb{R}$ and for all $f\in C_0^\infty(\mathbb{G}\backslash\{0\})$ families of inequalities

$$\int_{\mathbb{G}}|\mathcal{R}f(x)|^2dx\geq\int_{\mathbb{G}}\left(\alpha\mathcal{R}\phi(x)+\alpha(Q-1)\frac{\psi(x)}{|x|}-\alpha^2(\psi(x))^2\right)|f(x)|^2dx$$

and

$$
\begin{aligned}
&\int_{\mathbb{G}}|\mathcal{R}f(x)|^2dx\\
&\quad\geq\int_{\mathbb{G}}\left(-\alpha\mathcal{R}\phi(x)+\alpha(Q-1)\frac{\psi(x)}{|x|}-\alpha^2(\psi(x))^2-\frac{Q-1}{|x|^2}\right)|f(x)|^2dx.
\end{aligned}
$$

2. If we take $\phi(x)=|x|^{-a}$ and $\psi(x)=|x|^{-b}$ for $a,b\in\mathbb{R}$, then (2.35) implies that

$$\int_{\mathbb{G}}\frac{|\mathcal{R}f(x)|^2}{|x|^{2a}}dx\geq\alpha(Q-a-b-1)\int_{\mathbb{G}}\frac{|f(x)|^2}{|x|^{a+b+1}}dx-\alpha^2\int_{\mathbb{G}}\frac{|f(x)|^2}{|x|^{2b}}dx.$$

In the case when we take $b=a+1$, we get

$$\int_{\mathbb{G}}\frac{|\mathcal{R}f(x)|^2}{|x|^{2a}}dx\geq(\alpha(Q-2a-2)-\alpha^2)\int_{\mathbb{G}}\frac{|f(x)|^2}{|x|^{2a+2}}dx.$$

Then, by maximizing the constant $(\alpha(Q-2a-2)-\alpha^2)$ with respect to α we obtain the weighted Hardy inequality from Corollary 2.1.6, namely,

$$\int_{\mathbb{G}}\frac{|\mathcal{R}f(x)|^2}{|x|^{2a}}dx\geq\frac{(Q-2a-2)^2}{4}\int_{\mathbb{G}}\frac{|f(x)|^2}{|x|^{2a+2}}dx, \qquad (2.37)$$

for which it is known that the constant in (2.37) is sharp.

3. If we take $\phi(x) = |x|^{-a}(\log|x|)^c$ and $\psi(x) = |x|^{-b}(\log|x|)^d$ for $a, b, c, d \in \mathbb{R}$, then we obtain from (2.35) the inequality

$$\begin{aligned}
&\int_{\mathbb{G}} \frac{(\log|x|)^{2c}}{|x|^{2a}} |\mathcal{R}f(x)|^2 dx \\
&\geq \alpha \int_{\mathbb{G}} \left(\frac{(c+d)(\log|x|)^{c+d-1} + (Q-1-a-b)(\log|x|)^{c+d}}{|x|^{a+b+1}} \right) |f(x)|^2 dx \\
&- \alpha^2 \int_{\mathbb{G}} \frac{(\log|x|)^{2d}}{|x|^{2b}} |f(x)|^2 dx.
\end{aligned}$$

If we take $a = \frac{Q-2}{2}$, $b = \frac{Q}{2}$, $c = 1$ and $d = 0$, it follows that

$$\int_{\mathbb{G}} \frac{(\log|x|)^2}{|x|^{Q-2}} |\mathcal{R}f(x)|^2 dx \geq (\alpha - \alpha^2) \int_{\mathbb{G}} \frac{|f(x)|^2}{|x|^Q} dx.$$

After maximizing the above constant with respect to α we obtain the critical Hardy inequality

$$\int_{\mathbb{G}} \frac{(\log|x|)^2}{|x|^{Q-2}} |\mathcal{R}f(x)|^2 dx \geq \frac{1}{4} \int_{\mathbb{G}} \frac{|f(x)|^2}{|x|^Q} dx, \tag{2.38}$$

recovering the critical inequality in Theorem 2.2.4. There, it is shown that the constant $\frac{1}{4}$ in (2.38) is sharp.

4. We can refer to [GL17] for a thorough discussion of the factorization method, its history and different features. We also refer to [GP80] for obtaining the Hardy inequality and to [Ges84] for logarithmic refinements by this factorization method.

   The inequalities in Theorem 2.1.14 were obtained in [RY17] which we follow for the proof.

Proof of Theorem 2.1.14. Let us introduce the one-parameter differential expression

$$T_\alpha := \phi(x)\mathcal{R} + \alpha\psi(x).$$

One can readily calculate for the formal adjoint operator of T_α on $C_0^\infty(\mathbb{G}\backslash\{0\})$:

$$\begin{aligned}
&\int_{\mathbb{G}} \phi(x)\mathcal{R}f(x)\overline{g(x)}dx + \alpha \int_{\mathbb{G}} \psi(x)f(x)\overline{g(x)}dx \\
&= \int_0^\infty \int_{\wp} \phi(ry)\frac{d}{dr}(f(ry))\overline{g(ry)}r^{Q-1}d\sigma(y)dr + \alpha \int_{\mathbb{G}} \psi(x)f(x)\overline{g(x)}dx \\
&= -\int_0^\infty \int_{\wp} \phi(ry)f(ry)\overline{\frac{d}{dr}(g(ry))}r^{Q-1}d\sigma(y)dr \\
&- (Q-1)\int_0^\infty \int_{\wp} \phi(ry)f(ry)\overline{g(ry)}r^{Q-2}d\sigma(y)dr + \alpha \int_{\mathbb{G}} \psi(x)f(x)\overline{g(x)}dx
\end{aligned}$$

$$
\begin{aligned}
&= \int_{\mathbb{G}} f(x) \left(-\phi(x)\overline{\mathcal{R}g(x)}\right) dx - (Q-1)\int_{\mathbb{G}} f(x) \left(\frac{\phi(x)}{|x|}\overline{g(x)}\right) dx \\
&\quad - \int_{\mathbb{G}} f(x) \left(\mathcal{R}\phi(x)\overline{g(x)}\right) dx + \alpha \int_{\mathbb{G}} f(x) \left(\psi(x)\overline{g(x)}\right) dx.
\end{aligned}
$$

Thus, the formal adjoint operator of T_α has the form

$$
T_\alpha^+ := -\phi(x)\mathcal{R} - \frac{Q-1}{|x|}\phi(x) - \mathcal{R}\phi(x) + \alpha\psi(x),
$$

where $x \neq 0$. Then we have

$$
\begin{aligned}
(T_\alpha^+ T_\alpha f)(x) = &- \phi(x)\mathcal{R}(\phi(x)\mathcal{R}f(x)) - \alpha\phi(x)\mathcal{R}(f(x)\psi(x)) - \frac{Q-1}{|x|}(\phi(x))^2\mathcal{R}f(x) \\
&- \frac{\alpha(Q-1)}{|x|}\phi(x)\psi(x)f(x) - \phi(x)\mathcal{R}\phi(x)\mathcal{R}f(x) - \alpha\psi(x)f(x)\mathcal{R}\phi(x) \\
&+ \alpha\phi(x)\psi(x)\mathcal{R}f(x) + \alpha^2(\psi(x))^2 f(x).
\end{aligned}
$$

By the non-negativity of $T_\alpha^+ T_\alpha$, introducing polar coordinates $(r, y) = (|x|, \frac{x}{|x|}) \in (0, \infty) \times \wp$ on $\mathbb{G}$, where $\wp$ is the quasi-sphere as in (1.12), and using Proposition 1.2.10 one calculates

$$
\begin{aligned}
0 \leq \int_{\mathbb{G}} |(T_\alpha f)(x)|^2 dx &= \int_{\mathbb{G}} f(x)\overline{(T_\alpha^+ T_\alpha f)(x)} dx \\
&= \mathrm{Re} \int_{\mathbb{G}} f(x)\overline{(T_\alpha^+ T_\alpha f)(x)} dx =: I_1 + I_2 + I_3 + I_4 + I_5 + I_6,
\end{aligned}
\tag{2.39}
$$

where we set

$$
\begin{aligned}
I_1 &= -\mathrm{Re} \int_0^\infty \int_\wp f(ry)\phi(ry)\overline{\frac{d}{dr}\left(\phi(ry)\frac{d}{dr}(f(ry))\right)} r^{Q-1} d\sigma(y)dr, \qquad (2.40) \\
I_2 &= -\alpha\mathrm{Re} \int_0^\infty \int_\wp f(ry)\phi(ry)\overline{\frac{d}{dr}(f(ry)\psi(ry))} r^{Q-1} d\sigma(y)dr, \\
I_3 &= -(Q-1)\mathrm{Re} \int_0^\infty \int_\wp f(ry)\frac{(\phi(ry))^2 \overline{\frac{d}{dr}(f(ry))}}{r} r^{Q-1} d\sigma(y)dr, \\
I_4 &= -\alpha(Q-1)\int_{\mathbb{G}} \frac{\phi(x)\psi(x)}{|x|}|f(x)|^2 dx - \alpha\int_{\mathbb{G}} \mathcal{R}\phi(x)\psi(x)|f(x)|^2 dx \\
&\quad + \alpha^2 \int_{\mathbb{G}} (\psi(x))^2 |f(x)|^2 dx, \\
I_5 &= -\mathrm{Re} \int_0^\infty \int_\wp f(ry)\frac{d}{dr}(\phi(ry))\phi(ry)\overline{\frac{d}{dr}(f(ry))} r^{Q-1} d\sigma(y)dr
\end{aligned}
$$

and

$$
I_6 = \alpha\mathrm{Re} \int_0^\infty \int_\wp f(ry)\phi(ry)\psi(ry)\overline{\frac{d}{dr}(f(ry))} r^{Q-1} d\sigma(y)dr.
$$

Now we simplify the sum of the terms I_1, I_2, I_3, I_5 and I_6. By a direct calculation we obtain

$$
\begin{aligned}
I_1 = &(Q-1)\mathrm{Re}\int_0^\infty \int_\wp (\phi(ry))^2 f(ry)\overline{\frac{d}{dr}(f(ry))}r^{Q-2}d\sigma(y)dr \\
&+\mathrm{Re}\int_0^\infty \int_\wp (\phi(ry))^2 \frac{d}{dr}(f(ry))\overline{\frac{d}{dr}(f(ry))}r^{Q-1}d\sigma(y)dr \\
&+\mathrm{Re}\int_0^\infty \int_\wp \phi(ry)\frac{d}{dr}(\phi(ry))\overline{\frac{d}{dr}f(ry)}f(ry)r^{Q-1}d\sigma(y)dr \\
=&\frac{Q-1}{2}\int_0^\infty \int_\wp (\phi(ry))^2 \frac{d}{dr}|f(ry)|^2 r^{Q-2}d\sigma(y)dr \\
&+\int_0^\infty \int_\wp (\phi(ry))^2 \left|\frac{d}{dr}(f(ry))\right|^2 r^{Q-1}d\sigma(y)dr \\
&+\frac{1}{2}\int_0^\infty \int_\wp \phi(ry)\frac{d}{dr}(\phi(ry))\frac{d}{dr}|f(ry)|^2 r^{Q-1}d\sigma(y)dr \\
=&\int_0^\infty \int_\wp (\phi(ry))^2 \left|\frac{d}{dr}(f(ry))\right|^2 r^{Q-1}d\sigma(y)dr \\
&-\frac{(Q-1)(Q-2)}{2}\int_0^\infty \int_\wp (\phi(ry))^2 |f(ry)|^2 r^{Q-3}d\sigma(y)dr \\
&-(Q-1)\int_0^\infty \int_\wp \phi(ry)\frac{d}{dr}(\phi(ry))|f(ry)|^2 r^{Q-2}d\sigma(y)dr \\
&-\frac{1}{2}\int_0^\infty \int_\wp \left(\frac{d}{dr}(\phi(ry))\right)^2 |f(ry)|^2 r^{Q-1}d\sigma(y)dr \\
&-\frac{1}{2}\int_0^\infty \int_\wp \phi(ry)\frac{d^2}{dr^2}(\phi(ry))|f(ry)|^2 r^{Q-1}d\sigma(y)dr \\
&-\frac{Q-1}{2}\int_0^\infty \int_\wp \phi(ry)\frac{d}{dr}(\phi(ry))|f(ry)|^2 r^{Q-2}d\sigma(y)dr \\
=&\int_{\mathbb{G}} (\phi(x))^2|\mathcal{R}f(x)|^2dx - \frac{(Q-1)(Q-2)}{2}\int_{\mathbb{G}} \frac{(\phi(x))^2}{|x|^2}|f(x)|^2dx \\
&-(Q-1)\int_{\mathbb{G}} \frac{\phi(x)\mathcal{R}\phi(x)}{|x|}|f(x)|^2dx - \frac{1}{2}\int_{\mathbb{G}} (\mathcal{R}\phi(x))^2|f(x)|^2dx \\
&-\frac{1}{2}\int_{\mathbb{G}} \mathcal{R}^2\phi(x)\phi(x)|f(x)|^2dx - \frac{Q-1}{2}\int_{\mathbb{G}} \frac{\phi(x)\mathcal{R}\phi(x)}{|x|}|f(x)|^2dx.
\end{aligned}
$$

Now we calculate I_2 as follows,

$$
I_2 = -\alpha\mathrm{Re}\int_0^\infty \int_\wp \phi(ry)\psi(ry)f(ry)\overline{\frac{d}{dr}(f(ry))}r^{Q-1}d\sigma(y)dr
$$

$$
\begin{aligned}
&-\alpha\int_0^\infty\int_\wp \phi(ry)\frac{d}{dr}(\psi(ry))|f(ry)|^2 r^{Q-1}d\sigma(y)dr\\
=&-\frac{\alpha}{2}\int_0^\infty\int_\wp \phi(ry)\psi(ry)\frac{d}{dr}|f(ry)|^2 r^{Q-1}d\sigma(y)dr\\
&-\alpha\int_0^\infty\int_\wp \phi(ry)\frac{d}{dr}(\psi(ry))|f(ry)|^2 r^{Q-1}d\sigma(y)dr\\
=&-\alpha\int_0^\infty\int_\wp \phi(ry)\frac{d}{dr}(\psi(ry))|f(ry)|^2 r^{Q-1}d\sigma(y)dr\\
&+\frac{\alpha}{2}\int_0^\infty\int_\wp \psi(ry)\frac{d}{dr}(\phi(ry))|f(ry)|^2 r^{Q-1}d\sigma(y)dr\\
&+\frac{\alpha}{2}\int_0^\infty\int_\wp \phi(ry)\frac{d}{dr}(\psi(ry))|f(ry)|^2 r^{Q-1}d\sigma(y)dr\\
&+\frac{\alpha(Q-1)}{2}\int_0^\infty\int_\wp \phi(ry)\psi(ry)|f(ry)|^2 r^{Q-2}d\sigma(y)dr\\
=&\frac{\alpha}{2}\int_{\mathbb{G}}\psi(x)\mathcal{R}\phi(x)|f(x)|^2dx+\frac{\alpha}{2}\int_{\mathbb{G}}\phi(x)\mathcal{R}\psi(x)|f(x)|^2dx\\
&+\frac{\alpha(Q-1)}{2}\int_{\mathbb{G}}\frac{\phi(x)\psi(x)}{|x|}|f(x)|^2dx-\alpha\int_{\mathbb{G}}\phi(x)\mathcal{R}\psi(x)|f(x)|^2dx.
\end{aligned}
$$

For I_3, one has

$$
\begin{aligned}
I_3=&-(Q-1)\mathrm{Re}\int_0^\infty\int_\wp (\phi(ry))^2 f(ry)\overline{\frac{d}{dr}(f(ry))}r^{Q-2}d\sigma(y)dr\\
=&-\frac{Q-1}{2}\int_0^\infty\int_\wp (\phi(ry))^2\frac{d}{dr}|f(ry)|^2 r^{Q-2}d\sigma(y)dr\\
=&(Q-1)\int_0^\infty\int_\wp \phi(ry)\frac{d}{dr}(\phi(ry))|f(ry)|^2 r^{Q-2}d\sigma(y)dr\\
&+\frac{(Q-1)(Q-2)}{2}\int_0^\infty\int_\wp (\phi(ry))^2|f(ry)|^2 r^{Q-3}d\sigma(y)dr\\
=&(Q-1)\int_{\mathbb{G}}\frac{\phi(x)\mathcal{R}\phi(x)}{|x|}|f(x)|^2dx+\frac{(Q-1)(Q-2)}{2}\int_{\mathbb{G}}\frac{(\phi(x))^2}{|x|^2}|f(x)|^2dx.
\end{aligned}
$$

For I_5, we have

$$
\begin{aligned}
I_5=&-\mathrm{Re}\int_0^\infty\int_\wp f(ry)\frac{d}{dr}(\phi(ry))\phi(ry)\overline{\frac{d}{dr}(f(ry))}r^{Q-1}d\sigma(y)dr\\
=&-\frac{1}{2}\int_0^\infty\int_\wp \frac{d}{dr}(\phi(ry))\phi(ry)\frac{d}{dr}|f(ry)|^2 r^{Q-1}d\sigma(y)dr\\
=&\frac{1}{2}\int_0^\infty\int_\wp \phi(ry)\frac{d^2}{dr^2}(\phi(ry))|f(ry)|^2 r^{Q-1}d\sigma(y)dr
\end{aligned}
$$

$$
\begin{aligned}
&+\frac{1}{2}\int_0^\infty\int_\wp\left(\frac{d}{dr}(\phi(ry))\right)^2|f(ry)|^2r^{Q-1}d\sigma(y)dr\\
&+\frac{Q-1}{2}\int_0^\infty\int_\wp\frac{\phi(ry)\frac{d}{dr}(\phi(ry))}{r}|f(ry)|^2r^{Q-1}d\sigma(y)dr\\
=&\frac{1}{2}\int_{\mathbb{G}}\mathcal{R}^2\phi(x)\phi(x)|f(x)|^2dx+\frac{1}{2}\int_{\mathbb{G}}(\mathcal{R}\phi(x))^2|f(x)|^2dx\\
&+\frac{Q-1}{2}\int_{\mathbb{G}}\frac{\mathcal{R}\phi(x)\phi(x)}{|x|}|f(x)|^2dx.
\end{aligned}
$$

Finally, for I_6 we obtain

$$
\begin{aligned}
I_6=&\alpha\mathrm{Re}\int_0^\infty\int_\wp f(ry)\phi(ry)\psi(ry)\overline{\frac{d}{dr}(f(ry))}r^{Q-1}d\sigma(y)dr \qquad (2.41)\\
=&\frac{\alpha}{2}\int_0^\infty\int_\wp\phi(ry)\psi(ry)\frac{d}{dr}|f(ry)|^2r^{Q-1}d\sigma(y)dr\\
=&-\frac{\alpha}{2}\int_0^\infty\int_\wp\frac{d}{dr}(\phi(ry))\psi(ry)|f(ry)|^2r^{Q-1}d\sigma(y)dr\\
&-\frac{\alpha}{2}\int_0^\infty\int_\wp\frac{d}{dr}(\psi(ry))\phi(ry)|f(ry)|^2r^{Q-1}d\sigma(y)dr\\
&-\frac{\alpha(Q-1)}{2}\int_0^\infty\int_\wp\phi(ry)\psi(ry)\frac{|f(ry)|^2}{r}r^{Q-1}d\sigma(y)dr\\
=&-\frac{\alpha}{2}\int_{\mathbb{G}}\mathcal{R}\phi(x)\psi(x)|f(x)|^2dx-\frac{\alpha}{2}\int_{\mathbb{G}}\phi(x)\mathcal{R}\psi(x)|f(x)|^2dx\\
&-\frac{\alpha(Q-1)}{2}\int_{\mathbb{G}}\frac{\phi(x)\psi(x)}{|x|}|f(x)|^2dx.
\end{aligned}
$$

Putting (2.40)–(2.41) in (2.39), we obtain that

$$
\begin{aligned}
&\int_{\mathbb{G}}(\phi(x))^2|\mathcal{R}f(x)|^2dx\\
&-\alpha\int_{\mathbb{G}}\left(\phi(x)\mathcal{R}\psi(x)+\psi(x)\mathcal{R}\phi(x)+(Q-1)\frac{\phi(x)\psi(x)}{|x|}\right)|f(x)|^2dx\\
&+\alpha^2\int_{\mathbb{G}}(\psi(x))^2|f(x)|^2dx\geq 0,
\end{aligned}
$$

which implies (2.35).

Thus, we have obtained (2.35) using the non-negativity of $T_\alpha^+T_\alpha$. Now we can obtain (2.36) using the non-negativity of $T_\alpha T_\alpha^+$. Similar to the above we calculate

$$
(T_\alpha T_\alpha^+ f)(x)=-\phi(x)\mathcal{R}(\phi(x)\mathcal{R}f(x))+\alpha\phi(x)\mathcal{R}(f(x)\psi(x))-\frac{Q-1}{|x|}(\phi(x))^2\mathcal{R}f(x)
$$

$$
\begin{aligned}
&-\frac{\alpha(Q-1)}{|x|}\phi(x)\psi(x)f(x)-\phi(x)\mathcal{R}\phi(x)\mathcal{R}f(x)-\alpha\psi(x)f(x)\mathcal{R}\phi(x)\\
&-\alpha\phi(x)\psi(x)\mathcal{R}f(x)+\alpha^2(\psi(x))^2f(x)\\
&-(Q-1)\phi(x)f(x)\mathcal{R}\left(\frac{\phi(x)}{|x|}\right)-\phi(x)f(x)\mathcal{R}^2\phi(x).
\end{aligned}
$$

Using the non-negativity of $T_\alpha T_\alpha^+$ we get

$$
\begin{aligned}
0&\le\int_{\mathbb{G}}|(T_\alpha f)(x)|^2dx=\int_{\mathbb{G}}f(x)\overline{(T_\alpha T_\alpha^+ f)(x)}dx\\
&=\mathrm{Re}\int_{\mathbb{G}}f(x)\overline{(T_\alpha T_\alpha^+ f)(x)}dx=\widetilde{I}_1+\widetilde{I}_2+\widetilde{I}_3+\widetilde{I}_4+\widetilde{I}_5+\widetilde{I}_6,
\end{aligned}
$$

where

$$
\widetilde{I}_1=I_1,\ \widetilde{I}_2=-I_2,\ \widetilde{I}_3=I_3,\ \widetilde{I}_4=I_4+A,\ \widetilde{I}_5=I_5,\ \widetilde{I}_6=-I_6
$$

with

$$
A=-(Q-1)\int_{\mathbb{G}}\phi(x)|f(x)|^2\mathcal{R}\left(\frac{\phi(x)}{|x|}\right)dx-\int_{\mathbb{G}}\phi(x)\mathcal{R}^2\phi(x)|f(x)|^2dx.
$$

Taking into account these and (2.40)–(2.41), we obtain (2.36). □

To finish this section, we observe another version of weighted Hardy inequality with the radial derivative. Anticipating the material presented in later chapters, the proof will be based on the integral Hardy inequality from Theorem 5.1.1.

Theorem 2.1.16 (Weighted Hardy inequality for radial functions). *Let $\mathbb{G}$ be a homogeneous group of homogeneous dimension Q. Let $\phi>0$, $\psi>0$ be positive weight functions on $\mathbb{G}$ and let $1<p\le q<\infty$. Then there exists a positive constant $C>0$ such that*

$$
\left(\int_{\mathbb{G}}\phi(x)|f(x)|^qdx\right)^{1/q}\le C\left(\int_{\mathbb{G}}\psi(x)|\mathcal{R}f(x)|^pdx\right)^{1/p}\tag{2.42}
$$

holds for all radial functions f with $f(0)=0$ if and only if

$$
\sup_{R>0}\left(\int_{|x|\ge R}\phi(x)dx\right)^{1/q}\left(\int_0^R\left(\int_{\wp}r^{Q-1}\psi(ry)d\sigma(y)\right)^{1-p'}dr\right)^{1/p'}<\infty.\tag{2.43}
$$

Remark 2.1.17. In the Abelian case $\mathbb{G}=(\mathbb{R}^n,+)$ and $Q=n$, (2.42) was obtained in [DHK97] and in [Saw84]. On homogeneous groups it was observed in [RY18a] and we follow the proof there.

Proof of Theorem 2.1.16. For $r=|x|$, let us denote $\tilde{f}(r)=f(x)$. We also denote

$$
\Phi(r):=\int_{\wp}r^{Q-1}\phi(ry)d\sigma(y),\quad \Psi(r):=\int_{\wp}r^{Q-1}\psi(ry)d\sigma(y).
$$

Consequently, using $\tilde{f}(0)=0$, we have

$$\begin{aligned}\left(\int_{\mathbb{G}} \phi(x)|f(x)|^q dx\right)^{1/q} &= \left(\int_{\wp}\int_0^\infty r^{Q-1}\phi(ry)|\tilde{f}(r)|^q dr d\sigma(y)\right)^{1/q} \\ &= \left(\int_0^\infty \Phi(r)|\tilde{f}(r)|^q dr\right)^{1/q} \\ &= \left(\int_0^\infty \Phi(r)\left|\int_0^r \mathcal{R}\tilde{f}(r)dr\right|^q dr\right)^{1/q} \\ &\leq C\left(\int_0^\infty \Psi(r)\left|\mathcal{R}\tilde{f}(r)\right|^p dr\right)^{1/p} \\ &= C\left(\int_{\mathbb{G}} \psi(x)|\mathcal{R}f(x)|^p dx\right)^{1/p}\end{aligned}$$

if and only if the condition (2.43) holds by Theorem 5.1.1, namely by (5.2) and (5.3). □

2.2 Critical Hardy inequalities

In this section we discuss critical Hardy inequalities. The critical behaviour may be manifested with respect to different parameters. For example, one major critical case arises when we have $p=Q$ in (2.2). In this case, in the Euclidean case of $\mathbb{R}^n$ with $p=Q=n$ it is known that the Hardy inequality (2.1) fails for any constant, see, e.g., [ET99] and [IIO16a], and references therein. In such critical cases it is natural to expect the appearance of the logarithmic terms.

One version of such a critical case is the inequality

$$\sup_{R>0}\left\|\frac{f-f_R}{|x|\log\frac{R}{|x|}}\right\|_{L^Q(\mathbb{G})} \leq \frac{Q}{Q-1}\|\mathcal{R}f\|_{L^Q(\mathbb{G})}, \quad Q\geq 2, \tag{2.44}$$

for all $f\in C_0^\infty(\mathbb{G}\backslash\{0\})$, where we denote

$$f_R(x) := f\left(R\frac{x}{|x|}\right) \quad \text{for } x\in\mathbb{G} \text{ and } R>0.$$

In fact, this inequality is a special case (with $p=Q$) of the following more general family of critical inequalities derived in Theorem 2.2.1, namely,

$$\sup_{R>0}\left\|\frac{f-f_R}{|x|^{\frac{Q}{p}}\log\frac{R}{|x|}}\right\|_{L^p(\mathbb{G})} \leq \frac{p}{p-1}\left\|\frac{1}{|x|^{\frac{Q}{p}-1}}\mathcal{R}f\right\|_{L^p(\mathbb{G})}, \tag{2.45}$$

for all $1<p<\infty$. Here the constant $\frac{p}{p-1}$ in (2.45) is sharp but is in general unattainable.

The inequalities (2.45) are all critical with respect to their weight $|x|^{-\frac{Q}{p}}$ in L^p-space because its pth power gives $|x|^{-Q}$ which is the critical order for the integrability at zero and at infinity in $L^p(\mathbb{G})$.

Moreover, we show another type of a critical Hardy inequality on general homogeneous groups with the logarithm being on the right-hand side:

$$\left\| \frac{f}{|x|} \right\|_{L^Q(\mathbb{G})} \leq Q \|(\log |x|)\mathcal{R}f\|_{L^Q(\mathbb{G})}. \tag{2.46}$$

Furthermore, we also give improved versions of the Hardy inequality (2.1) on quasi-balls of homogeneous (Lie) groups, the so-called Hardy–Sobolev type inequalities.

2.2.1 Critical Hardy inequalities

First we discuss a family of generalized critical Hardy inequalities on the homogeneous group $\mathbb{G}$ mentioned in (2.45). In this section, $\mathbb{G}$ is a homogeneous group of homogeneous dimension $Q \geq 2$ and $|\cdot|$ is an arbitrary homogeneous quasi-norm on $\mathbb{G}$.

Theorem 2.2.1 (A family of critical Hardy inequalities). *Let $f \in C_0^\infty(\mathbb{G}\backslash\{0\})$ and denote $f_R(x) := f(R\frac{x}{|x|})$ for $x \in \mathbb{G}$ and $R > 0$. Then we have the inequalities*

$$\sup_{R>0} \left\| \frac{f - f_R}{|x|^{\frac{Q}{p}} \log \frac{R}{|x|}} \right\|_{L^p(\mathbb{G})} \leq \frac{p}{p-1} \left\| \frac{1}{|x|^{\frac{Q}{p}-1}} \mathcal{R}f \right\|_{L^p(\mathbb{G})}, \quad 1 < p < \infty, \tag{2.47}$$

where the constant $\frac{p}{p-1}$ is sharp. Moreover, denoting

$$u_R(x) := \frac{f(x) - f_R(x)}{|x|^{\frac{Q}{p}} \log \frac{R}{|x|}} \quad \textit{and} \quad v(x) := \frac{1}{|x|^{\frac{Q}{p}-1}} \mathcal{R}f(x),$$

for each $R > 0$ we have the following expression for the remainder:

$$\begin{aligned} \left\| \frac{f - f_R}{|x|^{\frac{Q}{p}} \log \frac{R}{|x|}} \right\|^p_{L^p(\mathbb{G})} &= \left(\frac{p}{p-1} \right)^p \left\| \frac{1}{|x|^{\frac{Q}{p}-1}} \mathcal{R}f \right\|^p_{L^p(\mathbb{G})} \\ &\quad - p \int_{\mathbb{G}} I(u_R, -\frac{p}{p-1}v) \left| \frac{p}{p-1}v + u_R \right|^2 dx, \end{aligned} \tag{2.48}$$

where I is defined by

$$I(u, g) := \left(\frac{1}{p}|g|^p + \left(1 - \frac{1}{p}\right) |u|^p - |u|^{p-2} \mathrm{Re}(u\overline{g}) \right) |u - g|^{-2} \geq 0, \quad u \neq g,$$

$$I(g, g) := \frac{p-1}{2} |g|^{p-2}.$$

Remark 2.2.2.

1. For $p = Q$ the inequality (2.47) becomes

$$\sup_{R>0}\left\|\frac{f-f_R}{|x|\log\frac{R}{|x|}}\right\|_{L^Q(\mathbb{G})}\leq\frac{Q}{Q-1}\|\mathcal{R}f\|_{L^Q(\mathbb{G})},\quad Q\geq 2,\tag{2.49}$$

which gives the mentioned critical estimate (2.44).

2. In the Euclidean isotropic case $\mathbb{G}=\mathbb{R}^n$ with the Euclidean norm $|\cdot|_E$, similar to Remark 2.1.2, Part 1, inequalities (2.47) yields inequalities for the usual gradient ∇ in $\mathbb{R}^n$ for all $1<p<\infty$:

$$\sup_{R>0}\left\|\frac{f-f_R}{|x|_E^{\frac{n}{p}}\log\frac{R}{|x|_E}}\right\|_{L^p(\mathbb{R}^n)}\leq\frac{p}{p-1}\left\|\frac{1}{|x|_E^{\frac{n}{p}-1}}\nabla f\right\|_{L^p(\mathbb{R}^n)},\tag{2.50}$$

where $f_R(x):=f(R\frac{x}{|x|_E})$ for $x\in\mathbb{R}^n$ and $R>0$. The critical inequality (2.49) becomes

$$\sup_{R>0}\left\|\frac{f-f_R}{|x|_E\log\frac{R}{|x|_E}}\right\|_{L^n(\mathbb{R}^n)}\leq\frac{n}{n-1}\|\nabla f\|_{L^n(\mathbb{R}^n)},\quad n\geq 2.\tag{2.51}$$

3. For the sharpness of the constant $\frac{p}{p-1}$ in (2.47) we refer to the Euclidean case with the Euclidean norm where this constant is known to be sharp but is in general unattainable. This was shown in [IIO16a]. In the case of general homogeneous groups this follows from the expression (2.48) for the remainder.

4. For $p=2$, in the Euclidean case $\mathbb{G}=\mathbb{R}^n$ with the Euclidean norm, the estimate (2.50) was shown in [MOW17a]. In principle, this case can be also obtained from [MOW15a, Theorem 1.1]. Inequality (2.51) in bounded domains of $\mathbb{R}^n$ was analysed in [II15].

Proof of Theorem 2.2.1. Using the polar decomposition in Proposition 1.2.10, a straightforward calculation shows

$$\begin{aligned}&\int_{B(0,R)}\frac{|f(x)-f_R(x)|^p}{|x|^Q|\log\frac{R}{|x|}|^p}dx\\&=\int_0^R\int_{\wp}\frac{|f(ry)-f(Ry)|^p}{r^Q\left(\log\frac{R}{r}\right)^p}r^{Q-1}d\sigma(y)dr\\&=\int_0^R\frac{d}{dr}\left(\frac{1}{p-1}\frac{1}{\left(\log\frac{R}{r}\right)^{p-1}}\int_{\wp}|f(ry)-f(Ry)|^pd\sigma(y)\right)dr\end{aligned}$$

$$-\frac{p}{p-1}\mathrm{Re}\int_0^R\left(\frac{1}{(\log\frac{R}{r})^{p-1}}\int_{\wp}|f(ry)-f(Ry)|^{p-2}(f(ry)-f(Ry))\overline{\frac{df(ry)}{dr}}d\sigma(y)\right)dr$$

$$=-\frac{p}{p-1}\mathrm{Re}\int_0^R\left(\frac{1}{(\log\frac{R}{r})^{p-1}}\int_{\wp}|f(ry)-f(Ry)|^{p-2}(f(ry)-f(Ry))\overline{\frac{df(ry)}{dr}}d\sigma(y)\right)dr,$$

where σ is the Borel measure on $\wp$ and the contribution on the boundary at $r = R$ vanishes due to the inequalities

$$|f(ry)-f(Ry)|\le C(R-r),\qquad \frac{R-r}{R}\le\log\frac{R}{r}.$$

By using the formula (1.30) we obtain

$$\int_{B(0,R)}\frac{|f(x)-f_R(x)|^p}{|x|^Q|\log\frac{R}{|x|}|^p}dx$$

$$=-\frac{p}{p-1}\mathrm{Re}\int_0^R\frac{1}{(\log\frac{R}{r})^{p-1}}\int_{\wp}|f(ry)-f(Ry)|^{p-2}(f(ry)-f(Ry))\overline{\frac{df(ry)}{dr}}d\sigma(y)dr$$

$$=-\frac{p}{p-1}\mathrm{Re}\int_{B(0,R)}\left|\frac{f(x)-f_R(x)}{|x|^{\frac{Q}{p}}\log\frac{R}{|x|}}\right|^{p-2}\frac{(f(x)-f_R(x))}{|x|^{\frac{Q}{p}}\log\frac{R}{|x|}}\frac{1}{|x|^{\frac{Q}{p}-1}}\overline{\frac{df(x)}{d|x|}}dx.$$

Similarly, one has

$$\int_{B^c(0,R)}\frac{|f(x)-f_R(x)|^p}{|x|^Q|\log\frac{R}{|x|}|^p}dx=\int_R^\infty\int_{\wp}\frac{|f(ry)-f(Ry)|^p}{r^Q(\log\frac{r}{R})^p}r^{Q-1}d\sigma(y)dr$$

$$=-\int_R^\infty\frac{d}{dr}\left(\frac{1}{p-1}\frac{1}{(\log\frac{r}{R})^{p-1}}\int_{\wp}|f(ry)-f(Ry)|^pd\sigma(y)\right)dr$$

$$+\frac{p}{p-1}\mathrm{Re}\int_R^\infty\left(\frac{1}{(\log\frac{r}{R})^{p-1}}\int_{\wp}|f(ry)-f(Ry)|^{p-2}(f(ry)-f(Ry))\overline{\frac{df(ry)}{dr}}d\sigma(y)\right)dr$$

$$=-\frac{p}{p-1}\mathrm{Re}\int_{B^c(0,R)}\left|\frac{f(x)-f_R(x)}{|x|^{\frac{Q}{p}}\log\frac{R}{|x|}}\right|^{p-2}\frac{(f(x)-f_R(x))}{|x|^{\frac{Q}{p}}\log\frac{R}{|x|}}\frac{1}{|x|^{\frac{Q}{p}-1}}\overline{\frac{df(x)}{d|x|}}dx.$$

This implies that

$$\int_{\mathbb{G}}\frac{|f(x)-f_R(x)|^p}{|x|^Q|\log\frac{R}{|x|}|^p}dx$$

$$=-\frac{p}{p-1}\mathrm{Re}\int_{\mathbb{G}}\left|\frac{f(x)-f_R(x)}{|x|^{\frac{Q}{p}}\log\frac{R}{|x|}}\right|^{p-2}\frac{(f(x)-f_R(x))}{|x|^{\frac{Q}{p}}\log\frac{R}{|x|}}\frac{1}{|x|^{\frac{Q}{p}-1}}\overline{\frac{df(x)}{d|x|}}dx$$

$$=\left(\frac{p}{p-1}\right)^p\|v\|^p_{L^p(\mathbb{G})}-p\int_{\mathbb{G}}I(u,-\frac{p}{p-1}v)\left|\frac{p}{p-1}v+u\right|^2dx,$$

with

$$u = \frac{f(x) - f_R(x)}{|x|^{\frac{Q}{p}} \log \frac{R}{|x|}}, \qquad v = \frac{1}{|x|^{\frac{Q}{p}-1}} \frac{df(x)}{d|x|},$$

and I is defined by

$$I(f,g) := \left(\frac{1}{p}|g|^p + \frac{1}{p'}|f|^p - |f|^{p-2}\operatorname{Re}(f\overline{g})\right)|f-g|^{-2} \geq 0,\ f \neq g,\ \frac{1}{p} + \frac{1}{p'} = 1,$$

$$I(g,g) := \frac{p-1}{2}|g|^{p-2}.$$

Thus, we establish

$$\|u\|^p_{L^p(\mathbb{G})} = \left(\frac{p}{p-1}\right)^p \|v\|^p_{L^p(\mathbb{G})} - p\int_{\mathbb{G}} I\left(u, -\frac{p}{p-1}v\right)\left|\frac{p}{p-1}v + u\right|^2 dx.$$

This proves the equality (2.48) and the inequality (2.47) since the last term is non-positive. □

We have the following consequence of Theorem 2.2.1:

Corollary 2.2.3 (Critical uncertainty type principles). *Let $1 < p < \infty$ and $f \in C_0^\infty(\mathbb{G}\backslash\{0\})$. Then for any $R > 0$ and $\frac{1}{p} + \frac{1}{q} = \frac{1}{2}$ with $q > 1$, we have*

$$\left\|\frac{1}{|x|^{\frac{Q}{p}-1}}\mathcal{R}f\right\|_{L^p(\mathbb{G})} \|f\|_{L^q(\mathbb{G})} \geq \frac{p-1}{p}\left\|\frac{f(f-f_R)}{|x|^{\frac{Q}{p}}\log\frac{R}{|x|}}\right\|_{L^2(\mathbb{G})} \tag{2.52}$$

and also

$$\left\|\frac{1}{|x|^{\frac{Q}{p}-1}}\mathcal{R}f\right\|_{L^p(\mathbb{G})} \left\|\frac{f-f_R}{|x|^{\frac{Q}{p'}}\log\frac{R}{|x|}}\right\|_{L^{p'}(\mathbb{G})} \geq \frac{p-1}{p}\left\|\frac{f-f_R}{|x|^{\frac{Q}{2}}\log\frac{R}{|x|}}\right\|^2_{L^2(\mathbb{G})} \tag{2.53}$$

for $\frac{1}{p} + \frac{1}{p'} = 1$.

Proof of Corollary 2.2.3. By Theorem 2.2.1 and Hölder's inequality we have

$$\left\|\frac{1}{|x|^{\frac{Q}{p}-1}}\mathcal{R}f\right\|_{L^p(\mathbb{G})} \|f\|_{L^q(\mathbb{G})} \geq \frac{p-1}{p}\left\|\frac{f-f_R}{|x|^{\frac{Q}{p}}\log\frac{R}{|x|}}\right\|_{L^p(\mathbb{G})} \|f\|_{L^q(\mathbb{G})}$$

$$= \frac{p-1}{p}\left(\int_{\mathbb{G}}\left|\frac{f-f_R}{|x|^{\frac{Q}{p}}\log\frac{R}{|x|}}\right|^{2\frac{p}{2}} dx\right)^{\frac{1}{2}\frac{2}{p}} \left(\int_{\mathbb{G}}|f|^{2\frac{q}{2}}dx\right)^{\frac{1}{2}\frac{2}{q}}$$

$$\geq \frac{p-1}{p}\left(\int_{\mathbb{G}}\left|\frac{f(f-f_R)}{|x|^{\frac{Q}{p}}\log\frac{R}{|x|}}\right|^2 dx\right)^{1/2} = \frac{p-1}{p}\left\|\frac{f(f-f_R)}{|x|^{\frac{Q}{p}}\log\frac{R}{|x|}}\right\|_{L^2(\mathbb{G})}.$$

The formula (2.52) is proved. The proof of (2.53) is similar so we can omit it. □

2.2.2 Another type of critical Hardy inequality

In this section we analyse another type of a critical Hardy inequality when the logarithmic term appears on the other side of the inequality. This extends inequality (2.38) that was obtained for $p = 2$ by a factorization method.

Theorem 2.2.4 (Critical Hardy inequality). *Let $\mathbb{G}$ be a homogeneous group of homogeneous dimension $Q \geq 1$ and let $|\cdot|$ be a homogeneous quasi-norm on $\mathbb{G}$. Let $1 < p < \infty$. Then for any complex-valued function $f \in C_0^\infty(\mathbb{G}\backslash\{0\})$ we have*

$$\int_{\mathbb{G}} \frac{|f(x)|^p}{|x|^Q} dx \leq p^p \int_{\mathbb{G}} \frac{|\log |x||^p}{|x|^{Q-p}} |\mathcal{R} f(x)|^p dx, \tag{2.54}$$

and the constant p^p in this inequality is sharp.

Remark 2.2.5.

1. Inequality (2.54) was mentioned in (2.46) as another type of a critical Hardy inequality in the critical case of $p = Q$, in which case we have

$$\int_{\mathbb{G}} \frac{|f(x)|^Q}{|x|^Q} dx \leq Q^Q \int_{\mathbb{G}} |(\log |x|) \mathcal{R} f(x)|^Q dx. \tag{2.55}$$

2. In the Euclidean case $\mathbb{G} = (\mathbb{R}^n, +)$ we have $Q = n$, so for any quasi-norm $|\cdot|$ on $\mathbb{R}^n$ the inequality (2.54) implies a new inequality with the optimal constant: For each $f \in C_0^\infty(\mathbb{R}^n\backslash\{0\})$, we have

$$\left\| \frac{f}{|x|} \right\|_{L^n(\mathbb{R}^n)} \leq n \left\| (\log |x|) \frac{x}{|x|} \cdot \nabla f \right\|_{L^n(\mathbb{R}^n)}. \tag{2.56}$$

   If we take now the standard Euclidean distance $|x|_E = \sqrt{x_1^2 + \cdots + x_n^2}$, it follows that we have

$$\left\| \frac{f}{|x|_E} \right\|_{L^n(\mathbb{R}^n)} \leq n \left\| (\log |x|_E) \nabla f \right\|_{L^n(\mathbb{R}^n)}, \tag{2.57}$$

   for all $f \in C_0^\infty(\mathbb{R}^n\backslash\{0\})$, where ∇ is the standard gradient in $\mathbb{R}^n$. The constants in the above inequalities are sharp.

3. The inequality (2.57) and its consequence are analogous to the critical Hardy inequality of Edmunds and Triebel [ET99] that they showed in $\mathbb{R}^n$ for the Euclidean norm $|\cdot|_E$ in bounded domains $B \subset \mathbb{R}^n$:

$$\left\| \frac{f(x)}{|x|_E (1 + \log \frac{1}{|x|_E})} \right\|_{L^n(B)} \leq \frac{n}{n-1} \|\nabla f\|_{L^n(B)}, \quad n \geq 2, \tag{2.58}$$

   with sharp constant $\frac{n}{n-1}$, which was also discussed in [AS06]. This inequality was also shown to be equivalent to the critical case of the Sobolev–Lorentz inequality. However, a different feature of (2.57) compared to (2.58) is that the logarithmic term enters the other side of the inequality. Inequalities of this type have been investigated in [RS16a].

Proof of Theorem 2.2.4. Let $R > 0$ be such that $\mathrm{supp} f \subset B(0, R)$. A direct calculation using the polar decomposition in Proposition 1.2.10 with integration by parts yields

$$\begin{aligned}\int_{B(0,R)} \frac{|f(x)|^p}{|x|^Q} dx &= \int_0^R \int_{\wp} |f(\delta_r(y))|^p r^{Q-1-Q} d\sigma(y) dr \\ &= -p \int_0^R \log r \mathrm{Re} \int_{\wp} |f(\delta_r(y))|^{p-2} f(\delta_r(y)) \overline{\frac{df(\delta_r(y))}{dr}} d\sigma(y) dr \\ &\leq p \int_{B(0,R)} \frac{|\mathcal{R} f(x)||f(x)|^{p-1}}{|x|^{Q-1}} |\log |x|| dx \\ &= p \int_{B(0,R)} \frac{|\mathcal{R} f(x)||\log |x||}{|x|^{\frac{Q}{p}-1}} \frac{|f(x)|^{p-1}}{|x|^{\frac{Q(p-1)}{p}}} dx,\end{aligned}$$

and by using the Hölder inequality, we obtain that

$$\int_{B(0,R)} \frac{|f(x)|^p}{|x|^Q} dx \leq p \left(\int_{B(0,R)} \frac{|\mathcal{R} f(x)|^p |\log |x||^p}{|x|^{Q-p}} dx \right)^{\frac{1}{p}} \left(\int_{B(0,R)} \frac{|f(x)|^p}{|x|^Q} dx \right)^{\frac{p-1}{p}},$$

which gives (2.54).

Now it remains to show the optimality of the constant, so we need to check the equality condition in the above Hölder inequality. Let us consider the test function

$$h(x) = \log |x|.$$

Thus, we have

$$\left(\frac{|\mathcal{R} h(x)||\log |x||}{|x|^{\frac{Q}{p}-1}} \right)^p = \left(\frac{|h(x)|^{p-1}}{|x|^{\frac{Q(p-1)}{p}}} \right)^{\frac{p}{p-1}},$$

which satisfies the equality condition in Hölder's inequality. This gives the optimality of the constant p^p in (2.54). □

As usual, the Hardy inequality implies the corresponding uncertainty principle:

Corollary 2.2.6 (Another type of critical uncertainty principle). *Let $\mathbb{G}$ be a homogeneous group of homogeneous dimension $Q \geq 2$. Let $|\cdot|$ be a homogeneous quasi-norm on $\mathbb{G}$. Then for each $f \in C_0^\infty(\mathbb{G} \backslash \{0\})$ we have*

$$\left(\int_{\mathbb{G}} |(\log |x|) \mathcal{R} f|^Q dx \right)^{\frac{1}{Q}} \left(\int_{\mathbb{G}} |x|^{\frac{Q}{Q-1}} |f|^{\frac{Q}{Q-1}} dx \right)^{\frac{Q-1}{Q}} \geq \frac{1}{Q} \int_{\mathbb{G}} |f|^2 dx. \quad (2.59)$$

Proof. From the inequality (2.54) we get

$$\left(\int_{B(0,R)} |(\log|x|)\mathcal{R}f|^Q\,dx\right)^{\frac{1}{Q}}\left(\int_{B(0,R)} |x|^{\frac{Q}{Q-1}}|f|^{\frac{Q}{Q-1}}\,dx\right)^{\frac{Q-1}{Q}}$$
$$\geq \frac{1}{Q}\left(\int_{B(0,R)} \frac{|f|^Q}{|x|^Q}\,dx\right)^{\frac{1}{Q}}\left(\int_{B(0,R)} |x|^{\frac{Q}{Q-1}}|f|^{\frac{Q}{Q-1}}\,dx\right)^{\frac{Q-1}{Q}} \geq \frac{1}{Q}\int_{B(0,R)} |f|^2 dx,$$

where we have used the Hölder inequality in the last line. This shows (2.59). □

2.2.3 Critical Hardy inequalities of logarithmic type

In this section we present yet another logarithmic type of critical Hardy inequalities on the homogeneous group $\mathbb{G}$ of homogeneous dimension $Q \geq 1$. As usual, let $|\cdot|$ be a homogeneous quasi-norm on $\mathbb{G}$.

Theorem 2.2.7 (Another family of logarithmic Hardy inequalities). *Let* $1 < \gamma < \infty$ *and* $\max\{1, \gamma-1\} < p < \infty$. *Then for all complex-valued functions* $f \in C_0^\infty(\mathbb{G}\backslash\{0\})$ *and all* $R > 0$ *we have the inequality*

$$\left\|\frac{f - f_R}{|x|^{\frac{Q}{p}}\left(\log\frac{R}{|x|}\right)^{\frac{\gamma}{p}}}\right\|_{L^p(\mathbb{G})} \leq \frac{p}{\gamma-1}\left\||x|^{\frac{p-Q}{p}}\left(\log\frac{R}{|x|}\right)^{\frac{p-\gamma}{p}}\mathcal{R}f\right\|_{L^p(\mathbb{G})}, \tag{2.60}$$

where $f_R(x) := f\left(R\frac{x}{|x|}\right)$, *and the constant* $\frac{p}{\gamma-1}$ *is sharp.*

Proof of Theorem 2.2.7. For a quasi-ball $B(0,R)$ we have using the polar decomposition in Proposition 1.2.10 that

$$\int_{B(0,R)} \frac{|f(x)-f_R(x)|^p}{|x|^Q\left|\log\frac{R}{|x|}\right|^\gamma}dx = \int_0^R\int_\wp \frac{|f(ry)-f(Ry)|^p}{r^Q\left(\log\frac{R}{r}\right)^\gamma}r^{Q-1}d\sigma(y)dr$$
$$= \int_0^R \frac{d}{dr}\left(\frac{1}{(\gamma-1)\left(\log\frac{R}{r}\right)^{\gamma-1}}\int_\wp |f(ry)-f(Ry)|^p d\sigma(y)\right)dr$$
$$-\frac{p}{\gamma-1}\mathrm{Re}\int_0^R\left(\log\frac{R}{r}\right)^{-\gamma+1}$$
$$\times\int_\wp |f(ry)-f(Ry)|^{p-2}(f(ry)-f(Ry))\overline{\frac{df(ry)}{dr}}d\sigma(y)dr$$
$$= -\frac{p}{\gamma-1}\mathrm{Re}\int_0^R\left(\log\frac{R}{r}\right)^{-\gamma+1}$$
$$\times\int_\wp |f(ry)-f(Ry)|^{p-2}(f(ry)-f(Ry))\overline{\frac{df(ry)}{dr}}d\sigma(y)dr,$$

where $p-\gamma+1 > 0$, so that the boundary term at $r = R$ vanishes due to inequalities

$$|f(ry) - f(Ry)| \le C(R-r), \qquad \log\frac{R}{r} \ge \frac{R-r}{R}.$$

Then by the Hölder inequality we get

$$\begin{aligned}
&\int_0^R \int_{\wp} \frac{|f(ry)-f(Ry)|^p}{r\left(\log\frac{R}{r}\right)^{\gamma}} d\sigma(y)dr \\
&= -\frac{p}{\gamma-1}\mathrm{Re}\int_0^R \left(\log\frac{R}{r}\right)^{-\gamma+1} \int_{\wp} |f(ry)-f(Ry)|^{p-2}(f(ry)-f(Ry))\overline{\frac{df(ry)}{dr}}d\sigma(y)dr \\
&\le \frac{p}{\gamma-1}\int_0^R \left(\log\frac{R}{r}\right)^{-\gamma+1} \int_{\wp} |f(ry)-f(Ry)|^{p-1}\left|\frac{df(ry)}{dr}\right| d\sigma(y)dr \\
&\le \frac{p}{\gamma-1}\left(\int_0^R \int_{\wp} \frac{|f(ry)-f(Ry)|^p}{r\left(\log\frac{R}{r}\right)^{\gamma}} d\sigma(y)dr\right)^{(p-1)/p} \\
&\quad\times\left(\int_0^R \int_{\wp} r^{p-1}\left(\log\frac{R}{r}\right)^{p-\gamma}\left|\frac{df(ry)}{dr}\right|^p d\sigma(y)dr\right)^{1/p}.
\end{aligned}$$

Thus, we obtain

$$\begin{aligned}
&\left(\int_{B(0,R)} \frac{|f(x)-f_R(x)|^p}{|x|^Q\left|\log\frac{R}{|x|}\right|^{\gamma}}dx\right)^{1/p} \\
&\le \frac{p}{\gamma-1}\left(\int_{B(0,R)} |x|^{p-Q}\left|\log\frac{R}{|x|}\right|^{p-\gamma} |\mathcal{R}f(x)|^p\,dx\right)^{1/p}.
\end{aligned} \tag{2.61}$$

Similarly, we have

$$\begin{aligned}
&\left(\int_{B^c(0,R)} \frac{|f(x)-f_R(x)|^p}{|x|^Q\left|\log\frac{R}{|x|}\right|^{\gamma}}dx\right)^{1/p} \\
&\le \frac{p}{\gamma-1}\left(\int_{B^c(0,R)} |x|^{p-Q}\left|\log\frac{R}{|x|}\right|^{p-\gamma} |\mathcal{R}f(x)|^p\,dx\right)^{1/p}.
\end{aligned} \tag{2.62}$$

The inequalities (2.61) and (2.62) imply (2.60). Furthermore, the optimality of the constant in (2.60) is proved exactly in the same way as in the Euclidean case (see [MOW15b, Section 3]). □

Corollary 2.2.8 (Another family of logarithmic uncertainty type principles). *Let $1 < p < \infty$ and $q > 1$ be such that $\frac{1}{p}+\frac{1}{q}=\frac{1}{2}$. Let $1 < \gamma < \infty$ and $\max\{1,\gamma-1\} < p < \infty$. Then for any $R > 0$ and $f \in C_0^{\infty}(\mathbb{G}\backslash\{0\})$ we have*

$$\left\| |x|^{\frac{p-Q}{p}}\left(\log\frac{R}{|x|}\right)^{(p-\gamma)/p}\mathcal{R}f\right\|_{L^p(\mathbb{G})} \|f\|_{L^q(\mathbb{G})} \ge \frac{\gamma-1}{p}\left\|\frac{f(f-f_R)}{|x|^{\frac{Q}{p}}(\log(R/|x|))^{\gamma/p}}\right\|_{L^2(\mathbb{G})}.$$

Moreover,

$$\left\| |x|^{\frac{p-Q}{p}} \left(\log \frac{R}{|x|}\right)^{\frac{p-\gamma}{p}} \mathcal{R}f \right\|_{L^p(\mathbb{G})} \left\| \frac{f - f_R}{|x|^{\frac{Q}{p'}} \left(\log \frac{R}{|x|}\right)^{2-\frac{\gamma}{p}}} \right\|_{L^{p'}(\mathbb{G})} \geq \frac{\gamma-1}{p} \left\| \frac{f-f_R}{|x|^{\frac{Q}{2}} \log \frac{R}{|x|}} \right\|^2_{L^2(\mathbb{G})} \tag{2.63}$$

holds for $\frac{1}{p} + \frac{1}{p'} = 1$.

Proof of Corollary 2.2.8. By (2.60), we have

$$\left\| |x|^{\frac{p-Q}{p}} \left(\log \frac{R}{|x|}\right)^{\frac{p-\gamma}{p}} \mathcal{R}f \right\|_{L^p(\mathbb{G})} \|f\|_{L^q(\mathbb{G})} \geq \frac{\gamma-1}{p} \left\| \frac{f-f_R}{|x|^{\frac{Q}{p}} \left(\log \frac{R}{|x|}\right)^{\frac{\gamma}{p}}} \right\|_{L^p(\mathbb{G})} \|f\|_{L^q(\mathbb{G})}$$

$$= \frac{\gamma-1}{p} \left(\int_{\mathbb{G}} \left| \frac{f(x)-f_R(x)}{|x|^{\frac{Q}{p}} \left(\log \frac{R}{|x|}\right)^{\frac{\gamma}{p}}} \right|^{2\frac{p}{2}} dx \right)^{\frac{1}{2}\frac{2}{p}} \left(\int_{\mathbb{G}} |f(x)|^{2\frac{q}{2}} dx \right)^{\frac{1}{2}\frac{2}{q}},$$

and using Hölder's inequality, we obtain

$$\left\| |x|^{\frac{p-Q}{p}} \left(\log \frac{R}{|x|}\right)^{\frac{p-\gamma}{p}} \mathcal{R}f \right\|_{L^p(\mathbb{G})} \|f\|_{L^q(\mathbb{G})}$$

$$\geq \frac{\gamma-1}{p} \left(\int_{\mathbb{G}} \left| \frac{f(x)(f(x)-f_R(x))}{|x|^{\frac{Q}{p}} \left(\log \frac{R}{|x|}\right)^{\gamma/p}} \right|^2 dx \right)^{1/2} = \frac{\gamma-1}{p} \left\| \frac{f(f-f_R)}{|x|^{\frac{Q}{p}} \left(\log \frac{R}{|x|}\right)^{\gamma/p}} \right\|_{L^2(\mathbb{G})}.$$

Similarly, one can prove (2.63). □

Remark 2.2.9. When $\gamma = p$, the statement in Theorem 2.2.7 appeared in [RS16a, Theorem 3.1] in the form

$$\left\| \frac{f-f_R}{|x|^{\frac{Q}{p}} \log \frac{R}{|x|}} \right\|_{L^p(\mathbb{G})} \leq \frac{p}{p-1} \left\| |x|^{\frac{p-Q}{p}} \mathcal{R}f \right\|_{L^p(\mathbb{G})}, \quad 1 < p < \infty, \tag{2.64}$$

for all $R > 0$. For general p and γ it was analysed in [RS16a]. In the Euclidean case with the Euclidean distance such inequalities have been analysed in [MOW15b, Section 3].

2.3 Remainder estimates

In this section we analyse the remainder estimates for L^p-weighted Hardy inequalities with sharp constants on homogeneous groups. In addition, other refined versions involving a distance and the critical case $p = Q = 2$ on the quasi-ball are discussed.

The analysis of remainder terms in Hardy inequalities has a long history initiated by Brézis and Nirenberg in [BN83], with subsequent works by Brézis and Lieb [BL85] for Hardy–Sobolev inequalities, Brézis and Vázquez in [BV97, Section 4]. Nowadays there is a lot of literature on this subject and this section will contain some further references on this subject.

2.3.1 Remainder estimates for L^p-weighted Hardy inequalities

Let $\mathbb{G}$ be a homogeneous group of homogeneous dimension $Q \geq 3$ and let $|\cdot|$ be a homogeneous quasi-norm on $\mathbb{G}$. We now present a family of remainder estimates for the weighted L^p-Hardy inequalities, with a freedom of choosing the parameter $b \in \mathbb{R}$.

Theorem 2.3.1 (Remainder estimates for L^p-weighted Hardy inequalities). *Let*

$$2 \leq p < Q, \quad -\infty < \alpha < \frac{Q-p}{p},$$

and let

$$\delta_1 = Q - p - \alpha p - \frac{Q+pb}{p},$$
$$\delta_2 = Q - p - \alpha p - \frac{bp}{p-1},$$

for $b \in \mathbb{R}$. Then for all complex-valued functions $f \in C_0^\infty(\mathbb{G}\backslash\{0\})$ we have

$$\begin{aligned} \int_{\mathbb{G}} \frac{|\mathcal{R}f(x)|^p}{|x|^{\alpha p}} dx - \left(\frac{Q-p-\alpha p}{p}\right)^p \int_{\mathbb{G}} \frac{|f(x)|^p}{|x|^{p(\alpha+1)}} dx \\ \geq C_p \frac{\left(\int_{\mathbb{G}} |f(x)|^p |x|^{\delta_1} dx\right)^p}{\left(\int_{\mathbb{G}} |f(x)|^p |x|^{\delta_2} dx\right)^{p-1}}, \end{aligned} \tag{2.65}$$

where the constant $C_p = c_p \left|\frac{Q(p-1)-pb}{p^2}\right|^p$ is sharp, with

$$c_p = \min_{0<t\leq 1/2} ((1-t)^p - t^p + pt^{p-1}). \tag{2.66}$$

Due to the positivity of the last remainder term, Theorem 2.3.1 implies the L^p-weighted Hardy inequalities with the radial derivative. In the case of the Euclidean norm, they reduce to the usual L^p-weighted Hardy estimates:

Remark 2.3.2 (L^p-weighted Hardy inequalities).

1. If we take $b = \frac{Q(p-1)}{p}$ we have $C_p = 0$, and then inequality (2.65) gives the L^p-weighted Hardy inequalities with sharp constant on $\mathbb{G}$:

$$\int_{\mathbb{G}} \frac{|\mathcal{R}f(x)|^p}{|x|^{\alpha p}} dx \geq \left(\frac{Q-p-\alpha p}{p}\right)^p \int_{\mathbb{G}} \frac{|f(x)|^p}{|x|^{p(\alpha+1)}} dx,$$
$$-\infty < \alpha < \frac{Q-p}{p}, \quad 2 \leq p < Q, \tag{2.67}$$

for all complex-valued functions $f \in C_0^\infty(\mathbb{G}\backslash\{0\})$. Such inequalities on homogeneous groups have been investigated in [RS17b], and their remainders have been analysed in [RSY18b].

2. If $\mathbb{G} = (\mathbb{R}^n, +)$ with $Q = n$, the inequality (2.67) gives the L^p-weighted Hardy inequalities with sharp constant for any quasi-norm on $\mathbb{R}^n$: For any complex-valued function $f \in C_0^\infty(\mathbb{R}^n\backslash\{0\})$ we obtain

$$\int_{\mathbb{R}^n} \left|\frac{x}{|x|} \cdot \nabla f(x)\right|^p |x|^{-\alpha p} dx \geq \left(\frac{n-p-\alpha p}{p}\right)^p \int_{\mathbb{R}^n} \frac{|f(x)|^p}{|x|^{p(\alpha+1)}} dx,$$

where $-\infty < \alpha < \frac{n-p}{p}$ and $2 \leq p < n$, and where ∇ is the standard gradient in $\mathbb{R}^n$. Now, if we take the Euclidean norm $|x|_E = \sqrt{x_1^2 + x_2^2 + \cdots + x_n^2}$, by using the Schwarz inequality we obtain the Euclidean form of the L^p-weighted Hardy inequalities with sharp constants:

$$\int_{\mathbb{R}^n} \frac{|\nabla f(x)|^p}{|x|_E^{\alpha p}} dx \geq \left(\frac{n-p-\alpha p}{p}\right)^p \int_{\mathbb{R}^n} \frac{|f(x)|^p}{|x|_E^{p(\alpha+1)}} dx,$$
$$-\infty < \alpha < \frac{n-p}{p}, \quad 2 \leq p < n, \tag{2.68}$$

for any complex-valued function $f \in C_0^\infty(\mathbb{R}^n\backslash\{0\})$.

3. Moreover, for any function $f \in C_0^\infty(\mathbb{R}^n\backslash\{0\})$ and for any $b \in \mathbb{R}$, we have

$$\int_{\mathbb{R}^n} \left|\frac{x}{|x|} \cdot \nabla f(x)\right|^p |x|^{-\alpha p} dx - \left(\frac{n-p-\alpha p}{p}\right)^p \int_{\mathbb{R}^n} \frac{|f(x)|^p}{|x|^{p(\alpha+1)}} dx$$
$$\geq C_p \frac{\left(\int_{\mathbb{R}^n} |f(x)|^p |x|^{\delta_1} dx\right)^p}{\left(\int_{\mathbb{R}^n} |f(x)|^p |x|^{\delta_2} dx\right)^{p-1}}, \quad 2 \leq p < n, \ -\infty < \alpha < \frac{n-p}{p}, \tag{2.69}$$

where ∇ is the standard gradient in $\mathbb{R}^n$. As in Part 2 above, by the Schwarz inequality with the usual Euclidean distance $|\cdot|_E$, we obtain

$$\int_{\mathbb{R}^n} \frac{|\nabla f(x)|^p}{|x|_E^{\alpha p}} dx - \left(\frac{n-p-\alpha p}{p}\right)^p \int_{\mathbb{R}^n} \frac{|f(x)|^p}{|x|_E^{p(\alpha+1)}} dx$$
$$\geq C_p \frac{\left(\int_{\mathbb{R}^n} |f(x)|^p |x|_E^{\delta_1} dx\right)^p}{\left(\int_{\mathbb{R}^n} |f(x)|^p |x|_E^{\delta_2} dx\right)^{p-1}}, \quad 2 \leq p < n, \ -\infty < \alpha < \frac{n-p}{p}, \tag{2.70}$$

for all complex-valued functions $f \in C_0^\infty(\mathbb{R}^n \backslash \{0\})$ and for any $b \in \mathbb{R}$, where the constant C_p is sharp.

4. In $\mathbb{R}^n$ a variant of inequality (2.65) is known for radially symmetric functions. Let $n \geq 3$, $2 \leq p < n$ and $-\infty < \alpha < \frac{n-p}{p}$. Let $N \in \mathbb{N}$, $t \in (0,1)$, $\gamma < \min\{1-t, \frac{p-N}{p}\}$ and $\delta = N - n + \frac{N}{1-t-\gamma}\left(\gamma + \frac{n-p-\alpha p}{p}\right)$. Then there exists a constant $C > 0$ such that the inequality
$$\int_{\mathbb{R}^n} \frac{|\nabla f|^p}{|x|_E^{\alpha p}} dx - \left(\frac{n-p-\alpha p}{p}\right)^p \int_{\mathbb{R}^n} \frac{|f|^p}{|x|_E^{p(\alpha+1)}} dx$$
$$\geq C \frac{\left(\int_{\mathbb{R}^n} |f|^{\frac{N}{1-t-\gamma}} |x|_E^{\delta} dx\right)^{\frac{p(1-t-\gamma)}{Nt}}}{\left(\int_{\mathbb{R}^n} |f|^p |x|_E^{-\alpha p} dx\right)^{\frac{1-t}{t}}}$$
holds for all radially symmetric functions $f \in W_{0,\alpha}^{1,p}(\mathbb{R}^n)$, $f \neq 0$, where $W_{0,\alpha}^{1,p}(\mathbb{R}^n)$ is an appropriate Sobolev type space. For $\alpha = 0$ this was shown by Sano and Takahashi [ST17] and then extended in [ST18a] for any $-\infty < \alpha < \frac{n-p}{p}$.

5. The constant c_p in (2.66) appears in view of the following result that will be also of use in the determination of this constant:

   Lemma 2.3.3 ([FS08]). *Let $p \geq 2$ and let a, b be real numbers. Then there exists $c_p > 0$ such that*
   $$|a-b|^p \geq |a|^p - p|a|^{p-2}ab + c_p|b|^p$$
   holds, where $c_p = \min_{0<t\leq 1/2}((1-t)^p - t^p + pt^{p-1})$ is sharp in this inequality.

6. Remainder estimates of different forms are possible. In general, it is known from Ghoussoub and Moradifam [GM08] that there are no strictly positive functions $V \in C^1(0,\infty)$ such that the inequality
$$\int_{\mathbb{R}^n} |\nabla f|^2 dx \geq \left(\frac{n-2}{2}\right)^2 \int_{\mathbb{R}^n} \frac{|f|^2}{|x|_E^2} dx + \int_{\mathbb{R}^n} V(|x|_E)|f|^2 dx$$
holds for all Sobolev space functions $f \in W^{1,2}(\mathbb{R}^n)$. At the same time, Cianchi and Ferone showed in [CF08] that for all $1 < p < n$ there exists a constant $C = C(p,n)$ such that
$$\int_{\mathbb{R}^n} |\nabla f|^p dx \geq \left(\frac{n-p}{p}\right)^p \int_{\mathbb{R}^n} \frac{|f|^p}{|x|_E^p} dx \,(1 + Cd_p(f)^{2p^*})$$
holds for all real-valued weakly differentiable functions f in $\mathbb{R}^n$ such that f and $|\nabla f| \in L^p(\mathbb{R}^n)$ go to zero at infinity, where
$$d_p(f) = \inf_{c\in\mathbb{R}} \frac{\|f - c|x|_E^{-\frac{n-p}{p}}\|_{L^{p^*,\infty}(\mathbb{R}^n)}}{\|f\|_{L^{p^*,p}(\mathbb{R}^n)}}$$

with $p^* = \frac{np}{n-p}$, and $L^{\tau,\sigma}(\mathbb{R}^n)$ is the Lorentz space for $0 < \tau \leq \infty$ and $1 \leq \sigma \leq \infty$. In the case of a bounded domain Ω, Wang and Willem [WW03] for $p = 2$ and Abdellaoui, Colorado and Peral [ACP05] for $1 < p < \infty$ investigated other expressions of remainders, see also [ST17] and [ST18a] for more details.

In the following proof we will rely on a useful feature that some estimates involving radial derivatives of the Euler operator can be proved first for radial functions, and then extended to non-radial ones by a more abstract argument, see Section 1.3.3.

Proof of Theorem 2.3.1. Let $f \in C_0^\infty(\mathbb{G}\backslash\{0\})$ be a radial function, then f can be represented as $f(x) = \widetilde{f}(|x|)$. By using Brézis–Vázquez' idea ([BV97]), we define

$$\widetilde{g}(r) := r^{\frac{Q-p-\alpha p}{p}} \widetilde{f}(r). \tag{2.71}$$

Since $\widetilde{f} = \widetilde{f}(r) \in C_0^\infty(0,\infty)$ and $\alpha < \frac{Q-p}{p}$, we have $\widetilde{g}(0) = 0$ and $\widetilde{g}(+\infty) = 0$. We set

$$g(x) := \widetilde{g}(|x|)$$

for $x \in \mathbb{G}$. Introducing polar coordinates $(r,y) = (|x|, \frac{x}{|x|}) \in (0,\infty) \times \wp$ on $\mathbb{G}$, by Proposition 1.2.10 we have

$$\begin{aligned}
J &:= \int_{\mathbb{G}} \frac{|\mathcal{R}f(x)|^p}{|x|^{\alpha p}} dx - \left(\frac{Q-p-\alpha p}{p}\right)^p \int_{\mathbb{G}} \frac{|f(x)|^p}{|x|^{p(\alpha+1)}} dx \\
&= |\wp| \int_0^\infty \left| \frac{d}{dr} \widetilde{f}(r) \right|^p r^{-\alpha p + Q - 1} dr - |\wp| \left(\frac{Q-p-\alpha p}{p}\right)^p \int_0^\infty |\widetilde{f}(r)|^p r^{-p(\alpha+1)+Q-1} dr \\
&= |\wp| \int_0^\infty \left| \left(\frac{Q-p-\alpha p}{p}\right) r^{-\frac{Q-\alpha p}{p}} \widetilde{g}(r) - r^{-\frac{Q-p-\alpha p}{p}} \frac{d}{dr} \widetilde{g}(r) \right|^p r^{Q-1-\alpha p} dr \\
&\quad - |\wp| \left(\frac{Q-p-\alpha p}{p}\right)^p \int_0^\infty |\widetilde{g}(r)|^p r^{-1} dr,
\end{aligned}$$

where $|\wp|$ is the $Q-1$-dimensional surface measure of the unit sphere. Here applying Lemma 2.3.3 to the integrand of the first term in the last expression above, we get

$$\begin{aligned}
&\left| \left(\frac{Q-p-\alpha p}{p}\right) r^{-\frac{Q-\alpha p}{p}} \widetilde{g}(r) - r^{-\frac{Q-p-\alpha p}{p}} \frac{d}{dr} \widetilde{g}(r) \right|^p r^{Q-1-\alpha p} \\
&\geq \left(\left(\frac{Q-p-\alpha p}{p}\right)^p r^{-Q+\alpha p} |\widetilde{g}(r)|^p \right) r^{Q-1-\alpha p} \\
&\quad - p \left(\frac{Q-p-\alpha p}{p}\right)^{p-1} |\widetilde{g}(r)|^{p-2} \widetilde{g}(r) \frac{d}{dr} \widetilde{g}(r) r^{-(\frac{Q-\alpha p}{p})(p-1)} r^{-(\frac{Q-p-\alpha p}{p})} r^{Q-1-\alpha p} \\
&\quad + c_p \left| \frac{d}{dr} \widetilde{g}(r) \right|^p r^{-Q+p+\alpha p} r^{Q-1-\alpha p}
\end{aligned}$$

$$= \left(\frac{Q-p-\alpha p}{p}\right)^p r^{-1}|\widetilde{g}(r)|^p - p\left(\frac{Q-p-\alpha p}{p}\right)^{p-1} |\widetilde{g}(r)|^{p-2}\widetilde{g}(r)\frac{d}{dr}\widetilde{g}(r)$$
$$+ c_p\left|\frac{d}{dr}\widetilde{g}(r)\right|^p r^{p-1}.$$

Since $\widetilde{g}(0) = \widetilde{g}(+\infty) = 0$ and $p \geq 2$, we note that

$$p\int_0^\infty |\widetilde{g}(r)|^{p-2}\widetilde{g}(r)\frac{d}{dr}\widetilde{g}(r)dr = \int_0^\infty \frac{d}{dr}(|\widetilde{g}(r)|^p)dr = 0.$$

This gives a so-called ground state representation of the Hardy difference J:

$$J \geq c_p|\wp|\int_0^\infty \left|\frac{d}{dr}\widetilde{g}(r)\right|^p r^{p-1}dr = c_p\int_{\mathbb{G}} |\mathcal{R}g(x)|^p|x|^{p-Q}dx. \tag{2.72}$$

Putting $a = \frac{Q-p}{p}$ in Lemma 2.3.3, we obtain for any $b \in \mathbb{R}$ that

$$\left|\frac{Q(p-1)-pb}{p^2}\right|\int_{\mathbb{G}} |g|^p|x|^{-\frac{Q+pb}{p}}dx$$
$$\leq \left(\int_{\mathbb{G}} |\mathcal{R}g|^p|x|^{p-Q}dx\right)^{\frac{1}{p}}\left(\int_{\mathbb{G}} |g|^p|x|^{-\frac{bp}{p-1}}dx\right)^{\frac{p-1}{p}}.$$

It gives the estimate

$$\begin{aligned} J &\geq c_p\int_{\mathbb{G}} |\mathcal{R}g(x)|^p|x|^{p-Q}dx \\ &\geq c_p\left|\frac{Q(p-1)-pb}{p^2}\right|^p \frac{\left(\int_{\mathbb{G}} |g|^p|x|^{-\frac{Q+pb}{p}}dx\right)^p}{\left(\int_{\mathbb{G}} |g|^p|x|^{-\frac{bp}{p-1}}dx\right)^{p-1}}. \end{aligned} \tag{2.73}$$

Taking into account that $g(x) = \widetilde{g}(|x|)$, $x \in \mathbb{G}$, and (2.71), one calculates

$$\begin{aligned} \int_{\mathbb{G}} |x|^{-\frac{Q+pb}{p}}|g(x)|^p dx &= |\wp|\int_0^\infty r^{Q-p-\alpha p}|\widetilde{f}(r)|^p r^{-\frac{Q+pb}{p}} r^{Q-1}dr \\ &= \int_{\mathbb{G}} |f(x)|^p|x|^{Q-p-\alpha p-\frac{Q+pb}{p}}dx = \int_{\mathbb{G}} |f(x)|^p|x|^{\delta_1}dx. \end{aligned}$$

On the other hand,

$$\begin{aligned} \int_{\mathbb{G}} |x|^{-\frac{bp}{p-1}}|g(x)|^p dx &= |\wp|\int_0^\infty r^{Q-p-\alpha p}|\widetilde{f}(r)|^p r^{-\frac{bp}{p-1}} r^{Q-1}dr \\ &= \int_{\mathbb{G}} |f(x)|^p|x|^{Q-p-\alpha p-\frac{bp}{p-1}}dx = \int_{\mathbb{G}} |f(x)|^p|x|^{\delta_2}dx. \end{aligned}$$

Putting these equalities into (2.73), we obtain

$$J \geq c_p \left| \frac{Q(p-1)-pb}{p^2} \right|^p \frac{\left(\int_{\mathbb{G}} |f(x)|^p |x|^{\delta_1} dx\right)^p}{\left(\int_{\mathbb{G}} |f(x)|^p |x|^{\delta_2} dx\right)^{p-1}}.$$

Now let us prove the statement for non-radial functions. For a non-radial function f we consider the radial one obtained as its spherical average with respect to the homogeneous quasi-norm $|\cdot|$:

$$U(r) := \left(\frac{1}{|\wp|} \int_{\wp} |f(ry)|^p d\sigma(y) \right)^{\frac{1}{p}}. \tag{2.74}$$

Using Hölder's inequality, we calculate

$$\frac{d}{dr}U(r) = \frac{1}{p}\left(\frac{1}{|\wp|} \int_{\wp} |f(ry)|^p d\sigma(y) \right)^{\frac{1}{p}-1} \frac{1}{|\wp|} \int_{\wp} p|f(ry)|^{p-2} f(ry) \overline{\frac{d}{dr} f(ry)} d\sigma(y)$$
$$\leq \left(\frac{1}{|\wp|} \int_{\wp} |f(ry)|^p d\sigma(y) \right)^{\frac{1}{p}-1} \frac{1}{|\wp|} \int_{\wp} |f(ry)|^{p-1} \left| \frac{d}{dr} f(ry) \right| d\sigma(y)$$
$$\leq \left(\frac{1}{|\wp|} \int_{\wp} |f(ry)|^p d\sigma(y) \right)^{\frac{1}{p}-1} \frac{1}{|\wp|} \left(\int_{\wp} \left| \frac{d}{dr} f(ry) \right|^p d\sigma(y) \right)^{\frac{1}{p}} \left(\int_{\wp} |f(ry)|^p d\sigma(y) \right)^{\frac{p-1}{p}}$$
$$= \left(\frac{1}{|\wp|} \int_{\wp} \left| \frac{d}{dr} f(ry) \right|^p d\sigma(y) \right)^{\frac{1}{p}}.$$

Here we note that since there exists the function $h(x) = e^{-|x|}$ which satisfies the equality

$$\left| \frac{d}{dr} h(x) \right|^p = (|h(x)|^{p-1})^{\frac{p}{p-1}},$$

the equality condition in the above Hölder inequality holds. Thus, we have

$$\frac{d}{dr}U(r) \leq \left(\frac{1}{|\wp|} \int_{\wp} \left| \frac{d}{dr} f(ry) \right|^p d\sigma(y) \right)^{\frac{1}{p}}.$$

It follows that

$$|\wp| \int_0^\infty \left| \frac{d}{dr} U(r) \right|^p r^{Q-1-\alpha p} dr \leq |\wp| \int_0^\infty \frac{1}{|\wp|} \int_{\wp} \left| \frac{d}{dr} f(ry) \right|^p r^{Q-1-\alpha p} d\sigma(y) dr$$
$$= \int_{\mathbb{G}} |\mathcal{R} f|^p \, |x|^{-\alpha p} dx,$$

that is,

$$\int_{\mathbb{G}} |\mathcal{R} U|^p \, |x|^{-\alpha p} dx \leq \int_{\mathbb{G}} |\mathcal{R} f|^p \, |x|^{-\alpha p} dx. \tag{2.75}$$

In view of (2.74), we obtain the equality

$$\int_{\mathbb{G}} |U(|x|)|^p |x|^\theta dx = |\wp| \int_0^\infty |U(r)|^p r^{\theta+Q-1} dr \tag{2.76}$$
$$= |\wp| \int_0^\infty \frac{1}{|\wp|} \int_\wp |f(ry)|^p d\sigma(y) r^{\theta+Q-1} dr = \int_{\mathbb{G}} |f(x)|^p |x|^\theta dx,$$

for any $\theta \in \mathbb{R}$. Then, it is easy to see that (2.75) and (2.76) imply that (2.65) holds also for all non-radial functions. □

2.3.2 Critical and subcritical Hardy inequalities

Here we discuss the relation between the critical and the subcritical Hardy inequalities on homogeneous groups. We formulate this relation for functions that are radially symmetric with respect to a homogeneous quasi-norm $|\cdot|$ on $\mathbb{G}$.

Proposition 2.3.4 (Critical and subcritical Hardy inequalities). *Let $\mathbb{G}$ be a homogeneous group of homogeneous dimension $Q \geq 3$ and let $\widetilde{\mathbb{G}}$ be a homogeneous group of homogeneous dimension $m \geq 2$, and assume that $Q \geq m+1$. Let $|\cdot|$ denote homogeneous quasi-norms on $\mathbb{G}$ and on $\widetilde{\mathbb{G}}$. Then for any non-negative radially symmetric function $g \in C_0^1(B^m(0,R)\backslash\{0\})$, there exists a non-negative radially symmetric function $f \in C_0^1(B^Q(0,1)\backslash\{0\})$ such that*

$$\int_{B^Q(0,1)} |\mathcal{R}f(x)|^m dx - \left(\frac{Q-m}{m}\right)^m \int_{B^Q(0,1)} \frac{|f(x)|^m}{|x|^m} dx$$
$$= \frac{|\wp|}{|\widetilde{\wp}|} \left(\frac{Q-m}{m-1}\right)^{m-1} \tag{2.77}$$
$$\times \left(\int_{B^m(0,R)} |\mathcal{R}g|^m dz - \left(\frac{m-1}{m}\right)^m \int_{B^m(0,R)} \frac{|g|^m}{|z|^m \left(\log \frac{Re}{|z|}\right)^m} dz \right)$$

holds true, where $|\wp|$ and $|\widetilde{\wp}|$ are $Q-1$- and $m-1$-dimensional surface measures of the unit sphere, respectively.

Proof of Proposition 2.3.4. Let $r = |x|$, $x \in \mathbb{G}$ and $s = |z|$, $z \in \widetilde{\mathbb{G}}$, where $\widetilde{\mathbb{G}}$ is a homogeneous group of homogeneous dimension m. Let us define a radial function $f = f(x) \in C_0^1(B^Q(0,1)\backslash\{0\})$ for a non-negative radial function $g = g(z) \in C_0^1(B^m(0,R)\backslash\{0\})$:

$$f(r) = g(s(r)),$$

where $s(r) = R\exp(1 - r^{-\frac{Q-m}{m-1}})$, that is,

$$r^{-\frac{Q-m}{m-1}} = \log \frac{Re}{s}, \quad s'(r) = \frac{Q-m}{m-1} r^{-\frac{Q-m}{m-1}-1} s(r).$$

Here we see that $s'(r) > 0$ for $r \in [0,1]$ and $s(0) = 0$, $s(1) = R$. Since $g(s) \equiv 0$ near $s = R$, we also note that $f \equiv 0$ near $r = 1$.

Then a direct calculation shows

$$
\begin{aligned}
&\int_{B^Q(0,1)} |\mathcal{R}f|^m dx - \left(\frac{Q-m}{m}\right)^m \int_{B^Q(0,1)} \frac{|f|^m}{|x|^m} dx \\
&= |\wp| \int_0^1 |f'(r)|^m r^{Q-1} dr - \left(\frac{Q-m}{m}\right)^m |\wp| \int_0^1 f^m(r) r^{Q-m-1} dr \\
&= |\wp| \int_0^R |g'(s)s'(r(s))|^m r^{Q-1}(s) \frac{ds}{s'(r(s))} \\
&\quad - \left(\frac{Q-m}{m}\right)^m |\wp| \int_0^R g^m(s) r^{Q-m-1}(s) \frac{ds}{s'(r(s))} \\
&= |\wp| \left(\frac{Q-m}{m-1}\right)^{m-1} \int_0^R |g'(s)|^m s^{m-1} ds \\
&\quad - \left(\frac{Q-m}{m}\right)^m \frac{m-1}{Q-m} |\wp| \int_0^R \frac{g^m(s)}{s\left(\log \frac{Re}{s}\right)^m} ds \\
&= \frac{|\wp|}{|\widetilde{\wp}|} \left(\frac{Q-m}{m-1}\right)^{m-1} \\
&\quad \times \left(\int_{B^m(0,R)} |\mathcal{R}g|^m dz - \left(\frac{m-1}{m}\right)^m \int_{B^m(0,R)} \frac{|g|^m}{|z|^m \left(\log \frac{Re}{|z|}\right)^m} dz \right),
\end{aligned}
$$

yielding (2.77). □

2.3.3 A family of Hardy–Sobolev type inequalities on quasi-balls

Let $\mathbb{G}$ be a homogeneous group of homogeneous dimension $Q \geq 3$. It will be convenient to denote the dilations by $\delta_r(x) = rx$ in the following formulations. Here we discuss another type of Hardy–Sobolev inequalities for functions supported in balls of radius R. As usual, we denote by $B(0, R)$ a quasi-ball of radius R around 0 with respect to the quasi-norm $|\cdot|$.

Theorem 2.3.5 (Another type of Hardy inequalities for $Q \geq 3$). *For each $f \in C_0^\infty(B(0,R)\backslash\{0\})$ and any homogeneous quasi-norm $|\cdot|$ on $\mathbb{G}$ we have*

$$
\left(\int_{B(0,R)} \frac{1}{|x|^2} \left| f(x) - f\left(\frac{\delta_R(x)}{|x|}\right) \right|^2 dx \right)^{1/2} \leq \frac{2}{Q-2} \left(\int_{B(0,R)} |\mathcal{R}f|^2 dx \right)^{1/2}, \tag{2.78}
$$

and

$$
\begin{aligned}
\left(\int_{B(0,R)} \frac{1}{|x|^2} |f(x)|^2 dx \right)^{\frac{1}{2}} \leq \left(\frac{Q}{Q-2}\right)^{\frac{1}{2}} \frac{1}{R} \left(\int_{B(0,R)} |f(x)|^2 dx \right)^{\frac{1}{2}} \\
+ \frac{2}{Q-2} \left(1 + \left(\frac{Q}{Q-2}\right)^{\frac{1}{2}} \right) \left(\int_{B(0,R)} |\mathcal{R}f|^2 dx \right)^{\frac{1}{2}}.
\end{aligned} \tag{2.79}
$$

Remark 2.3.6.

1. Theorem 2.3.5 could have been formulated for functions $f \in C_0^\infty(\mathbb{G}\backslash\{0\})$ choosing $R > 0$ such that $\operatorname{supp} f \subset B(0, R)$. The introduction of R into the notation is essential here since the dilated function $f\left(\frac{\delta_R(x)}{|x|}\right)$ appears in the inequality (2.78).
2. In the Euclidean setting with the Euclidean norm inequalities in Theorem 2.3.5 have been studied in [MOW13b].

In the case $Q = 2$ we have the following inequalities:

Theorem 2.3.7 (Another type of critical Hardy inequality for $Q = 2$). *Let $\mathbb{G}$ be a homogeneous group of homogeneous dimension $Q = 2$. Then for each $f \in C_0^\infty(B(0, R)\backslash\{0\})$ and any homogeneous quasi-norm $|\cdot|$ on $\mathbb{G}$ we have*

$$\left(\int_{B(0,R)} \frac{1}{|x|^2 \left|\log \frac{R}{|x|}\right|^2} \left|f(x) - f\left(\frac{\delta_R(x)}{|x|}\right)\right|^2 dx\right)^{1/2} \leq 2\left(\int_{B(0,R)} |\mathcal{R}f|^2 dx\right)^{1/2}, \tag{2.80}$$

and

$$\left(\int_{B(0,R)} \frac{|f(x)|^2}{|x|^2 \left(1 + \left|\log \frac{R}{|x|}\right|^2\right)^2} dx\right)^{1/2} \tag{2.81}$$
$$\leq \frac{\sqrt{2}}{R}\left(\int_{B(0,R)} |f(x)|^2\, dx\right)^{1/2} + 2\left(1+\sqrt{2}\right)\left(\int_{B(0,R)} |\mathcal{R}f|^2 dx\right)^{1/2}.$$

Proof of Theorem 2.3.5. By the polar decomposition from Proposition 1.2.10, we write $(r, y) = (|x|, \frac{x}{|x|}) \in (0, \infty) \times \wp$ on $\mathbb{G}$, where $\wp$ is the unit quasi-sphere, so that

$$\begin{aligned}
&\int_{B(0,R)} \frac{1}{|x|^2} \left|f(x) - f\left(\frac{\delta_R(x)}{|x|}\right)\right|^2 dx \\
&= \int_0^R \int_\wp |f(\delta_r(y)) - f(\delta_R(y))|^2 r^{Q-3} d\sigma(y) dr \\
&= \frac{1}{Q-2} r^{Q-2} \int_\wp |f(\delta_r(y)) - f(\delta_R(y))|^2 d\sigma(y) \Bigg|_{r=0}^{r=R} \\
&\quad - \frac{1}{Q-2} \int_0^R r^{Q-2} \left(\frac{d}{dr} \int_\wp |f(\delta_r(y)) - f(\delta_R(y))|^2 d\sigma(y)\right) dr \\
&= -\frac{2}{Q-2} \int_0^R r^{Q-2} \mathrm{Re} \int_\wp (f(\delta_r(y)) - f(\delta_R(y))) \overline{\frac{df(\delta_r(y))}{dr}} d\sigma(y) dr.
\end{aligned}$$

Now using Schwarz' inequality, we obtain

$$\int_{B(0,R)} \frac{1}{|x|^2} \left| f(x) - f\left(\frac{\delta_R(x)}{|x|}\right)\right|^2 dx$$

$$\leq \frac{2}{Q-2}\left(\int_0^R \int_{\wp} |f(\delta_r(y)) - f(\delta_R(y))|^2 r^{Q-3} d\sigma(y) dr\right)^{1/2}$$

$$\times \left(\int_0^R \int_{\wp} \left|\frac{df(\delta_r(y))}{dr}\right|^2 r^{Q-1} d\sigma(y) dr\right)^{1/2}$$

$$= \frac{2}{Q-2}\left(\int_{B(0,R)} \frac{1}{|x|^2}\left|f(x) - f\left(\frac{\delta_R(x)}{|x|}\right)\right|^2 dx\right)^{1/2} \left(\int_{B(0,R)} |\mathcal{R}f|^2 dx\right)^{1/2}.$$

This implies that

$$\left(\int_{B(0,R)} \frac{1}{|x|^2}\left|f(x) - f\left(\frac{\delta_R(x)}{|x|}\right)\right|^2 dx\right)^{1/2} \leq \frac{2}{Q-2}\left(\int_{B(0,R)} |\mathcal{R}f|^2 dx\right)^{1/2},$$

that is, the inequality (2.78) is proved. The triangle inequality gives

$$\left(\int_{B(0,R)} \frac{1}{|x|^2}|f|^2 dx\right)^{\frac{1}{2}} = \left(\int_{B(0,R)} \frac{1}{|x|^2}\left|f(x) - f\left(\frac{\delta_R(x)}{|x|}\right) + f\left(\frac{\delta_R(x)}{|x|}\right)\right|^2 dx\right)^{\frac{1}{2}}$$

$$\leq \left(\int_{B(0,R)} \frac{1}{|x|^2}\left|f(x) - f\left(\frac{\delta_R(x)}{|x|}\right)\right|^2 dx\right)^{\frac{1}{2}} + \left(\int_{B(0,R)} \frac{1}{|x|^2}\left|f\left(\frac{\delta_R(x)}{|x|}\right)\right|^2 dx\right)^{\frac{1}{2}}. \tag{2.82}$$

Moreover, we have

$$\left(\int_{B(0,R)} \frac{1}{|x|^2}\left|f\left(\frac{\delta_R(x)}{|x|}\right)\right|^2 dx\right)^{\frac{1}{2}} = \left(\int_0^R \int_{\wp} |f(\delta_R(y))|^2 r^{Q-3} d\sigma(y) dr\right)^{\frac{1}{2}}$$

$$= \left(\frac{R^{Q-2}}{Q-2}\int_{\wp} |f(\delta_R(y))|^2 d\sigma(y)\right)^{\frac{1}{2}}$$

$$= \left(\frac{R^{Q-2}}{Q-2}\frac{Q}{R^Q}\int_0^R \int_{\wp} |f(\delta_R(y))|^2 r^{Q-1} d\sigma(y) dr\right)^{\frac{1}{2}}$$

$$= \left(\frac{Q}{Q-2}\right)^{\frac{1}{2}} \frac{1}{R}\left(\int_{B(0,R)} \left|f\left(\frac{\delta_R(x)}{|x|}\right)\right|^2 dx\right)^{\frac{1}{2}}$$

$$\leq \left(\frac{Q}{Q-2}\right)^{\frac{1}{2}} \frac{1}{R}\left(\left(\int_{B(0,R)} \left|f\left(\frac{\delta_R(x)}{|x|}\right) - f(x)\right|^2 dx\right)^{\frac{1}{2}} + \left(\int_{B(0,R)} |f|^2 dx\right)^{\frac{1}{2}}\right)$$

$$\leq \left(\frac{Q}{Q-2}\right)^{\frac{1}{2}} \left(\int_{B(0,R)} \frac{1}{|x|^2} \left| f\left(\frac{\delta_R(x)}{|x|}\right) - f(x) \right|^2 dx\right)^{\frac{1}{2}}$$
$$+ \left(\frac{Q}{Q-2}\right)^{\frac{1}{2}} \frac{1}{R} \left(\int_{B(0,R)} |f|^2 dx\right)^{\frac{1}{2}},$$

thus,

$$\left(\int_{B(0,R)} \frac{1}{|x|^2} \left| f\left(\frac{\delta_R(x)}{|x|}\right) \right|^2 dx\right)^{1/2}$$
$$\leq \left(\frac{Q}{Q-2}\right)^{\frac{1}{2}} \left(\int_{B(0,R)} \frac{1}{|x|^2} \left| f\left(\frac{\delta_R(x)}{|x|}\right) - f(x) \right|^2 dx\right)^{1/2} \tag{2.83}$$
$$+ \left(\frac{Q}{Q-2}\right)^{\frac{1}{2}} \frac{1}{R} \left(\int_{B(0,R)} |f|^2 dx\right)^{1/2}.$$

Combining (2.83) with (2.82) we arrive at

$$\left(\int_{B(0,R)} \frac{1}{|x|^2} |f|^2 dx\right)^{1/2}$$
$$\leq \left(1 + \left(\frac{Q}{Q-2}\right)^{1/2}\right) \left(\int_{B(0,R)} \frac{1}{|x|^2} \left| f(x) - f\left(\frac{\delta_R(x)}{|x|}\right) \right|^2 dx\right)^{1/2}$$
$$+ \left(\frac{Q}{Q-2}\right)^{1/2} \frac{1}{R} \left(\int_{B(0,R)} |f|^2 dx\right)^{1/2}.$$

Now by using (2.78) we arrive at (2.79). □

Proof of Theorem 2.3.7. By using polar coordinates $(r, y) = (|x|, \frac{x}{|x|}) \in (0, \infty) \times \wp$ on $\mathbb{G}$, where $\wp$ is the unit quasi-sphere, one calculates

$$\int_{B(0,R)} \frac{1}{|x|^2 |\log(R/|x|)|^2} \left| f(x) - f\left(\frac{\delta_R(x)}{|x|}\right) \right|^2 dx$$
$$= \int_0^R \int_{\wp} |f(\delta_r(y)) - f(\delta_R(y))|^2 \frac{1}{r\,(\log(R/r))^2} d\sigma(y) dr$$
$$= \frac{1}{\log(R/r)} \int_{\wp} |f(\delta_r(y)) - f(\delta_R(y))|^2 d\sigma(y) \Bigg|_{r=0}^{r=R}$$
$$- \int_0^R \frac{1}{\log(R/r)} \left(\frac{d}{dr} \int_{\wp} |f(\delta_r(y)) - f(\delta_R(y))|^2 d\sigma(y)\right) dr$$

$$= -2\int_0^R \frac{1}{\log(R/r)} \mathrm{Re} \int_{\wp} (f(\delta_r(y)) - f(\delta_R(y))) \overline{\frac{df(\delta_r(y))}{dr}} d\sigma(y)dr.$$

Here we have used the fact that

$$\log(R/r) = \log\left(1 + \left(\frac{R}{r} - 1\right)\right) \geq \frac{R}{r} - 1 = \frac{R-r}{r},$$

and that

$$|f(\delta_r(y)) - f(\delta_R(y))|^2 \leq C|R-r|^2.$$

Using Schwarz's inequality we obtain

$$\int_{B(0,R)} \frac{1}{|x|^2|\log(R/|x|)|^2} \left| f(x) - f\left(\frac{\delta_R(x)}{|x|}\right)\right|^2 dx$$

$$\leq 2\left(\int_0^R \int_{\wp} \frac{1}{r\left(\log(R/r)\right)^2} |f(\delta_r(y)) - f(\delta_R(y))|^2 d\sigma(y)dr\right)^{\frac{1}{2}}$$

$$\times \left(\int_0^R \int_{\wp} \left|\frac{df(\delta_r(y))}{dr}\right|^2 r d\sigma(y)dr\right)^{\frac{1}{2}}$$

$$= 2\left(\int_{B(0,R)} \frac{1}{|x|^2|\log(R/|x|)|^2} \left|f(x) - f\left(\frac{\delta_R(x)}{|x|}\right)\right|^2 dx\right)^{\frac{1}{2}} \left(\int_{B(0,R)} |\mathcal{R}f|^2 dx\right)^{\frac{1}{2}}.$$

It completes the proof of (2.80). To show (2.81) we calculate

$$\left(\int_{B(0,R)} \frac{1}{|x|^2\left(1+|\log(R/|x|)|\right)^2} |f(x)|^2 dx\right)^{1/2}$$

$$\leq \left(\int_{B(0,R)} \frac{1}{|x|^2\left(1+|\log(R/|x|)|\right)^2} \left|f(x) - f\left(\frac{\delta_R(x)}{|x|}\right)\right|^2 dx\right)^{1/2} \tag{2.84}$$

$$+ \left(\int_{B(0,R)} \frac{1}{|x|^2\left(1+|\log(R/|x|)|\right)^2} \left|f\left(\frac{\delta_R(x)}{|x|}\right)\right|^2 dx\right)^{1/2}.$$

Moreover, we have

$$\left(\int_{B(0,R)} \frac{1}{|x|^2\left(1+|\log(R/|x|)|\right)^2} \left|f\left(\frac{\delta_R(x)}{|x|}\right)\right|^2 dx\right)^{\frac{1}{2}}$$

$$= \left(\int_0^R \int_{\wp} \frac{1}{r\left(1+|\log(R/r)|\right)^2} |f(\delta_R(y))|^2 d\sigma(y)dr\right)^{\frac{1}{2}}$$

$$
\begin{aligned}
&= \left(\int_0^R \frac{1}{r\left(1+|\log(R/r)|\right)^2} dr \int_{\wp} |f(\delta_R(y))|^2 d\sigma(y) \right)^{\frac{1}{2}} \\
&= \left(\frac{1}{1+|\log(R/r)|} \Bigg|_0^R \int_{\wp} |f(\delta_R(y))|^2 d\sigma(y) \right)^{\frac{1}{2}} \\
&= \left(\int_{\wp} |f(\delta_R(y))|^2 d\sigma(y) \right)^{\frac{1}{2}} = \left(\frac{2}{R^2} \int_0^R \int_{\wp} |f(\delta_R(y))|^2 r d\sigma(y) dr \right)^{\frac{1}{2}} \\
&= \left(\frac{2}{R^2} \right)^{\frac{1}{2}} \left(\int_{B(0,R)} \left| f\left(\frac{\delta_R(x)}{|x|} \right) \right|^2 dx \right)^{\frac{1}{2}} \\
&\leq \frac{\sqrt{2}}{R} \left(\left(\int_{B(0,R)} \left| f\left(\frac{\delta_R(x)}{|x|} \right) - f(x) \right|^2 dx \right)^{\frac{1}{2}} + \left(\int_{B(0,R)} |f(x)|^2 dx \right)^{\frac{1}{2}} \right) \\
&\leq \sqrt{2} \left(\int_{B(0,R)} \frac{1}{|x|^2\left(1+|\log(R/|x|)|\right)^2} \left| f\left(\frac{\delta_R(x)}{|x|} \right) - f(x) \right|^2 dx \right)^{\frac{1}{2}} \\
&+ \frac{\sqrt{2}}{R} \left(\int_{B(0,R)} |f(x)|^2 dx \right)^{\frac{1}{2}},
\end{aligned}
$$

where we use the simple inequality

$$
\frac{1}{R^2} \leq \frac{1}{r^2(1+\log(R/r))^2}, \quad r \in (0,R).
$$

Therefore, we obtain

$$
\begin{aligned}
&\left(\int_{B(0,R)} \frac{1}{|x|^2\left(1+|\log(R/|x|)|\right)^2} \left| f\left(\frac{\delta_R(x)}{|x|} \right) \right|^2 dx \right)^{\frac{1}{2}} \\
&\quad \leq \sqrt{2} \left(\int_{B(0,R)} \frac{1}{|x|^2\left(1+|\log(R/|x|)|\right)^2} \left| f\left(\frac{\delta_R(x)}{|x|} \right) - f(x) \right|^2 dx \right)^{\frac{1}{2}} \\
&\quad + \frac{\sqrt{2}}{R} \left(\int_{B(0,R)} |f(x)|^2 dx \right)^{\frac{1}{2}}.
\end{aligned} \tag{2.85}
$$

Combining (2.85) with (2.84) we arrive at

$$
\left(\int_{B(0,R)} \frac{1}{|x|^2\left(1+|\log(R/|x|)|\right)^2} |f(x)|^2 dx \right)^{\frac{1}{2}}
$$

$$\leq (1+\sqrt{2})\left(\int_{B(0,R)} \frac{1}{|x|^2\left(1+|\log(R/|x|)|\right)^2}\left|f(x)-f\left(\frac{\delta_R(x)}{|x|}\right)\right|^2 dx\right)^{\frac{1}{2}}$$
$$+\frac{\sqrt{2}}{R}\left(\int_{B(0,R)}|f(x)|^2 dx\right)^{\frac{1}{2}}.$$

Finally, using (2.80) we obtain (2.81). □

2.3.4 Improved Hardy inequalities on quasi-balls

For $p=Q=2$ and any homogeneous quasi-norm $|\cdot|$ on $\mathbb{G}$ we have the following refinement of Theorem 2.3.7 with an estimate for a remainder.

Theorem 2.3.8 (Remainder estimate in critical Hardy inequality for $Q=2$). *We have*

$$\int_{B^2(0,R)}|\mathcal{R}f|^2 dx-\frac{1}{4}\int_{B^2(0,R)}\frac{|f(|x|)|^2}{|x|^2\left(\log\frac{R}{|x|}\right)^2}dx \tag{2.86}$$
$$\geq \frac{4}{R^2|\widehat{\sigma}|}\sup_{\nu>0}\nu^{-4}\left|\int_{B^2(0,R)} f(|x|)\frac{\nu-\log\frac{R}{|x|}}{|x|\left(\log\frac{R}{|x|}\right)^{\frac{1}{2}}}\left(\frac{R}{|x|}\right)^{1-\frac{1}{\nu}}dx\right|^2,\ \forall\nu>0,$$

for all real-valued radial functions $f\in C_0^\infty(B^2(0,R)\backslash\{0\})$, where $B^2(0,R)$ and $|\widehat{\sigma}|$ are 2-dimensional quasi-ball with radius R and 1-dimensional surface measure of the unit sphere, respectively.

Proof of Theorem 2.3.8. Let us define the new function $g=g(x)$ on $B^2(0,R)$ as

$$g(r):=\left(\log\frac{R}{r}\right)^{-\frac{1}{2}} f(r),\ r=|x|,\ x\in B^2(0,R).$$

One calculates

$$I_1:=\int_{B^2(0,R)}|\mathcal{R}f|^2 dx-\frac{1}{4}\int_{B^2(0,R)}\frac{|f|^2}{|x|^2\left(\log\frac{R}{|x|}\right)^2}dx$$
$$=|\widehat{\sigma}|\int_0^R (f'(r))^2 r dr-\frac{|\widehat{\sigma}|}{4}\int_0^R\frac{|f(r)|^2}{r^2\left(\log\frac{R}{r}\right)^2}rdr$$
$$=|\widehat{\sigma}|\int_0^R\left(-\frac{1}{2}\left(\log\frac{R}{r}\right)^{-\frac{1}{2}}\frac{g(r)}{r}+\left(\log\frac{R}{r}\right)^{\frac{1}{2}}g'(r)\right)^2 rdr-\frac{|\widehat{\sigma}|}{4}\int_0^R\frac{|g(r)|^2}{r\log\frac{R}{r}}dr$$
$$=-|\widehat{\sigma}|\int_0^R g(r)g'(r)dr+|\widehat{\sigma}|\int_0^R|g'(r)|^2 r\log\frac{R}{r}dr$$

$$= -\frac{|\widehat{\sigma}|}{2}\int_0^R (g^2(r))'dr + \frac{R^2|\widehat{\sigma}|}{4}\left(\int_0^R |g'(r)|^2\frac{4}{R^2}r\log\frac{R}{r}dr\right),$$

and applying $g(0) = g(R) = 0$, we obtain

$$I_1 = \frac{R^2|\widehat{\sigma}|}{4}\left(\int_0^R |g'(r)|^2\frac{4}{R^2}r\log\frac{R}{r}dr\right).$$

Taking into account that $\frac{4}{R^2}\int_0^R r\log\frac{R}{r}dr = 1$ and using Hölder's inequality, we get

$$\begin{aligned}\left(\int_0^R |g'(r)|^2\frac{4}{R^2}r\log\frac{R}{r}dr\right)^{\frac{1}{2}} &= \left(\int_0^R |g'(r)|^2\frac{4}{R^2}r\log\frac{R}{r}dr\right)^{\frac{1}{2}}\left(\frac{4}{R^2}\int_0^R r\log\frac{R}{r}dr\right)^{\frac{1}{2}}\\ &\geq \int_0^R |g'(r)|\frac{4}{R^2}r\log\frac{R}{r}dr.\end{aligned}$$

It follows that

$$I_1 \geq \frac{R^2|\widehat{\sigma}|}{4}\left(\int_0^R |g'(r)|\frac{4}{R^2}r\log\frac{R}{r}dr\right)^2 \geq \frac{4|\widehat{\sigma}|}{R^2}\left|\int_0^R g'(r)r\log\frac{R}{r}dr\right|^2. \tag{2.87}$$

Using $g(0) = g(R) = 0$, we have

$$\begin{aligned}\int_0^R g'(r)r\log\frac{R}{r}dr &= -\int_0^R g(r)\left(\log\frac{R}{r} - 1\right)dr = \int_0^R f(r)\frac{1-\log\frac{R}{r}}{\left(\log\frac{R}{r}\right)^{\frac{1}{2}}}dr\\ &= \frac{1}{|\widehat{\sigma}|}\int_{B^2(0,R)} f(x)\frac{1-\log\frac{R}{|x|}}{|x|\left(\log\frac{R}{|x|}\right)^{\frac{1}{2}}}dx.\end{aligned}$$

Putting this in (2.87), we arrive at

$$I_1 \geq \frac{4}{R^2|\widehat{\sigma}|}\left|\int_{B^2(0,R)} f(x)\frac{1-\log\frac{R}{|x|}}{|x|\left(\log\frac{R}{|x|}\right)^{\frac{1}{2}}}dx\right|^2.$$

Now we note that I_1 is invariant under the scaling $f \longmapsto f_\nu(r) = \nu^{-\frac{1}{2}}f(R^{1-\nu}r^\nu)$. Then setting

$$I_2(f) := \int_{B^2(0,R)} f(x)\frac{1-\log\frac{R}{|x|}}{|x|\left(\log\frac{R}{|x|}\right)^{\frac{1}{2}}}dx = |\widehat{\sigma}|\int_0^R f(r)\frac{1-\log\frac{R}{r}}{\left(\log\frac{R}{r}\right)^{\frac{1}{2}}}dr,$$

one obtains

$$I_1 \geq \frac{4}{R^2|\widehat{\sigma}|}|I_2(f_\nu)|^2 \tag{2.88}$$

for any $\nu > 0$. On the other hand, we have

$$I_2(f_\nu) = |\widehat{\sigma}| \int_0^R f_\nu(r) \frac{1-\log\frac{R}{r}}{\left(\log\frac{R}{r}\right)^{\frac{1}{2}}} dr = |\widehat{\sigma}|\nu^{-\frac{1}{2}} \int_0^R f(r^\nu R^{1-\nu}) \frac{1-\log\frac{R}{r}}{\left(\log\frac{R}{r}\right)^{\frac{1}{2}}} dr.$$

Using a change of variable $s = r^\nu R^{1-\nu}$, $dr = \frac{1}{\nu}\left(\frac{R}{s}\right)^{\frac{\nu-1}{\nu}} ds$, one calculates

$$\begin{aligned} I_2(f_\nu) &\geq |\widehat{\sigma}|\nu^{-\frac{1}{2}} \int_0^R f(s) \frac{1-\log\left(\frac{R}{s}\right)^{\frac{1}{\nu}}}{\left(\log\left(\frac{R}{s}\right)^{\frac{1}{\nu}}\right)^{\frac{1}{2}}} \frac{1}{\nu}\left(\frac{R}{s}\right)^{1-\frac{1}{\nu}} ds \\ &= |\widehat{\sigma}|\nu^{-2} \int_0^R f(s) \frac{\nu - \log\frac{R}{s}}{\left(\log\frac{R}{s}\right)^{\frac{1}{2}}} \left(\frac{R}{s}\right)^{1-\frac{1}{\nu}} ds \\ &= \nu^{-2} \int_{B^2(0,R)} f(x) \frac{\nu - \log\frac{R}{|x|}}{|x|\left(\log\frac{R}{|x|}\right)^{\frac{1}{2}}} \left(\frac{R}{|x|}\right)^{1-\frac{1}{\nu}} dx. \end{aligned} \tag{2.89}$$

The estimates (2.88) and (2.89) imply (2.86). □

Theorem 2.3.9 (Remainder estimate in critical Hardy inequality). *Let $\mathbb{G}$ be a homogeneous group of homogeneous dimension $Q \geq 2$. Let $|\cdot|$ be a homogeneous quasi-norm on $\mathbb{G}$. Let $q > 0$ be such that*

$$\alpha = \alpha(q, L) := \frac{Q-1}{Q}q + L + 2 \leq Q,$$

for $-1 < L < Q-2$. Then for all real-valued positive non-increasing radial functions $u \in C_0^\infty(B(0,R))$ we have

$$\begin{aligned} &\int_{B(0,R)} |\mathcal{R}u|^Q dx - \left(\frac{Q-1}{Q}\right)^Q \int_{B(0,R)} \frac{|u(x)|^Q}{|x|^Q\left(\log\frac{Re}{|x|}\right)^Q} dx \\ &\geq |\wp|^{1-\frac{Q}{q}} C^{\frac{Q}{q}} \left(\int_{B(0,R)} \frac{|u(x)|^q}{|x|^Q\left(\log\frac{Re}{|x|}\right)^\alpha} dx\right)^{\frac{Q}{q}}, \end{aligned} \tag{2.90}$$

where $|\wp|$ is the measure of the unit quasi-sphere in $\mathbb{G}$ and

$$\begin{aligned} C^{-1} = C(L,Q,q)^{-1} &:= \int_0^1 s^L \left(\log\frac{1}{s}\right)^{\frac{Q-1}{Q}q} ds \\ &= (L+1)^{-\left(\frac{Q-1}{Q}q+1\right)} \Gamma\left(\frac{Q-1}{Q}q+1\right), \end{aligned}$$

where $\Gamma(\cdot)$ is the Gamma function.

Proof of Theorem 2.3.9. As in previous proofs we set

$$v(s) = \left(\log \frac{Re}{r}\right)^{-\frac{Q-1}{Q}} u(r), \quad \text{where} \quad r = |x|, s = s(r) = \left(\log \frac{Re}{r}\right)^{-1},$$

$$s'(r) = \frac{s(r)}{r \log \frac{Re}{r}} \geq 0.$$

We have $v(0) = v(1) = 0$ since $u(R) = 0$ and, moreover,

$$u'(r) = -\left(\frac{Q-1}{Q}\right)\left(\log \frac{Re}{r}\right)^{-\frac{1}{Q}} \frac{v(s(r))}{r} + \left(\log \frac{Re}{r}\right)^{\frac{Q-1}{Q}} v'(s(r))s'(r) \leq 0.$$

It is straightforward to calculate that

$$\begin{aligned}
I &:= \int_{B(0,R)} |\mathcal{R}u|^Q dx - \left(\frac{Q-1}{Q}\right)^Q \int_{B(0,R)} \frac{|u|^Q}{|x|^Q \left(\log \frac{Re}{|x|}\right)^Q} dx \\
&= |\wp| \int_0^R |u'(r)|^Q r^{Q-1} dr - \left(\frac{Q-1}{Q}\right)^Q |\wp| \int_0^R \frac{|u(r)|^Q}{r \left(\log \frac{Re}{r}\right)^Q} dr \\
&= |\wp| \int_0^R \left(\frac{Q-1}{Q}\left(\log \frac{Re}{r}\right)^{-\frac{1}{Q}} \frac{v(s(r))}{r} - \left(\log \frac{Re}{r}\right)^{\frac{Q-1}{Q}} v'(s(r))s'(r)\right)^Q r^{Q-1} dr \\
&\quad - \left(\frac{Q-1}{Q}\right)^Q |\wp| \int_0^R \frac{|u(r)|^Q}{r \left(\log \frac{Re}{r}\right)^Q} dr.
\end{aligned}$$

By applying the third relation in Lemma 2.4.2 with

$$a = \frac{Q-1}{Q}\left(\log \frac{Re}{r}\right)^{-\frac{1}{Q}} \frac{v(s(r))}{r} \quad \text{and} \quad b = \left(\log \frac{Re}{r}\right)^{\frac{Q-1}{Q}} v'(s(r))s'(r),$$

and dropping $a^Q \geq 0$ as well as using the boundary conditions $v(0) = v(1) = 0$, we get

$$\begin{aligned}
I \geq & -|\wp| Q \left(\frac{Q-1}{Q}\right)^{Q-1} \int_0^R v(s(r))^{Q-1} v'(s(r)) s'(r) dr \\
& + |\wp| \int_0^R |v'(s(r))|^Q (s'(r))^Q \left(r \log \frac{Re}{r}\right)^{Q-1} dr \\
= & -|\wp| Q \left(\frac{Q-1}{Q}\right)^{Q-1} \int_0^R v(s(r))^{Q-1} v'(s(r)) s'(r) dr \\
& + |\wp| \int_0^R |v'(s(r))|^Q \frac{1}{r^Q \left(\log \frac{Re}{r}\right)^{2Q}} \left(r \log \frac{Re}{r}\right)^{Q-1} dr
\end{aligned} \tag{2.91}$$

$$
\begin{aligned}
&= -|\wp|Q\left(\frac{Q-1}{Q}\right)^{Q-1}\int_0^R v(s(r))^{Q-1}v'(s(r))s'(r)dr \\
&\quad + |\wp|\int_0^R |v'(s(r))|^Q s(r)^{Q-1}s'(r)dr \\
&= -|\wp|Q\left(\frac{Q-1}{Q}\right)^{Q-1}\int_0^1 v(s)^{Q-1}v'(s)ds + |\wp|\int_0^1 |v'(s)|^Q s^{Q-1}ds \\
&= |\wp|\int_0^1 |v'(s)|^Q s^{Q-1}ds.
\end{aligned}
$$

Moreover, by using the inequality

$$
\begin{aligned}
|v(s)| = \left|\int_s^1 v'(t)dt\right| &= \left|\int_s^1 v'(t)t^{\frac{Q-1}{Q}-\frac{Q-1}{Q}}dt\right| \\
&\leq \left(\int_0^1 |v'(t)|^Q t^{Q-1}dt\right)^{\frac{1}{Q}}\left(\log\frac{1}{s}\right)^{\frac{Q-1}{Q}},
\end{aligned}
$$

we obtain

$$
\int_0^1 |v(s)|^q s^L ds \leq \left(\int_0^1 |v'(s)|^Q s^{Q-1}ds\right)^{\frac{q}{Q}}\int_0^1 s^L\left(\log\frac{1}{s}\right)^{\frac{Q-1}{Q}q}ds
$$

for $-1 < L < Q-2$. Thus, we have

$$
\int_0^1 |v'(s)|^Q s^{Q-1}ds \geq C^{\frac{q}{Q}}\left(\int_0^1 |v(s)|^q s^L ds\right)^{\frac{Q}{q}}. \tag{2.92}
$$

Now it follows from (2.91) and (2.92) that

$$
\begin{aligned}
I &\geq |\wp|C^{\frac{Q}{q}}\left(\int_0^1 |v(s)|^q s^L ds\right)^{\frac{Q}{q}} = |\wp|C^{\frac{Q}{q}}\left(\int_0^R \frac{|u(r)|^q}{r\left(\log\frac{Re}{r}\right)^{\alpha}}dr\right)^{\frac{Q}{q}} \\
&= |\wp|^{1-\frac{Q}{q}}C^{\frac{Q}{q}}\left(\int_0^R \frac{|u(x)|^q}{|x|^Q\left(\log\frac{Re}{|x|}\right)^{\alpha}}dx\right)^{\frac{Q}{q}},
\end{aligned}
$$

where $\alpha = \alpha(q, L) = \frac{Q-1}{Q}q + L + 2$. The proof is complete. □

2.4 Stability of Hardy inequalities

Expressions for the remainder in an estimate in terms of the distance function to the set of extremisers are sometimes called the stability estimates in the literature. The purpose of this section is to discuss such estimates for the Hardy inequalities. Several results of such a type have been established in the Euclidean space $\mathbb{R}^n$ in [San18, ST17, ST18a, ST15, ST16]. In the following section our presentation on homogeneous groups follows [RS18].

2.4.1 Stability of Hardy inequalities for radial functions

Let $\mathbb{G}$ be a homogeneous group of homogeneous dimension $Q \geq 3$ and let $|\cdot|$ be a homogeneous quasi-norm on $\mathbb{G}$. We note that although sharp constants in the Hardy inequalities are not achieved the formal extremisers of such inequalities are given by homogeneous functions. To this end let us denote

$$f_\alpha(x) := |x|^{-\frac{Q-p-\alpha p}{p}} \tag{2.93}$$

for $-\infty < \alpha < \frac{Q-p}{p}$, and

$$d_R(f, g) := \left(\int_{\mathbb{G}} \frac{|f(x) - g(x)|^p}{\left|\log \frac{R}{|x|}\right|^p |x|^{p(\alpha+1)}} dx \right)^{\frac{1}{p}} \tag{2.94}$$

for functions f, g for which the integral in (2.94) is finite. We start with the case of radially symmetric functions.

Theorem 2.4.1 (Stability of Hardy inequalities for radially symmetric functions). *Let $\mathbb{G}$ be a homogeneous group of homogeneous dimension Q and let $|\cdot|$ be a homogeneous quasi-norm on $\mathbb{G}$. Let*

$$2 \leq p < Q \quad and \quad -\infty < \alpha < \frac{Q-p}{p}.$$

Then for all radial complex-valued functions $f \in C_0^\infty(\mathbb{G}\backslash\{0\})$ we have

$$\begin{aligned} \int_{\mathbb{G}} \frac{|\mathcal{R} f(x)|^p}{|x|^{\alpha p}} dx - \left(\frac{Q-p-\alpha p}{p}\right)^p \int_{\mathbb{G}} \frac{|f(x)|^p}{|x|^{p(\alpha+1)}} dx \\ \geq c_p \left(\frac{p-1}{p}\right)^p \sup_{R>0} d_R(f, c_f(R) f_\alpha)^p. \end{aligned} \tag{2.95}$$

where $c_f(R) := R^{\frac{Q-p-\alpha p}{p}} \widetilde{f}(R)$ with $f(x) = \widetilde{f}(r)$, $|x| = r$, $\mathcal{R} := \frac{d}{d|x|}$ is the radial derivative, c_p is defined in Lemma 2.3.3, i.e.,

$$c_p = \min_{0<t\leq 1/2} ((1-t)^p - t^p + pt^{p-1}),$$

and f_α and $d_R(\cdot, \cdot)$ are defined in (2.93) and (2.94), respectively.

Proof of Theorem 2.4.1. Since $p \geq 2$, as in (2.72) in the proof of Theorem 2.3.1, we have

$$J(f) = \int_{\mathbb{G}} \frac{|\mathcal{R}f|^p}{|x|^{\alpha p}} dx - \left(\frac{Q-p-\alpha p}{p}\right)^p \int_{\mathbb{G}} \frac{|f|^p}{|x|^{p(\alpha+1)}} dx$$
$$\geq c_p |\wp| \int_0^\infty \left|\frac{d}{dr}\widetilde{g}\right|^p r^{p-1} dr = c_p \int_{\mathbb{G}} \left|\frac{d}{dr} g\right|^p |x|^{p-Q} dx.$$

By Corollary 2.2.8 with $\gamma = p$, we obtain

$$J(f) \geq c_p \int_{\mathbb{G}} |\mathcal{R}g|^p |x|^{p-Q} dx \geq c_p \left(\frac{p-1}{p}\right)^p \int_{\mathbb{G}} \frac{\left|g(x) - g(\frac{Rx}{|x|})\right|^p}{\left|\log \frac{R}{|x|}\right|^p |x|^Q} dx$$
$$= c_p \left(\frac{p-1}{p}\right)^p \int_{\mathbb{G}} \frac{\left||x|^{\frac{Q-p-\alpha p}{p}} f(x) - R^{\frac{Q-p-\alpha p}{p}} f(\frac{Rx}{|x|})\right|^p}{\left|\log \frac{R}{|x|}\right|^p |x|^Q} dx$$

for any $R > 0$. Here using $f(x) = \widetilde{f}(r)$, $r = |x|$, we can estimate

$$J(f) \geq c_p \left(\frac{p-1}{p}\right)^p \int_{\mathbb{G}} \frac{\left|f(x) - R^{\frac{Q-p-\alpha p}{p}} \widetilde{f}(R) |x|^{-\frac{Q-p-\alpha p}{p}}\right|^p}{\left|\log \frac{R}{|x|}\right|^p |x|^{p(\alpha+1)}} dx$$
$$= c_p \left(\frac{p-1}{p}\right)^p \int_{\mathbb{G}} \frac{\left|f(x) - c_f(R) |x|^{-\frac{Q-p-\alpha p}{p}}\right|^p}{\left|\log \frac{R}{|x|}\right|^p |x|^{p(\alpha+1)}} dx,$$

yielding (2.95). □

2.4.2 Stability of Hardy inequalities for general functions

Here we discuss the stability of L^p-Hardy inequalities for functions which do not have to be radially symmetric. Before discussing the stability estimates for non-radial functions let us recall the following relations.

Lemma 2.4.2. *Let $a, b \in \mathbb{R}$. Then*

(i) *We have*

$$|a-b|^p - |a|^p \geq -p|a|^{p-2}ab, \quad p \geq 1.$$

(ii) *There exists a constant $C = C(p) > 0$ such that*

$$|a-b|^p - |a|^p \geq -p|a|^{p-2}ab + C|b|^p, \quad p \geq 2.$$

(iii) *If $a \geq 0$ and $a - b \geq 0$, then*

$$(a-b)^p + pa^{p-1}b - a^p \geq |b|^p, \quad p \geq 2.$$

Proof of Lemma 2.4.2. One has (see, e.g., [Lin90]) the inequality

$$|d|^p \geq |c|^p + p|c|^{p-2}c(d-c), \quad p \geq 1,$$

which is the condition for the convexity of the function $|x|^p$. Now, taking $a = c$ and $d = a - b$ we get the first relation. Furthermore, consider the function

$$g(s) := |1-s|^p - |s|^p + p|s|^{p-2}s, \quad s \in \mathbb{R}.$$

To show (ii), it is sufficient to prove that $g(s) \geq C > 0$, and then take $s = \frac{c}{b}$. Let $s \geq 1$. Then for $p \geq 2$ we have

$$g'(s) = p((t-1)^{p-1} - t^{p-1}) + p(p-1)s^{p-2} = p(p-1)(s^{p-2} - t^{p-2}) \geq 0,$$

by the mean value theorem for the function x^{p-2}, $x \geq 0$. Thus, we have

$$g(s) \geq g(1) = p - 1$$

for all $s \geq 1$. Similarly, we get $g(s) \geq g(0) = 1$ for all $s \leq 1$. Let now $0 \leq s \leq 1$. Setting

$$C_p := \min_{0 \leq t \leq 1} f(t),$$

we assume that $C_p = f(s_0)$ for some $0 \leq s_0 \leq 1$. The first relation implies that $C_p \geq 0$. If $C_p = 0$, then we have $f(s_0) = 0$ and $\frac{s_0-1}{p} f'(s_0) = 0$ which implies $s_0 = 0$, contradicting that $f(0) = 1$. Therefore, we have $g(s) \geq C_p > 0$. This proves (ii). The Taylor formula yields that

$$(a-b)^p + pa^{p-1}b - a^p \geq |b|^p = p(p-1)b^2 \int_0^1 (1-t)(a-\tau b)^{p-2} d\tau.$$

Thus, if $b \leq 0$, then $a - \tau b \geq \tau|b|$, which implies (iii) in this case. On the other hand, if $0 \leq b$, then $a - \tau b \geq (1-\tau)|b|$ which implies (iii) in this case as well. □

To formulate the following result, for $R > 0$, let us set

$$d_H(u; R) := \left(\int_{\mathbb{G}} \frac{\left| u(x) - R^{\frac{Q-p}{p}} u\left(R\frac{x}{|x|}\right) |x|^{-\frac{Q-p}{p}} \right|^p}{|x|^p | \log \frac{R}{|x|}|^p} dx \right)^{\frac{1}{p}}.$$

Then we have the following stability property.

Theorem 2.4.3 (Stability of Hardy inequalities for general real-valued functions). *Let $\mathbb{G}$ be a homogeneous group of homogeneous dimension Q and let $|\cdot|$ be a homogeneous quasi-norm on $\mathbb{G}$. Let $2 \leq p < Q$. Then there exists a constant $C > 0$ such that for all real-valued functions $u \in C_0^\infty(\mathbb{G})$ we have*

$$\int_{\mathbb{G}} |\mathcal{R}u|^p \, dx - \left(\frac{Q-p}{p}\right)^p \int_{\mathbb{G}} \frac{|u|^p}{|x|^p} dx \geq C_p \sup_{R>0} d_H^p(u; R). \tag{2.96}$$

Proof of Theorem 2.4.3. Let $x = (r, y) = (|x|, \frac{x}{|x|}) \in (0, \infty) \times \wp$ on $\mathbb{G}$, where $\wp$ is the unit quasi-sphere $\wp := \{x \in \mathbb{G} : |x| = 1\}$, and

$$v(ry) := r^{\frac{Q-p}{p}} u(ry),$$

where $u \in C_0^\infty(\mathbb{G})$. It follows that $v(0) = 0$ and that $\lim_{r\to\infty} v(ry) = 0$ for $y \in \wp$ since u is compactly supported. Using the polar decomposition from Proposition 1.2.10 and integrating by parts, we get

$$\begin{aligned} D :&= \int_{\mathbb{G}} |\mathcal{R}u|^p \, dx - \left(\frac{Q-p}{p}\right)^p \int_{\mathbb{G}} \frac{|u|^p}{|x|^p} dx \\ &= \int_\wp \int_0^\infty \left| -\frac{\partial}{\partial r} u(ry) \right|^p r^{Q-1} - \left(\frac{Q-p}{p}\right)^p |u(ry)|^p r^{Q-p-1} drdy \\ &= \int_\wp \int_0^\infty \left| \frac{Q-p}{p} r^{-\frac{Q}{p}} v(ry) - r^{-\frac{Q-p}{p}} \frac{\partial}{\partial r} v(ry) \right|^p r^{Q-1} \\ &\quad - \left(\frac{Q-p}{p}\right)^p |v(ry)|^p r^{-1} drdy. \end{aligned}$$

Now using relation (ii) in Lemma 2.4.2 with the choice $a = \frac{Q-p}{p} r^{-\frac{Q}{p}} v(ry)$ and $b = r^{-\frac{Q-p}{p}} \frac{\partial}{\partial r} v(ry)$, and using the fact that $\int_0^\infty |v|^{p-2} v \left(\frac{\partial}{\partial r} v\right) dr = 0$, we obtain

$$\begin{aligned} D &\geq \int_\wp \int_0^\infty -p \left(\frac{Q-p}{p}\right)^{p-1} |v(ry)|^{p-2} v(ry) \frac{\partial}{\partial r} v(ry) \\ &\quad + C \left| \frac{\partial}{\partial r} v(ry) \right|^p r^{p-1} drdy \\ &= C \int_{\mathbb{G}} |x|^{p-Q} \, |\mathcal{R}v|^p \, dx. \end{aligned} \tag{2.97}$$

Finally, combining (2.97) and Remark 2.2.9, we arrive at

$$\begin{aligned} D &\geq C_p \int_{\mathbb{G}} \frac{|v(x) - v(R\frac{x}{|x|})|^p}{|x|^Q |\log \frac{R}{|x|}|^p} dx = C \int_\wp \int_0^\infty \frac{|v(ry) - v(Ry)|^p}{r \left|\log \frac{R}{r}\right|^p} drdy \\ &= C_p \int_\wp \int_0^\infty \frac{|u(ry) - R^{\frac{Q-p}{p}} u(Ry) r^{-\frac{Q-p}{p}}|^p}{r^{1+p-Q} |\log \frac{R}{r}|^p} drdy, \end{aligned}$$

for any $R > 0$. This proves the desired result. □

2.4.3 Stability of critical Hardy inequality

Here we discuss the stability of the critical Hardy inequality, i.e., the L^p-Hardy inequality in the case $p = Q$. For this, let us denote

$$f_{T,R}(x) := T^{\frac{Q-1}{Q}} u\left(Re^{-\frac{1}{T}} \frac{x}{|x|}\right) \left(\log \frac{R}{|x|}\right)^{\frac{Q-1}{Q}}.$$

We also introduce the following 'distance' function:

$$d_{cH}(u;T,R) := \left(\int_{B(0,R)} \frac{|u(x) - f_{T,R}(x)|^Q}{|x|^Q \left|\log \frac{R}{|x|}\right|^Q \left|T \log \frac{R}{|x|}\right|^Q} dx \right)^{\frac{1}{Q}}, \tag{2.98}$$

for some parameter $T > 0$, functions u and $f_{T,R}$ for which the integral in (2.98) is finite.

Theorem 2.4.4 (Stability of critical Hardy inequality). *Let $\mathbb{G}$ be a homogeneous group of homogeneous dimension Q and let $|\cdot|$ be a homogeneous quasi-norm on $\mathbb{G}$. Then there exists a constant $C > 0$ such that for all real-valued functions $u \in C_0^\infty(B(0,R))$ we have*

$$\begin{aligned} \int_{B(0,R)} |\mathcal{R}u(x)|^Q \, dx - \left(\frac{Q-1}{Q}\right)^Q \int_{B(0,R)} \frac{|u(x)|^Q}{|x|^Q (\log \frac{R}{|x|})^Q} dx \\ \geq C_p \sup_{T>0} d_{cH}^Q(u;T,R). \end{aligned} \tag{2.99}$$

Proof of Theorem 2.4.4. With polar coordinates $(r,y) = (|x|, \frac{x}{|x|}) \in (0,\infty) \times \wp$ on $\mathbb{G}$, where $\wp$ is the sphere as in (1.12), we have $u(x) = u(ry) \in C_0^\infty(B(0,R))$. In addition, let us set

$$v(sy) := \left(\log \frac{R}{r}\right)^{-\frac{Q-1}{Q}} u(ry), \quad y \in \wp,$$

where

$$s = s(r) := \left(\log \frac{R}{r}\right)^{-1}.$$

Since $u \in C_0^\infty(B(0,R))$ we have $v(0) = 0$ and v has a compact support. Moreover, it is straightforward that

$$\frac{\partial}{\partial r} u(ry) = -\left(\frac{Q-1}{Q}\right) \left(\log \frac{R}{r}\right)^{-\frac{1}{Q}} \frac{v(sy)}{r} + \left(\log \frac{R}{r}\right)^{\frac{Q-1}{Q}} \frac{\partial}{\partial s} v(sy) s'(r).$$

A direct calculation using the polar decomposition in Proposition 1.2.10 gives

$$\begin{aligned} S &:= \int_{B(0,R)} |\mathcal{R}u|^Q \, dx - \left(\frac{Q-1}{Q}\right)^Q \int_{B(0,R)} \frac{|u|^Q}{|x|^Q \left(\log \frac{R}{|x|}\right)^Q} dx \\ &= \int_\wp \left(\int_0^R \left|\frac{\partial}{\partial r} u(ry)\right|^Q r^{Q-1} - \left(\frac{Q-1}{Q}\right)^Q \frac{|u(ry)|^Q}{r \left(\log \frac{R}{r}\right)^Q} dr \right) dy \end{aligned}$$

$$= \int_{\wp}\int_0^R \left(\left| \left(\frac{Q-1}{Q}\right)\left(r\log\frac{R}{r}\right)^{-\frac{1}{Q}} v(sy) + \left(r\log\frac{R}{r}\right)^{\frac{Q-1}{Q}} \frac{\partial}{\partial s} v(sy)s'(r) \right|^Q \right.$$
$$\left. - \left(\frac{Q-1}{Q}\right)^Q \frac{|v(sy)|^Q}{r\log\frac{R}{r}} \right) drdy.$$

Now by applying relation (ii) from Lemma 2.4.2 with the choice

$$a = \frac{Q-1}{Q}\left(r\log\frac{R}{r}\right)^{-\frac{1}{Q}} v(sy) \quad \text{and} \quad b = \left(r\log\frac{R}{r}\right)^{\frac{Q-1}{Q}} \frac{\partial}{\partial s} v(sy)s'(r),$$

and by using the properties that $v(0) = 0$ and $\lim_{r\to\infty} v(ry) = 0$, we obtain

$$S \geq \int_{\wp}\int_0^R \left(-Q\left(\frac{Q-1}{Q}\right)^{Q-1} |v(sy)|^{Q-2} v(sy) \frac{\partial}{\partial s} v(sy)s'(r) \right.$$
$$\left. + C\left|\frac{\partial}{\partial s} v(sy)\right|^Q (s'(r))^Q \left(r\log\frac{R}{r}\right)^{Q-1} \right) drdy$$
$$= \int_{\wp}\int_0^R \left(-Q\left(\frac{Q-1}{Q}\right)^{Q-1} |v(sy)|^{Q-2} v(sy) \frac{\partial}{\partial s} v(sy)s'(r) \right.$$
$$\left. + C\left|\frac{\partial}{\partial s} v(sy)\right|^Q \frac{1}{r^Q\left(\log\frac{R}{r}\right)^{2Q}} \left(r\log\frac{R}{r}\right)^{Q-1} \right) drdy$$
$$= \int_{\wp}\int_0^R \left(-Q\left(\frac{Q-1}{Q}\right)^{Q-1} |v(sy)|^{Q-2} v(sy) \frac{\partial}{\partial s} v(sy)s'(r) \right.$$
$$\left. + C\left|\frac{\partial}{\partial s} v(sy)\right|^Q \frac{1}{\left(\log\frac{R}{r}\right)^{Q-1}} s'(r) \right) drdy$$
$$= \int_{\wp}\int_0^R \left(-Q\left(\frac{Q-1}{Q}\right)^{Q-1} |v(sy)|^{Q-2} v(sy) \frac{\partial}{\partial s} v(s) \right.$$
$$\left. + C\left|\frac{\partial}{\partial s} v(sy)\right|^Q s^{Q-1} \right) dsdy$$
$$= C\int_{\mathbb{G}} |\mathcal{R}v|^Q \, dx,$$

that is,

$$S \geq C\int_{\mathbb{G}} |\mathcal{R}v|^Q \, dx. \tag{2.100}$$

According to Remark 2.2.9 for $v \in C_0^\infty(\mathbb{G}\backslash\{0\})$ with $p = Q$ and (2.100), it follows that

$$S \geq C_p \int_{\mathbb{G}} \frac{|v(x) - v(T\frac{x}{|x|})|^Q}{|x|^Q |\log\frac{T}{|x|}|^Q} dx = C_p \int_{\wp}\int_0^\infty \frac{|v(sy) - v(Ty)|^Q}{s|\log\frac{T}{s}|^Q} dsdy$$

$$= C_p \int_{\wp} \int_0^R \frac{\left| (\log \frac{R}{r})^{-\frac{Q-1}{Q}} u(ry) - T^{\frac{Q-1}{Q}} u(Re^{-\frac{1}{T}} y) \right|^Q}{r(\log \frac{R}{r}) |\log(T \log \frac{R}{r})|^Q} dr dy$$

$$= C_p \int_{\wp} \int_0^R \frac{\left| u(ry) - T^{\frac{Q-1}{Q}} u(Re^{-\frac{1}{T}} y)(\log \frac{R}{r})^{\frac{Q-1}{Q}} \right|^Q}{r(\log \frac{R}{r})^Q |\log(T \log \frac{R}{r})|^Q} dr dy.$$

Thus, we arrive at

$$S \geq C_p \int_{B(0,R)} \frac{\left| u(x) - T^{\frac{Q-1}{Q}} u\left(Re^{-\frac{1}{T}} \frac{x}{|x|}\right) \left(\log \frac{R}{|x|}\right)^{\frac{Q-1}{Q}} \right|^Q}{|x|^Q \left|\log \frac{R}{|x|}\right|^Q \left|\log\left(T \log \frac{R}{|x|}\right)\right|^Q} dx$$

for all $T > 0$. This completes the proof of Theorem 2.4.4. □

3

Understanding Rellich, Caffarelli–Kohn–Nirenberg, and Sobolev Type Inequalities

This chapter is devoted to other functional inequalities usually associated to the Hardy inequalities. These include Rellich and Caffarell–Kohn–Nirenberg inequalities. We also discuss different aspects of this analysis such as their stability, higher-order inequalities, their weighted and extended versions.

3.1 Rellich inequality

In general, the Rellich type inequalities have the following form

$$\int_{\mathbb{R}^n} \frac{|f(x)|^p}{|x|_E^\alpha} dx \leq C \int_{\mathbb{R}^n} \frac{|\Delta f(x)|^p}{|x|_E^\beta} dx$$

for certain constants α, β and p. The classical result by Rellich appearing at the 1954 ICM in Amsterdam [Rel56] was the inequality

$$\left\| \frac{f}{|x|_E^2} \right\|_{L^2(\mathbb{R}^n)} \leq \frac{4}{n(n-4)} \|\Delta f\|_{L^2(\mathbb{R}^n)}, \quad n \geq 5. \tag{3.1}$$

To find analogues of (3.1) on the homogeneous groups is an interesting question. The first obstacle to it is that there is neither stratification nor gradation on general homogeneous groups, so there may be no homogeneous left invariant hypoelliptic differential operators on $\mathbb{G}$ at all, to formulate an expression of the form of the right-hand side of (3.1).

However, similar to the results on the Hardy type inequalities before, the inequality (3.1) can be expressed in terms of the radial derivative

$$\partial_r = \frac{x}{|x|_E} \cdot \nabla,$$

taking the form

$$\left\| \frac{f}{|x|_E^2} \right\|_{L^2(\mathbb{R}^n)} \leq \frac{4}{n(n-4)} \left\| \partial_r^2 f + \frac{n-1}{|x|_E} \partial_r f \right\|_{L^2(\mathbb{R}^n)}, \quad n \geq 5. \tag{3.2}$$

It is well known that in the spherical coordinates on $\mathbb{R}^n$ the Laplacian $\Delta_{\mathbb{R}^n}$ decomposes in the radial and spherical parts as

$$\Delta_{\mathbb{R}^n} f = \frac{\partial^2 f}{\partial r^2} + \frac{n-1}{r} \frac{\partial f}{\partial r} + \frac{1}{r^2} \Delta_{\mathbb{S}^{n-1}} f.$$

So, the operator on the right-hand side in (3.2) is precisely the radial part of the Laplacian on $\mathbb{R}^n$, which can be also expressed as

$$\frac{\partial^2 f}{\partial r^2} + \frac{n-1}{r} \frac{\partial f}{\partial r} = \frac{1}{r^{n-1}} \frac{\partial}{\partial r} \left(r^{n-1} \frac{\partial f}{\partial r} \right). \tag{3.3}$$

Although there is no analogue of the Laplacian on general homogeneous groups, using the radial operator $\mathcal{R}$ from Section 1.3, the expression (3.3) on a homogeneous group of homogeneous dimension Q and homogeneous quasi-norm $|\cdot|$ makes perfect sense in the form of

$$\mathcal{R}^2 f + \frac{Q-1}{|x|} \mathcal{R} f = \frac{1}{|x|^{Q-1}} \mathcal{R} \left(|x|^{Q-1} \mathcal{R} f \right). \tag{3.4}$$

One aim of this section is to show that the Rellich inequality in the form (3.2) extends to general homogeneous groups using the radial operator $\mathcal{R}$, taking the form:

$$\left\| \frac{f}{|x|^2} \right\|_{L^2(\mathbb{G})} \leq \frac{4}{Q(Q-4)} \left\| \mathcal{R}^2 f + \frac{Q-1}{|x|} \mathcal{R} f \right\|_{L^2(\mathbb{G})}, \quad Q \geq 5, \tag{3.5}$$

for all complex-valued functions $f \in C_0^\infty(\mathbb{G} \backslash \{0\})$. The operator on the right-hand side of (3.5) is thus an analogue of the radial part of the Laplacian on $\mathbb{R}^n$.

Thus, in the sequel we will be frequently use the *Rellich type operator* appearing in the right-hand side of (3.8) which we may denote by

$$\tilde{\mathcal{R}} f := \mathcal{R}^2 f + \frac{Q-1}{|x|} \mathcal{R} f. \tag{3.6}$$

Similarly to the Hardy type inequalities from Chapter 2, the expression on the right-hand side of (3.5) appears to be natural since there is no analogue of homogeneous Laplacian or sub-Laplacian on general homogeneous groups to extend (3.1). Moreover, even on a group where such operators exist, this would usually give a refinement of those inequalities since derivatives only in one direction appear.

3.1.1 Rellich type inequalities in L^2

In this section we prove the Rellich type inequality (3.5) and its weighted version. This will be a corollary of the following identity relating different expressions involving the radial derivatives and radially symmetric weights.

Theorem 3.1.1 (Identity leading to Rellich inequality). *Let $\mathbb{G}$ be a homogeneous group of homogeneous dimension $Q \geq 5$. Then for arbitrary homogeneous quasi-norm $|\cdot|$ on $\mathbb{G}$ and every complex-valued function $f \in C_0^\infty(\mathbb{G}\backslash\{0\})$ we have the identity*

$$\left\|\mathcal{R}^2 f + \frac{Q-1}{|x|}\mathcal{R}f + \frac{Q(Q-4)}{4|x|^2}f\right\|^2_{L^2(\mathbb{G})} + \frac{Q(Q-4)}{2}\left\|\frac{1}{|x|}\mathcal{R}f + \frac{Q-4}{2|x|^2}f\right\|^2_{L^2(\mathbb{G})}$$
$$= \left\|\mathcal{R}^2 f + \frac{Q-1}{|x|}\mathcal{R}f\right\|^2_{L^2(\mathbb{G})} - \left(\frac{Q(Q-4)}{4}\right)^2\left\|\frac{f}{|x|^2}\right\|^2_{L^2(\mathbb{G})}. \tag{3.7}$$

Since the left-hand side of (3.7) is non-negative, this implies the following Rellich type inequality on $\mathbb{G}$ for $Q \geq 5$:

Corollary 3.1.2 (Rellich type inequality in $L^2(\mathbb{G})$). *For all complex-valued functions $f \in C_0^\infty(\mathbb{G}\backslash\{0\})$ we have*

$$\left\|\frac{f}{|x|^2}\right\|_{L^2(\mathbb{G})} \leq \frac{4}{Q(Q-4)}\left\|\mathcal{R}^2 f + \frac{Q-1}{|x|}\mathcal{R}f\right\|_{L^2(\mathbb{G})}, \quad Q \geq 5, \tag{3.8}$$

where the constant $\frac{4}{Q(Q-4)}$ is sharp and it is attained if and only if $f = 0$.

Remark 3.1.3. Let us show that the constant $\frac{4}{Q(Q-4)}$ in (3.8) is sharp and is never attained unless $f = 0$. Indeed, if the equality in (3.8) is attained, it follows that both terms on the left-hand side of (3.7) must be zero. In particular, it means that

$$\frac{1}{|x|}\mathcal{R}f + \frac{Q-4}{2|x|^2}f = 0, \tag{3.9}$$

and hence

$$\mathbb{E}f = -\frac{Q-4}{2}f.$$

In view of Proposition 1.3.1, Part (i), the function f must be positively homogeneous of order $-\frac{Q-4}{2}$ which is impossible since $f \in C_0^\infty(\mathbb{G}\backslash\{0\})$ unless $f = 0$, so that the constant is not attained unless $f = 0$.

Furthermore, the first term in (3.7) must be also zero, and by using (3.9) this is equivalent to

$$\mathcal{R}^2 f + \frac{Q-2}{2|x|}\mathcal{R}f = 0$$

which means that $\mathcal{R}f$ is positively homogeneous of order $-\frac{Q-2}{2}$. Thus, by taking an approximation of homogeneous functions f of order $-\frac{Q-4}{2}$, we have that $\mathcal{R}f$ is homogeneous of order $-\frac{Q-2}{2}$, so that the left-hand side of (3.7) converges to zero. Therefore, the constant $\frac{4}{Q(Q-4)}$ in (3.8) is sharp.

Thus, in view of Remark 3.1.3, the statement of Corollary 3.1.2 follows from the identity (3.7), which we will now prove.

Proof of Theorem 3.1.1. Introducing polar coordinates

$$(r,y)=\left(|x|,\frac{x}{|x|}\right)\in(0,\infty)\times\wp$$

on $\mathbb{G}$, using the polar decomposition formula from Proposition 1.2.10, as well as integrating by parts we obtain

$$\begin{aligned}
\int_{\mathbb{G}}\frac{|f(x)|^2}{|x|^4}dx&=\int_0^\infty\int_\wp\frac{|f(ry)|^2}{r^4}r^{Q-1}d\sigma(y)dr\\
&=-\frac{2}{Q-4}\mathrm{Re}\int_0^\infty r^{Q-4}\int_\wp f(ry)\overline{\frac{df(ry)}{dr}}d\sigma(y)dr\\
&=\frac{2}{(Q-3)(Q-4)}\mathrm{Re}\int_0^\infty r^{Q-3}\int_\wp\left(\left|\frac{df(ry)}{dr}\right|^2+f(ry)\overline{\frac{d^2f(ry)}{dr^2}}\right)d\sigma(y)dr\\
&=\frac{2}{(Q-3)(Q-4)}\left(\left\|\frac{1}{|x|}\mathcal{R}f\right\|^2_{L^2(\mathbb{G})}+\mathrm{Re}\int_{\mathbb{G}}\frac{f(x)}{|x|^2}\overline{\mathcal{R}^2f(x)}dx\right).
\end{aligned}\tag{3.10}$$

For the first term, using identity (2.16) with $\alpha=1$, we have identity (2.17), i.e.,

$$\left\|\frac{1}{|x|}\mathcal{R}f\right\|^2_{L^2(\mathbb{G})}=\left(\frac{Q-4}{2}\right)^2\left\|\frac{f}{|x|^2}\right\|^2_{L^2(\mathbb{G})}+\left\|\frac{1}{|x|}\mathcal{R}f+\frac{Q-4}{2|x|^2}f\right\|^2_{L^2(\mathbb{G})}.\tag{3.11}$$

For the second term a direct calculation shows

$$\begin{aligned}
&\mathrm{Re}\int_{\mathbb{G}}\frac{f(x)}{|x|^2}\overline{\mathcal{R}^2f(x)}dx\\
&=\mathrm{Re}\int_{\mathbb{G}}\frac{f(x)}{|x|^2}\overline{\left(\mathcal{R}^2f(x)+\frac{Q-1}{|x|}\mathcal{R}f(x)\right)}dx-(Q-1)\mathrm{Re}\int_{\mathbb{G}}\frac{f(x)}{|x|^3}\overline{\mathcal{R}f(x)}dx\\
&=\mathrm{Re}\int_{\mathbb{G}}\frac{f(x)}{|x|^2}\overline{\left(\mathcal{R}^2f(x)+\frac{Q-1}{|x|}\mathcal{R}f(x)\right)}dx\\
&\quad-\frac{Q-1}{2}\int_0^\infty r^{Q-4}\int_\wp\frac{d|f(ry)|^2}{dr}d\sigma(y)dr\\
&=\mathrm{Re}\int_{\mathbb{G}}\frac{f(x)}{|x|^2}\overline{\left(\mathcal{R}^2f(x)+\frac{Q-1}{|x|}\mathcal{R}f(x)\right)}dx\\
&\quad+\frac{(Q-1)(Q-4)}{2}\int_0^\infty r^{Q-5}\int_\wp|f(ry)|^2d\sigma(y)dr
\end{aligned}$$

$$
\begin{aligned}
&= \operatorname{Re} \int_{\mathbb{G}} \frac{f(x)}{|x|^2} \overline{\left(\mathcal{R}^2 f(x) + \frac{Q-1}{|x|}\mathcal{R}f(x)\right)} dx \\
&\quad + \frac{(Q-1)(Q-4)}{2} \left\| \frac{f}{|x|^2} \right\|^2_{L^2(\mathbb{G})}.
\end{aligned} \tag{3.12}
$$

Combining (3.11) and (3.12) with (3.10) we arrive at

$$
\begin{aligned}
\left\| \frac{f}{|x|^2} \right\|^2_{L^2(\mathbb{G})} &= \frac{2}{(Q-3)(Q-4)} \left(\left(\frac{Q-4}{2}\right)^2 \left\| \frac{f}{|x|^2} \right\|^2_{L^2(\mathbb{G})} \right. \\
&+ \left\| \frac{1}{|x|}\mathcal{R}f + \frac{Q-4}{2|x|^2} f \right\|^2_{L^2(\mathbb{G})} \\
&+ \left. \operatorname{Re} \int_{\mathbb{G}} \frac{f(x)}{|x|^2} \overline{\left(\mathcal{R}^2 f(x) + \frac{Q-1}{|x|}\mathcal{R}f(x)\right)} dx + \frac{(Q-1)(Q-4)}{2} \left\| \frac{f}{|x|^2} \right\|^2_{L^2(\mathbb{G})} \right).
\end{aligned}
$$

Collecting same terms, this gives

$$
\begin{aligned}
0 &= \frac{2Q}{(Q-3)4} \left\| \frac{f}{|x|^2} \right\|^2_{L^2(\mathbb{G})} \\
&+ \frac{2}{(Q-3)(Q-4)} \left(\operatorname{Re} \int_{\mathbb{G}} \frac{f(x)}{|x|^2} \overline{\left(\mathcal{R}^2 f(x) + \frac{Q-1}{|x|}\mathcal{R}f(x)\right)} dx \right. \\
&+ \left. \left\| \frac{1}{|x|}\mathcal{R}f + \frac{Q-4}{2|x|^2} f \right\|^2_{L^2(\mathbb{G})} \right),
\end{aligned}
$$

that is,

$$
\begin{aligned}
&\frac{Q(Q-4)}{4} \left\| \frac{f}{|x|^2} \right\|^2_{L^2(\mathbb{G})} \\
&= -\operatorname{Re} \int_{\mathbb{G}} \frac{f(x)}{|x|^2} \overline{\left(\mathcal{R}^2 f(x) + \frac{Q-1}{|x|}\mathcal{R}f(x)\right)} dx - \left\| \frac{1}{|x|}\mathcal{R}f + \frac{Q-4}{2|x|^2} f \right\|^2_{L^2(\mathbb{G})}.
\end{aligned}
$$

Multiplying both sides by $\frac{4}{Q(Q-4)}$ and simplifying we obtain

$$
\begin{aligned}
\operatorname{Re} \int_{\mathbb{G}} \frac{f(x)}{|x|^2} & \left(\overline{\frac{f(x)}{|x|^2} + \frac{4}{Q(Q-4)} \left(\mathcal{R}^2 f(x) + \frac{Q-1}{|x|}\mathcal{R}f(x)\right)} \right) dx \\
&= -\frac{4}{Q(Q-4)} \left\| \frac{1}{|x|}\mathcal{R}f + \frac{Q-4}{2|x|^2} f \right\|^2_{L^2(\mathbb{G})}.
\end{aligned} \tag{3.13}
$$

On the other hand, we also have

$$
2\operatorname{Re} \int_{\mathbb{G}} \frac{f(x)}{|x|^2} \left(\overline{\frac{f(x)}{|x|^2} + \frac{4}{Q(Q-4)} \left(\mathcal{R}^2 f(x) + \frac{Q-1}{|x|}\mathcal{R}f(x)\right)} \right) dx
$$

$$
\begin{aligned}
&= \left\| \frac{f}{|x|^2} + \frac{4}{Q(Q-4)} \left(\mathcal{R}^2 f + \frac{Q-1}{|x|} \mathcal{R} f \right) \right\|^2_{L^2(\mathbb{G})} + \left\| \frac{f}{|x|^2} \right\|^2_{L^2(\mathbb{G})} \\
&\quad - \left\| \frac{4}{Q(Q-4)} \left(\mathcal{R}^2 f + \frac{Q-1}{|x|} \mathcal{R} f \right) \right\|^2_{L^2(\mathbb{G})} . \qquad (3.14)
\end{aligned}
$$

From (3.13) and (3.14) we obtain

$$
\begin{aligned}
&- \frac{8}{Q(Q-4)} \left\| \frac{1}{|x|} \mathcal{R} f + \frac{Q-4}{2|x|^2} f \right\|^2_{L^2(\mathbb{G})} \\
&= \left(\frac{4}{Q(Q-4)} \right)^2 \left\| \frac{Q(Q-4)}{4} \frac{f}{|x|^2} + \mathcal{R}^2 f + \frac{Q-1}{|x|} \mathcal{R} f \right\|^2_{L^2(\mathbb{G})} + \left\| \frac{f}{|x|^2} \right\|^2_{L^2(\mathbb{G})} \\
&\quad - \left(\frac{4}{Q(Q-4)} \right)^2 \left\| \mathcal{R}^2 f + \frac{Q-1}{|x|} \mathcal{R} f \right\|^2_{L^2(\mathbb{G})} ,
\end{aligned}
$$

thus,

$$
\begin{aligned}
&\left\| \mathcal{R}^2 f + \frac{Q-1}{|x|} \mathcal{R} f + \frac{Q(Q-4)}{4|x|^2} f \right\|^2_{L^2(\mathbb{G})} + \frac{Q(Q-4)}{2} \left\| \frac{1}{|x|} \mathcal{R} f + \frac{Q-4}{2|x|^2} f \right\|^2_{L^2(\mathbb{G})} \\
&\qquad = \left\| \mathcal{R}^2 f + \frac{Q-1}{|x|} \mathcal{R} f \right\|^2_{L^2(\mathbb{G})} - \left(\frac{Q(Q-4)}{4} \right)^2 \left\| \frac{f}{|x|^2} \right\|^2_{L^2(\mathbb{G})} .
\end{aligned}
$$

This gives identity (3.7). ☐

In the Euclidean setting of $\mathbb{R}^n$ the equalities of the type of Theorem 3.1.1 were analysed in [MOW17b]. Theorem 3.1.1 was obtained in [RS17b] and it can be extended to all $\alpha \in \mathbb{R}$. We present these results next, following [Ngu17], also correcting relevant statements. We will be always using the notation

$$
\tilde{\mathcal{R}} f := \mathcal{R}^2 f + \frac{Q-1}{|x|} \mathcal{R} f. \qquad (3.15)
$$

Theorem 3.1.4 (Weighted L^2-Rellich inequalities). *Let $\mathbb{G}$ be a homogeneous group of homogeneous dimension $Q \geq 5$ and let $|\cdot|$ be a homogeneous quasi-norm on $\mathbb{G}$. Then we have the following properties:*

(1) *For any $\alpha \in \mathbb{R}$, for all complex-valued functions $f \in C_0^\infty(\mathbb{G}\backslash\{0\})$ we have the identity*

$$
\begin{aligned}
\left\| \frac{\tilde{\mathcal{R}} f}{|x|^\alpha} \right\|^2_{L^2(\mathbb{G})} &= C_\alpha^2 \left\| \frac{f}{|x|^{\alpha+2}} \right\|^2_{L^2(\mathbb{G})} + \left\| \frac{\tilde{\mathcal{R}} f}{|x|^\alpha} + C_\alpha \frac{f}{|x|^{\alpha+2}} \right\|^2_{L^2(\mathbb{G})} \\
&\quad + 2C_\alpha \left\| \frac{\mathcal{R} f}{|x|^{1+\alpha}} + \frac{Q-4-2\alpha}{2|x|^{\alpha+2}} f \right\|^2_{L^2(\mathbb{G})} , \qquad (3.16)
\end{aligned}
$$

where

$$C_\alpha = \frac{(Q+2\alpha)(Q-4-2\alpha)}{4}. \tag{3.17}$$

(2) *For any $\alpha \in (-Q/2, (Q-4)/2)$, we have the following weighted Rellich inequality for all complex-valued functions $f \in C_0^\infty(\mathbb{G}\backslash\{0\})$,*

$$C_\alpha \left\| \frac{f}{|x|^{\alpha+2}} \right\|_{L^2(\mathbb{G})} \leq \left\| \frac{\tilde{\mathcal{R}} f}{|x|^\alpha} \right\|_{L^2(\mathbb{G})}, \tag{3.18}$$

where the constant $C_\alpha > 0$ as in (3.17) is sharp and it is attained if and only if $f = 0$.

Proof of Theorem 3.1.4. Part (2) is an immediate consequence of Part (1), so we prove Part (1) now.

We can assume that $C_\alpha \neq 0$ since otherwise (3.16) trivially holds. Moreover, we can assume for the proof below that $Q-3-2\alpha \neq 0$. If this is zero, the statement (3.16) would follow by continuity from the same identity for other α's.

As in the proof of Theorem 3.1.1, we calculate

$$\begin{aligned}
\int_{\mathbb{G}} \frac{|f(x)|^2}{|x|^{4+2\alpha}} dx &= \int_0^\infty r^{Q-5-2\alpha} \int_{\wp} |f(ry)|^2 d\sigma(y) dr \\
&= \frac{1}{Q-4-2\alpha} \int_0^\infty (r^{Q-4-2\alpha})' \int_{\wp} |f(ry)|^2 d\sigma(y) dr \\
&= -\frac{2}{Q-4-2\alpha} \mathrm{Re} \int_0^\infty r^{Q-4-2\alpha} \int_{\wp} f(ry) \overline{\mathcal{R} f(ry)} d\sigma(y) dr \\
&= -\frac{2}{(Q-4-2\alpha)(Q-3-2\alpha)} \mathrm{Re} \int_0^\infty (r^{Q-3-2\alpha})' \int_{\wp} f(ry) \overline{\mathcal{R} f(ry)} d\sigma(y) dr \\
&= \frac{2}{(Q-4-2\alpha)(Q-3-2\alpha)} \mathrm{Re} \int_0^\infty r^{Q-3-2\alpha} \\
&\quad \times \int_{\wp} \left(|\mathcal{R} f(ry)|^2 + f(ry) \overline{\mathcal{R}^2 f(ry)} \right) d\sigma(y) dr \\
&= \frac{2}{(Q-4-2\alpha)(Q-3-2\alpha)} \mathrm{Re} \int_{\mathbb{G}} \left(\frac{|\mathcal{R} f(x)|^2}{|x|^{2+2\alpha}} + \frac{f(x) \overline{\mathcal{R}^2 f(x)}}{|x|^{2+2\alpha}} \right) dx.
\end{aligned} \tag{3.19}$$

By Theorem 3.1.1 we have

$$\int_{\mathbb{G}} \frac{|\mathcal{R} f|^2}{|x|^{2+2\alpha}} dx = \frac{(Q-4-2\alpha)^2}{4} \int_{\mathbb{G}} \frac{|f|^2}{|x|^{4+2\alpha}} dx + \int_{\mathbb{G}} \left| \frac{\mathcal{R} f}{|x|^{1+\alpha}} + \frac{Q-4-2\alpha}{2|x|^{2+\alpha}} f \right|^2 dx. \tag{3.20}$$

By integration by parts we obtain

$$\mathrm{Re} \int_{\mathbb{G}} \frac{f \overline{\mathcal{R}^2 f}}{|x|^{2+2\alpha}} dx = \mathrm{Re} \int_{\mathbb{G}} \frac{f \overline{\tilde{\mathcal{R}} f}}{|x|^{2+2\alpha}} dx - (Q-1) \mathrm{Re} \int_{\mathbb{G}} \frac{f \overline{\mathcal{R} f}}{|x|^{3+2\alpha}} dx \tag{3.21}$$

$$= \mathrm{Re}\int_{\mathbb{G}} \frac{f\overline{\tilde{\mathcal{R}}f}}{|x|^{2+2\alpha}}dx + \frac{(Q-1)(Q-4-2\alpha)}{2}\int_{\mathbb{G}} \frac{|f|^2}{|x|^{4+2\alpha}}dx.$$

Plugging (3.20) and (3.21) into (3.19) we get

$$\int_{\mathbb{G}} \frac{|f|^2}{|x|^{4+2\alpha}}dx = -\frac{1}{C_\alpha}\mathrm{Re}\int_{\mathbb{G}} \frac{f\overline{\tilde{\mathcal{R}}f}}{|x|^{2+2\alpha}}dx - \frac{1}{C_\alpha}\int_{\mathbb{G}}\left|\frac{\mathcal{R}f}{|x|^{1+\alpha}} + \frac{Q-4-2\alpha}{2|x|^{2+\alpha}}f\right|^2 dx$$

$$= \frac{1}{2C_\alpha^2}\int_{\mathbb{G}} \frac{|\tilde{\mathcal{R}}f|^2}{|x|^{2\alpha}}dx + \frac{1}{2}\int_{\mathbb{G}} \frac{|f|^2}{|x|^{2\alpha+4}}dx - \frac{1}{2C_\alpha^2}\int_{\mathbb{G}}\left|\frac{\tilde{\mathcal{R}}f}{|x|^{\alpha}} + C_\alpha\frac{f}{|x|^{\alpha+2}}\right|^2 dx$$

$$-\frac{1}{C_\alpha}\int_{\mathbb{G}}\left|\frac{\mathcal{R}f}{|x|^{\alpha+1}} + \frac{Q-4-2\alpha}{2|x|^{\alpha+2}}f\right|^2 dx,$$

which gives equality (3.16).

The inequality (3.18) is a straightforward consequence of (3.16) since $C_\alpha > 0$ for $\alpha \in (-Q/2, (Q-4)/2)$. Let us now show that this constant is sharp. For this, we consider the test function $g(r) := r^{-(Q-4-2\alpha)/2}$ which can be approximated by smooth functions. For example, let $\eta \in C_0^\infty(\mathbb{R})$ be such that $\eta = 1$ on $(-1,1)$ and $\eta = 0$ on $\mathbb{R}\backslash(-2,2)$. For any $\epsilon > 0$, define $f_\epsilon(r) := (1-\eta(r/\epsilon))r^{-\frac{Q-4-2\alpha}{2}}\eta(\epsilon r)$, then

$$\lim_{\epsilon\to 0} f_\epsilon(r) = g(r).$$

We have

$$\frac{\left\|\frac{\tilde{\mathcal{R}}g}{|x|^\alpha}\right\|^2_{L^2(\mathbb{G})}}{\left\|\frac{g}{|x|^{\alpha+2}}\right\|^2_{L^2(\mathbb{G})}} = \frac{\int_{\mathbb{G}} \frac{1}{|x|^{2\alpha}}\left|\mathcal{R}^2 g(|x|) + \frac{Q-1}{|x|}\mathcal{R}g(|x|)\right|^2 dx}{\int_{\mathbb{G}} \frac{|g(|x|)|^2}{|x|^{4+2\alpha}}dx} = C_\alpha^2,$$

which shows the sharpness of C_α. If for some function f, there is equality in (3.18), then from (3.16) we must have

$$\mathcal{R}f + \frac{Q-4-2\alpha}{2|x|}f = 0.$$

In terms of the Euler operator this can be equivalently expressed by

$$\mathbb{E}f = -\frac{Q-4-2\alpha}{2}f.$$

By Proposition 1.3.1 this implies that f is positively homogeneous of order $-(Q-4-2\alpha)/2$, that is, there exists function $h : \wp \to \mathbb{C}$ such that

$$f(x) = |x|^{-(Q-4-2\alpha)/2}h(x/|x|).$$

Since $f(x)/|x|^{2+\alpha}$ is in $L^2(\mathbb{G})$, we must have $h = 0$ on $\wp$ and, consequently, $f = 0$ on $\wp$. □

3.1.2 Rellich type inequalities in L^p

In this section we describe the L^p-versions of the L^2-properties presented in Section 3.1.1, where we have followed the proofs in [RS17b]. In this section we follow [Ngu17]. We start with the identity analogous to that in Theorem 3.1.1.

Theorem 3.1.5 (Identity leading to L^p-Rellich inequality). *Let $\mathbb{G}$ be a homogeneous group of homogeneous dimension Q and let $|\cdot|$ be a homogeneous quasi-norm on $\mathbb{G}$. Let $1<p<Q$ and $\alpha\in\mathbb{R}$. Then for all complex-valued functions $f\in C_0^\infty(\mathbb{G}\backslash\{0\})$ we have*

$$\int_{\mathbb{G}}\frac{1}{|x|^{p\alpha}}\left|\mathcal{R}f+\frac{Q-1}{|x|}f\right|^p dx=\frac{|Q+p'\alpha|^p}{(p')^p}\int_{\mathbb{G}}\frac{|f|^p}{|x|^{p(1+\alpha)}}dx \tag{3.22}$$
$$+p\int_{\mathbb{G}}\frac{1}{|x|^{p\alpha}}R_p\left(\frac{Q+p'\alpha}{p'}\frac{f}{|x|},\mathcal{R}f+\frac{Q-1}{|x|}f\right)dx,$$

where $p'=p/(p-1)$ and R_p is as in (2.22).

Proof of Theorem 3.1.5. First, a direct calculation shows

$$\begin{aligned}\int_{\mathbb{G}}\frac{|f|^p}{|x|^{p(1+\alpha)}}dx&=\int_0^\infty r^{Q-p(1+\alpha)-1}\int_{\wp}|f(ry)|^p d\sigma dr\\
&=\frac{1}{Q-p(1+\alpha)}\int_0^\infty(r^{Q-p(1+\alpha)})'\int_{\wp}|f(ry)|^p d\sigma dr\\
&=-\frac{p}{Q-p(1+\alpha)}\mathrm{Re}\int_0^\infty r^{Q-p(1+\alpha)}\int_{\wp}|f(ry)|^{p-2}f(ry)\overline{\mathcal{R}f(ry)}d\sigma(y)dr\\
&=-\frac{p}{Q-p(1+\alpha)}\mathrm{Re}\int_{\mathbb{G}}\frac{|f|^{p-2}f\overline{\mathcal{R}f}}{|x|^{p(1+\alpha)-1}}dx\\
&=-\frac{p}{Q-p(1+\alpha)}\left(\mathrm{Re}\int_{\mathbb{G}}\frac{|f|^{p-2}f}{|x|^{(p-1)(1+\alpha)}}\frac{\overline{\mathcal{R}f+\frac{Q-1}{|x|}f}}{|x|^\alpha}-(Q-1)\int_{\mathbb{G}}\frac{|f|^p}{|x|^{p(1+\alpha)}}dx\right).\end{aligned}$$

This implies further equalities,

$$\begin{aligned}\int_{\mathbb{G}}\frac{|f|^p}{|x|^{p(1+\alpha)}}dx&=\frac{p'}{Q+p'\alpha}\mathrm{Re}\int_{\mathbb{G}}\frac{|f|^{p-2}f}{|x|^{(p-1)(1+\alpha)}}\frac{\overline{\mathcal{R}f+\frac{Q-1}{|x|}f}}{|x|^\alpha}dx\\
&=\frac{p-1}{p}\int_{\mathbb{G}}\frac{|f|^p}{|x|^{p(1+\alpha)}}dx+\frac{1}{p}\frac{(p')^p}{|Q+p'\alpha|^p}\left|\mathcal{R}f+\frac{Q-1}{|x|}f\right|^2dx\\
&\quad-\int_{\mathbb{G}}R_p\left(\frac{f}{|x|^{1+\alpha}},\frac{p'}{Q+p'\alpha}\frac{\mathcal{R}f+\frac{Q-1}{|x|}f}{|x|^\alpha}\right)dx,\end{aligned}$$

which proves (3.22). □

Since the remainder term on the right-hand side of (3.22) is non-negative, this implies the following L^p-version of the Rellich type inequality on $\mathbb{G}$:

Corollary 3.1.6 (Rellich type inequality in $L^p(\mathbb{G})$). *Let $\mathbb{G}$ be a homogeneous group of homogeneous dimension Q and let $|\cdot|$ be a homogeneous quasi-norm on $\mathbb{G}$. Let $1 < p < Q$ and $\alpha \in \mathbb{R}$. Then for all complex-valued functions $f \in C_0^\infty(\mathbb{G}\backslash\{0\})$ we have*

$$\left\| \frac{1}{|x|^\alpha} \left(\mathcal{R}f + \frac{Q-1}{|x|} f \right) \right\|_{L^p(\mathbb{G})} \geq \frac{|Q+p'\alpha|}{p'} \left\| \frac{f}{|x|^{1+\alpha}} \right\|_{L^p(\mathbb{G})},$$

where the constant $\frac{4}{Q(Q-4)}$ is sharp and it is attained if and only if $f = 0$.

As a consequence of Theorem 3.1.5 we have

Corollary 3.1.7. *Let $\mathbb{G}$ be a homogeneous group of homogeneous dimension Q. Let $|\cdot|$ be any homogeneous quasi-norm on $\mathbb{G}$ and $1 < p < Q/2$. Then for any complex-valued $f \in C_0^\infty(\mathbb{G}\backslash\{0\})$, we have*

$$\int_{\mathbb{G}} \frac{|\tilde{\mathcal{R}}f|^p}{|x|^{p\alpha}} dx = \frac{|Q+p'\alpha|^p}{(p')^p} \int_{\mathbb{G}} \frac{|\mathcal{R}f|^p}{|x|^{p(1+\alpha)}} dx + p \int_{\mathbb{G}} \frac{1}{|x|^{p\alpha}} R_p \left(\frac{Q+p'\alpha}{p'} \frac{\mathcal{R}f}{|x|}, \tilde{\mathcal{R}}f \right) dx \tag{3.23}$$

for any $\alpha \in \mathbb{R}$. Here R_p is as in (2.22). As a consequence, we obtain the following weighted L^p-Rellich type inequality

$$\frac{|Q+p'\alpha|}{p'} \left\| \frac{\mathcal{R}f}{|x|^{(1+\alpha)}} \right\|_{L^p(\mathbb{G})} \leq \left\| \frac{\tilde{\mathcal{R}}f}{|x|^\alpha} \right\|_{L^p(\mathbb{G})} \tag{3.24}$$

for any $\alpha \in \mathbb{R}$ and any complex-valued function $f \in C_0^\infty(\mathbb{G}\backslash\{0\})$. Moreover, the inequality (3.24) is sharp and equality holds if and only if $f = 0$.

Proof of Corollary 3.1.7. The equality (3.23) is exactly (3.22) with f being replaced by $\mathcal{R}f$. The inequality (3.24) follows immediately from (3.23) by dropping the non-negative remainder term on the right-hand side of (3.23). The sharpness of (3.24) is proved by using approximations of the function $r^{-(Q'p(1+\alpha))/p}$. If the equality occurs in (3.24) for some function f, then by (3.23) we must have

$$\tilde{\mathcal{R}}f = \frac{Q+p'\alpha}{p'|x|} \mathcal{R}f,$$

which is equivalent to

$$\mathcal{R}^2 f + \frac{Q-p(1+\alpha)}{p|x|} \mathcal{R}f = 0.$$

This can be also expressed as

$$\mathbb{E}(\mathcal{R}f) = -\frac{Q-p(1+\alpha)}{p} \mathcal{R}f.$$

Hence $\mathcal{R}f$ is positively homogeneous of degree $-(Q'p(1+\alpha))/p$, which forces $\mathcal{R}f = 0$ since $\mathcal{R}f/|x|^{1+\alpha}$ is in $L^p(\mathbb{G})$. Thus, we get $f = 0$. □

Theorem 3.1.8 (Another weighted L^p-Rellich inequality). *Let $\mathbb{G}$ be a homogeneous group of homogeneous dimension Q. Let $1<p<Q/2$. Let $|\cdot|$ be a homogeneous quasi-norm on $\mathbb{G}$. Then we have the following properties.*

(1) *For all $\alpha\in\mathbb{R}$ and for all complex-valued functions $f\in C_0^\infty(\mathbb{G}\backslash\{0\})$ we have the identity*

$$\left\|\frac{\tilde{\mathcal{R}}f}{|x|^\alpha}\right\|^p_{L^p(\mathbb{G})}=|C_{p,\alpha}|^p\int_{\mathbb{G}}\frac{|f|^p}{|x|^{p(2+\alpha)}}dx+p\int_{\mathbb{G}}\frac{1}{|x|^{p\alpha}}R_p\left(C_{p,\alpha}\frac{f}{|x|^2},-\tilde{\mathcal{R}}f\right)dx$$
$$+p|C_{p,\alpha}|^{p-2}C_{p,\alpha}(p-1)\int_{\mathbb{G}}\frac{|f|^{p-2}}{|x|^{p(2+\alpha)-2}}\left|\mathcal{R}|f|+\frac{Q-p(2+\alpha)}{p|x|}|f|\right|^2dx$$
$$+p|C_{p,\alpha}|^{p-2}C_{p,\alpha}\int_{\mathbb{G}}\frac{|f|^{p-4}(\operatorname{Im}(f\overline{\mathcal{R}f}))^2}{|x|^{p(2+\alpha)-2}}dx. \tag{3.25}$$

Here R_p is as in (2.22) and

$$C_{p,\alpha}=\frac{(Q-2p-p\alpha)(Q+p'\alpha)}{pp'},\quad \frac{1}{p}+\frac{1}{p'}=1. \tag{3.26}$$

(2) *For any $\alpha\in(-(p-1)Q/p,(Q-2p)/p)$ and all complex-valued functions $f\in C_0^\infty(\mathbb{G}\backslash\{0\})$ we have*

$$C_{p,\alpha}\left\|\frac{f}{|x|^{2+\alpha}}\right\|_{L^p(\mathbb{G})}\le\left\|\frac{\tilde{\mathcal{R}}f}{|x|^\alpha}\right\|_{L^p(\mathbb{G})}. \tag{3.27}$$

Moreover, the constant $C_{p,\alpha}>0$ as in (3.26) is sharp and equality in (3.27) holds if and only if $f=0$.

Proof of Theorem 3.1.8. Inequality (3.27) in Part (2) follows from Part (1) in view of the positivity of the constant $C_{p,\alpha}>0$ under the corresponding conditions on α, so we prove Part (1) now. For the argument below we may assume that $C_{p,\alpha}\neq 0$, and that the constant on the left-hand side in the following estimate is non-zero. Then a direct calculation using Proposition 1.2.10 and integration by parts gives

$$\frac{(Q-p(2+\alpha))(Q-p(2+\alpha)+1)}{p}\int_{\mathbb{G}}\frac{|f(x)|^p}{|x|^{p(2+\alpha)}}dx$$
$$=\frac{(Q-p(2+\alpha))(Q-p(2+\alpha)+1)}{p}\int_0^\infty r^{Q-p(2+\alpha)-1}\int_\wp|f(ry)|^pd\sigma(y)dr$$
$$=\frac{(Q-p(2+\alpha)+1)}{p}\int_0^\infty(r^{Q-p(2+\alpha)})'\int_\wp|f(ry)|^pd\sigma(y)dr$$
$$=-(Q-p(2+\alpha)+1)\operatorname{Re}\int_0^\infty r^{Q-p(2+\alpha)}\int_\wp|f(ry)|^{p-2}f(ry)\overline{\mathcal{R}f(ry)}d\sigma(y)dr$$

$$
\begin{aligned}
&= -\mathrm{Re}\int_0^\infty (r^{Q-p(2+\alpha)+1})' \int_{\wp} |f(ry)|^{p-2} f(ry)\overline{\mathcal{R}f(ry)}d\sigma(y)dr \\
&= \mathrm{Re}\int_0^\infty r^{Q-p(2+\alpha)+1} \int_{\wp} (p-2)|f(ry)|^{p-4}(\mathrm{Re}(f(ry)\overline{\mathcal{R}f(ry)}))^2 \\
&\quad + |f(ry)|^{p-2}|\mathcal{R}f(ry)|^2 + |f(ry)|^{p-2}f(ry)\overline{\mathcal{R}^2 f(ry)}d\sigma(y)dr \\
&= \mathrm{Re}\int_{\mathbb{G}} \frac{1}{|x|^{p(2+\alpha)-2}} \left((p-2)|f|^{p-4}(\mathrm{Re}(f\overline{\mathcal{R}f}))^2 + |f|^{p-2}|\mathcal{R}f|^2 + |f|^{p-2}f\overline{\mathcal{R}^2 f}\right) dx \\
&= \mathrm{Re}\int_{\mathbb{G}} \frac{|f|^{p-2}f}{|x|^{(p-1)(2+\alpha)}} \frac{\overline{\tilde{\mathcal{R}}f}}{|x|^\alpha} dx - (Q-1)\mathrm{Re}\int_{\mathbb{G}} \frac{|f|^{p-2}f\overline{\mathcal{R}f}}{|x|^{p(2+\alpha)-1}} dx \\
&\quad + (p-1)\int_{\mathbb{G}} \frac{|f|^{p-4}(\mathrm{Re}(f\overline{\mathcal{R}f}))^2}{|x|^{p(2+\alpha)-2}} dx + \int_{\mathbb{G}} \frac{|f|^{p-4}(\mathrm{Im}(f\overline{\mathcal{R}f}))^2}{|x|^{p(2+\alpha)-2}} dx \\
&= \mathrm{Re}\int_{\mathbb{G}} \frac{|f|^{p-2}f}{|x|^{(p-1)(2+\alpha)}} \frac{\overline{\tilde{\mathcal{R}}f}}{|x|^\alpha} dx + \frac{(Q-1)(Q-p(2+\alpha))}{p}\int_{\mathbb{G}} \frac{|f|^p}{|x|^{p(2+\alpha)}} dx \\
&\quad + \frac{4(p-1)}{p^2}\int_{\mathbb{G}} \frac{(\mathcal{R}(|f|^{p/2}))^2}{|x|^{p(2+\alpha)-2}} dx + \int_{\mathbb{G}} \frac{|f|^{p-4}(\mathrm{Im}(f\overline{\mathcal{R}f}))^2}{|x|^{p(2+\alpha)-2}} dx. \qquad (3.28)
\end{aligned}
$$

By using Theorem 3.1.1 for $|f|^{\frac{p}{2}}$, we obtain

$$
\begin{aligned}
\int_{\mathbb{G}} \frac{(\mathcal{R}(|f|^{p/2}))^2}{|x|^{p(2+\alpha)-2}} dx &= \frac{(Q-p(2+\alpha))^2}{4}\int_{\mathbb{G}} \frac{|f|^p}{|x|^{p(2+\alpha)}} dx \\
&\quad + \int_{\mathbb{G}} \frac{1}{|x|^{p(2+\alpha)-2}} \left|\mathcal{R}|f|^{p/2} + \frac{Q-p(2+\alpha)}{2|x|}|f|^{p/2}\right|^2 dx.
\end{aligned}
$$

Plugging this in (3.28) we get

$$
\begin{aligned}
&\int_{\mathbb{G}} \frac{|f(x)|^p}{|x|^{p(2+\alpha)}} dx \\
&= -\frac{1}{C_{p,\alpha}}\mathrm{Re}\int_{\mathbb{G}} \frac{|f|^{p-2}f}{|x|^{(p-1)(2+\alpha)}} \frac{\overline{\tilde{\mathcal{R}}f}}{|x|^\alpha} dx - \frac{1}{C_{p,\alpha}}\int_{\mathbb{G}} \frac{|f|^{p-4}(\mathrm{Im}(f\overline{\mathcal{R}f}))^2}{|x|^{p(2+\alpha)-2}} dx \\
&\quad - \frac{1}{C_{p,\alpha}}\frac{4(p-1)}{p^2}\int_{\mathbb{G}} \frac{1}{|x|^{p(2+\alpha)-2}} \left|\mathcal{R}|f|^{\frac{p}{2}} + \frac{Q-p(2+\alpha)}{2|x|}|f|^{\frac{p}{2}}\right|^2 dx \\
&= \frac{p-1}{p}\int_{\mathbb{G}} \frac{|f|^p}{|x|^{p(2+\alpha)}} dx + \frac{1}{p}\frac{1}{|C_{p,\alpha}|^p}\int_{\mathbb{G}} \frac{|\tilde{\mathcal{R}}f|^p}{|x|^{p\alpha}} dx - \int_{\mathbb{G}} R_p\left(\frac{f}{|x|^{2+\alpha}}, -\frac{1}{C_{p,\alpha}}\frac{\tilde{\mathcal{R}}f}{|x|^\alpha}\right) dx \\
&\quad - \frac{4(p-1)}{p^2 C_{p,\alpha}}\int_{\mathbb{G}} \frac{1}{|x|^{p(2+\alpha)-2}} \left|\mathcal{R}|f|^{\frac{p}{2}} + \frac{Q-p(2+\alpha)}{2|x|}|f|^{\frac{p}{2}}\right|^2 dx \\
&\quad - \frac{1}{C_{p,\alpha}}\int_{\mathbb{G}} \frac{|f|^{p-4}(\mathrm{Im}(f\overline{\mathcal{R}f}))^2}{|x|^{p(2+\alpha)-2}} dx. \qquad (3.29)
\end{aligned}
$$

The equality (3.25) now follows from (3.29) and the equality

$$\mathcal{R}(|f|^{\frac{p}{2}}) = \frac{p}{2}|f|^{\frac{p}{2}-1}\mathcal{R}(|f|).$$

Part (2). Clearly, inequality (3.27) follows from equality (3.25). As in the proof of Theorem 3.1.4, the sharpness of (3.27) follows by considering appropriate approximations of the function $r^{-(Q-p(2+\alpha))/p}$. Moreover, if we have equality in (3.27) for some function f, then in view of Part (1) we must have

$$\mathcal{R}(|f|) + \frac{Q-p(2+\alpha)}{p|x|}|f| = 0.$$

This means that

$$\mathbb{E}(|f|) = -\frac{Q-p(2+\alpha)}{p}|f|.$$

By Proposition 1.3.1 the function $|f|$ must be positively homogeneous of degree $-(Q-p(2+\alpha))/p$. Since $|f|/|x|^{2+\alpha}$ is in $L^p(\mathbb{G})$ we must have $f=0$. □

3.1.3 Stability of Rellich type inequalities

The method used in the previous section also allows one to obtain the following stability property for Rellich type inequalities. We present such a result following [RS18].

Theorem 3.1.9 (Stability of Rellich type inequalities). *Let $\mathbb{G}$ be a homogeneous group of homogeneous dimension Q. Let $|\cdot|$ be a homogeneous quasi-norm on $\mathbb{G}$ and let $p \geq 1$. Let $k \geq 2$, $k \in \mathbb{N}$, be such that $kp < Q$. Then for all real-valued radial functions $u \in C_0^\infty(\mathbb{G})$ we have*

$$\int_{\mathbb{G}} \frac{|\tilde{\mathcal{R}}u|^p}{|x|^{(k-2)p}}dx - K_{k,p}^p \int_{\mathbb{G}} \frac{|u|^p}{|x|^{kp}}dx \geq C \sup_{R>0} \int_{\mathbb{G}} \frac{\left||u(x)|^{\frac{p-2}{2}}u(x) - R^{\frac{Q-kp}{2}}|u(R)|^{\frac{p-2}{2}}u(R)|x|^{-\frac{Q-kp}{2}}\right|^2}{|x|^{kp}\left|\log\frac{R}{|x|}\right|^2}dx, \tag{3.30}$$

where

$$\tilde{\mathcal{R}}f = \mathcal{R}^2 f + \frac{Q-1}{|x|}\mathcal{R}f$$

and

$$K_{k,p} = \frac{(Q-kp)[(k-2)p+(p-1)Q]}{p^2}. \tag{3.31}$$

Proof of Theorem 3.1.9. For $k \geq 2$, $k \in \mathbb{N}$, and $kp < Q$, as in the assumptions of the theorem, let us denote

$$v(r) := r^{\frac{Q-kp}{p}} u(r), \quad \text{where} \quad r \in [0, \infty). \tag{3.32}$$

In particular, we have $v(0) = 0$ and $v(\infty) = 0$. We then calculate as follows:

$$\begin{aligned}
-\tilde{\mathcal{R}}u =& -\mathcal{R}^2\left(r^{\frac{kp-Q}{p}}v(r)\right) - \frac{Q-1}{r}\mathcal{R}\left(r^{\frac{kp-Q}{p}}v(r)\right) \\
=& -\mathcal{R}\left(\frac{kp-Q}{p}r^{\frac{kp-Q}{p}-1}v(r) + r^{\frac{kp-Q}{p}}\mathcal{R}v(r)\right) \\
& - \frac{Q-1}{r}\frac{kp-Q}{p}r^{\frac{kp-Q}{p}-1}v(r) - \frac{Q-1}{r}r^{\frac{kp-Q}{p}}\mathcal{R}v(r) \\
=& -\frac{kp-Q}{p}\left(\frac{kp-Q}{p}-1\right)r^{\frac{kp-Q}{p}-2}v(r) - \frac{kp-Q}{p}r^{\frac{kp-Q}{p}-1}\mathcal{R}v(r) \\
& - \frac{kp-Q}{p}r^{\frac{kp-Q}{p}-1}\mathcal{R}v(r) - r^{\frac{kp-Q}{p}}\mathcal{R}^2v(r) \\
& - \frac{Q-1}{r}\frac{kp-Q}{p}r^{\frac{kp-Q}{p}-1}v(r) - \frac{Q-1}{r}r^{\frac{kp-Q}{p}}\mathcal{R}v(r) \\
=& -r^{\frac{kp-Q}{p}-2}\left(\frac{(kp-Q)(kp-Q-p)}{p^2} + \frac{(Q-1)(kp-Q)}{p}\right)v(r) \\
& - r^{\frac{kp-Q}{p}-2}r^2\left(\mathcal{R}^2v(r) + \frac{1}{r}\left(\frac{2(kp-Q)}{p} + (Q-1)\right)\mathcal{R}v(r)\right) \\
=& \, r^{k-2-\frac{Q}{p}}(K_{k,p}v(r) - r^2\tilde{\mathcal{R}}_kv(r)),
\end{aligned}$$

where $K_{k,p}$ is as in (3.31), and where we denote

$$\tilde{\mathcal{R}}_kf := \mathcal{R}^2f + \frac{2k + \frac{Q(p-2)}{p} - 1}{r}\mathcal{R}f.$$

By using the first inequality in Lemma 2.4.2 with $a = K_{k,p}v(r)$ and $b = r^2\tilde{\mathcal{R}}_kv(r)$, and the fact that $\int_0^\infty |v|^{p-2}vv'dr = 0$ in view of $v(0) = 0$ and $v(\infty) = 0$, we obtain

$$\begin{aligned}
J :=& \int_{\mathbb{G}} \frac{|\tilde{\mathcal{R}}u|^p}{|x|^{(k-2)p}}dx - K_{k,p}^p\int_{\mathbb{G}}\frac{|u|^p}{|x|^{kp}}dx \\
=& |\wp|\int_0^\infty |-\tilde{\mathcal{R}}u(r)|^pr^{Q-1-(k-2)p}dr - K_{k,p}^p|\wp|\int_0^\infty |u(r)|^pr^{Q-kp-1}dr \\
=& |\wp|\int_0^\infty \left(|K_{k,p}v(r) - r^2\tilde{\mathcal{R}}_kv(r)|^p - (K_{k,p}v(r))^p\right)r^{-1}dr \\
\geq& -p|\wp|K_{k,p}^{p-1}\int_0^\infty |v|^{p-2}v\tilde{\mathcal{R}}_kvrdr
\end{aligned}$$

$$= -p|\wp|K_{k,p}^{p-1}\int_0^\infty |v|^{p-2}v\left(v''+\frac{2k+\frac{Q(p-2)}{p}-1}{r}v'\right)rdr$$
$$= -p|\wp|K_{k,p}^{p-1}\int_0^\infty |v|^{p-2}vv''rdr.$$

On the other hand, we have

$$\begin{aligned}
-\int_0^\infty |v|^{p-2}vv''rdr &= (p-1)\int_0^\infty |v|^{p-2}(v')^2rdr+\int_0^\infty |v|^{p-2}vv'dr\\
&= (p-1)\int_0^\infty |v|^{p-2}(v')^2rdr\\
&= \frac{4(p-1)}{p^2}\int_0^\infty \left(\frac{p-2}{2}\right)^2 |v|^{p-2}(v')^2dr\\
&\quad+\frac{4(p-1)}{p^2}\int_0^\infty (p-2)|v|^{p-2}(v')^2+|v|^{p-2}(v')^2rdr\\
&= \frac{4(p-1)}{p^2}\int_0^\infty \left(\left(|v|^{\frac{p-2}{2}}\right)'v+|v|^{\frac{p-2}{2}}v'\right)^2 rdr\\
&= \frac{4(p-1)}{p^2}\int_0^\infty |(|v|^{\frac{p-2}{2}}v)'|^2rdr\\
&= \frac{4(p-1)}{|\wp_2|p^2}\int_{\mathbb{G}_2}\left|\mathcal{R}(|v|^{\frac{p-2}{2}}v)\right|^2dx,
\end{aligned}$$

where $\mathbb{G}_2$ is a homogeneous group of homogeneous degree 2 and $|\wp_2|$ is the measure of the corresponding unit 2-quasi-ball. By using Remark 2.2.9 for $|v|^{\frac{p-2}{2}}v\in C_0^\infty(\mathbb{G}_2\backslash\{0\})$ in $p=Q=2$ case, and combining the above equalities, we obtain

$$\begin{aligned}
J &\geq C_1\int_{\mathbb{G}_2}\frac{\left||v(x)|^{\frac{p-2}{2}}v(x)-|v(R\frac{x}{|x|})|^{\frac{p-2}{2}}v(R\frac{x}{|x|})\right|^2}{|x|^2\left|\log\frac{R}{|x|}\right|^2}dx\\
&= C_1\int_0^\infty \frac{\left||v(r)|^{\frac{p-2}{2}}v(r)-|v(R)|^{\frac{p-2}{2}}v(R)\right|^2}{r\left|\log\frac{R}{r}\right|^2}dr\\
&= C_1\int_0^\infty \frac{\left||u(r)|^{\frac{p-2}{2}}u(r)-R^{\frac{Q-kp}{2}}|u(R)|^{\frac{p-2}{2}}u(R)r^{-\frac{Q-kp}{2}}\right|^2}{r^{1-Q+kp}\left|\log\frac{R}{r}\right|^2}dr
\end{aligned}$$

for any $R>0$. That is, we have

$$J\geq C\sup_{R>0}\int_{\mathbb{G}}\frac{\left||u(x)|^{\frac{p-2}{2}}u(x)-R^{\frac{Q-kp}{2}}|u(R)|^{\frac{p-2}{2}}u(R)|x|^{-\frac{Q-kp}{2}}\right|^2}{|x|^{kp}\left|\log\frac{R}{|x|}\right|^2}dx.$$

The proof is complete. $\square$

3.1.4 Higher-order Hardy–Rellich inequalities

In this section we show that by iterating the already established weighted Hardy inequalities we get inequalities of higher order. An interesting feature is that we also obtain the exact formula for the remainder which yields the sharpness of the constants as well. That is, when $p = 2$, one can iterate the exact representation formulae of the remainder that we obtained in Theorem 3.1.1. It implies higher-order remainder equalities that can be then also used to argue the sharpness of the constant.

Theorem 3.1.10 (Higher-order Hardy–Rellich identities and inequalities). *Let $\mathbb{G}$ be a homogeneous group of homogeneous dimension $Q \geq 3$. Let $\alpha \in \mathbb{R}$ and $k \in \mathbb{N}$ be such that*

$$\prod_{j=0}^{k-1} \left| \frac{Q-2}{2} - (\alpha + j) \right| \neq 0.$$

Then for all complex-valued functions $f \in C_0^\infty(\mathbb{G}\backslash\{0\})$ we have

$$\left\| \frac{f}{|x|^{k+\alpha}} \right\|_{L^2(\mathbb{G})} \leq \left[\prod_{j=0}^{k-1} \left| \frac{Q-2}{2} - (\alpha + j) \right| \right]^{-1} \left\| \frac{1}{|x|^\alpha} \mathcal{R}^k f \right\|_{L^2(\mathbb{G})}, \tag{3.33}$$

where the constant above is sharp, and is attained if and only if $f = 0$.

Moreover, for all $k \in \mathbb{N}$ and $\alpha \in \mathbb{R}$, the following identity holds:

$$\begin{aligned} \left\| \frac{1}{|x|^\alpha} \mathcal{R}^k f \right\|^2_{L^2(\mathbb{G})} &= \left[\prod_{j=0}^{k-1} \left(\frac{Q-2}{2} - (\alpha + j) \right)^2 \right] \left\| \frac{f}{|x|^{k+\alpha}} \right\|^2_{L^2(\mathbb{G})} \\ &\quad + \sum_{l=1}^{k-1} \left[\prod_{j=0}^{l-1} \left(\frac{Q-2}{2} - (\alpha + j) \right)^2 \right] \\ &\quad \times \left\| \frac{1}{|x|^{l+\alpha}} \mathcal{R}^{k-l} f + \frac{Q - 2(l+1+\alpha)}{2|x|^{l+1+\alpha}} \mathcal{R}^{k-l-1} f \right\|^2_{L^2(\mathbb{G})} \\ &\quad + \left\| \frac{1}{|x|^\alpha} \mathcal{R}^k f + \frac{Q-2-2\alpha}{2|x|^{1+\alpha}} \mathcal{R}^{k-1} f \right\|^2_{L^2(\mathbb{G})}. \end{aligned} \tag{3.34}$$

Remark 3.1.11. Let us point out some special cases of inequality (3.33).

1. For $k = 1$ inequality (3.33) gives the weighted L^2-Hardy inequalities from Corollary 2.1.6.
2. In particular, for $k = 1$ and $\alpha = 0$, inequality (3.33) gives the L^2-Hardy inequality, i.e., (2.2) in the case of $p = 2$.
3. For $k = 2$, inequality (3.33) can be thought of as a (weighted) Hardy–Rellich type inequality, while for larger k this corresponds to higher-order (weighted) Rellich inequalities.

Proof of Theorem 3.1.10. For any $\alpha \in \mathbb{R}$ and $Q \geq 3$ let us iterate the identity

$$\begin{aligned} \left\| \frac{1}{|x|^{\alpha}} \mathcal{R} f \right\|^2_{L^2(\mathbb{G})} &= \left(\frac{Q-2}{2} - \alpha \right)^2 \left\| \frac{f}{|x|^{\alpha+1}} \right\|^2_{L^2(\mathbb{G})} \\ &+ \left\| \frac{1}{|x|^{\alpha}} \mathcal{R} f + \frac{Q-2(\alpha+1)}{2|x|^{\alpha+1}} f \right\|^2_{L^2(\mathbb{G})}, \end{aligned} \tag{3.35}$$

given in (2.16), as follows. First, replacing f in (3.35) by $\mathcal{R}f$ we have

$$\begin{aligned} \left\| \frac{1}{|x|^{\alpha}} \mathcal{R}^2 f \right\|^2_{L^2(\mathbb{G})} &= \left(\frac{Q-2}{2} - \alpha \right)^2 \left\| \frac{\mathcal{R} f}{|x|^{\alpha+1}} \right\|^2_{L^2(\mathbb{G})} \\ &+ \left\| \frac{1}{|x|^{\alpha}} \mathcal{R}^2 f + \frac{Q-2(\alpha+1)}{2|x|^{\alpha+1}} \mathcal{R} f \right\|^2_{L^2(\mathbb{G})}. \end{aligned} \tag{3.36}$$

Furthermore, replacing α by $\alpha+1$, (3.35) implies that

$$\begin{aligned} \left\| \frac{1}{|x|^{\alpha+1}} \mathcal{R} f \right\|^2_{L^2(\mathbb{G})} &= \left(\frac{Q-2}{2} - (\alpha+1) \right)^2 \left\| \frac{f}{|x|^{\alpha+2}} \right\|^2_{L^2(\mathbb{G})} \\ &+ \left\| \frac{1}{|x|^{\alpha+1}} \mathcal{R} f + \frac{Q-2(\alpha+2)}{2|x|^{\alpha+2}} f \right\|^2_{L^2(\mathbb{G})}. \end{aligned}$$

Combination of this with (3.36) gives

$$\begin{aligned} \left\| \frac{1}{|x|^{\alpha}} \mathcal{R}^2 f \right\|^2_{L^2(\mathbb{G})} &= \left(\frac{Q-2}{2} - \alpha \right)^2 \left(\frac{Q-2}{2} - (\alpha+1) \right)^2 \left\| \frac{f}{|x|^{\alpha+2}} \right\|^2_{L^2(\mathbb{G})} \\ &+ \left(\frac{Q-2}{2} - \alpha \right)^2 \left\| \frac{1}{|x|^{\alpha+1}} \mathcal{R} f + \frac{Q-2(\alpha+2)}{2|x|^{\alpha+2}} f \right\|^2_{L^2(\mathbb{G})} \\ &+ \left\| \frac{1}{|x|^{\alpha}} \mathcal{R}^2 f + \frac{Q-2-2\alpha}{2|x|^{\alpha+1}} \mathcal{R} f \right\|^2_{L^2(\mathbb{G})}. \end{aligned}$$

By using this iteration process further, we eventually arrive at the family of identities

$$\begin{aligned} \left\| \frac{1}{|x|^{\alpha}} \mathcal{R}^k f \right\|^2_{L^2(\mathbb{G})} &= \left[\prod_{j=0}^{k-1} \left(\frac{Q-2}{2} - (\alpha+j) \right)^2 \right] \left\| \frac{f}{|x|^{k+\alpha}} \right\|^2_{L^2(\mathbb{G})} \\ &+ \sum_{l=1}^{k-1} \left[\prod_{j=0}^{l-1} \left(\frac{Q-2}{2} - (\alpha+j) \right)^2 \right] \\ &\times \left\| \frac{1}{|x|^{l+\alpha}} \mathcal{R}^{k-l} f + \frac{Q-2(l+1+\alpha)}{2|x|^{l+1+\alpha}} \mathcal{R}^{k-l-1} f \right\|^2_{L^2(\mathbb{G})} \end{aligned}$$

$$+\left\|\frac{1}{|x|^{\alpha}}\mathcal{R}^{k}f+\frac{Q-2-2\alpha}{2|x|^{1+\alpha}}\mathcal{R}^{k-1}f\right\|^{2}_{L^{2}(\mathbb{G})}, \quad k=1,2,\ldots,$$

which give (3.34).

Now, by dropping positive terms, these identities imply that

$$\left\|\frac{1}{|x|^{\alpha}}\mathcal{R}^{k}f\right\|^{2}_{L^{2}(\mathbb{G})}\geq C_{k,Q}\left\|\frac{f}{|x|^{k+\alpha}}\right\|^{2}_{L^{2}(\mathbb{G})},$$

where

$$C_{k,Q}=\prod_{j=0}^{k-1}\left(\frac{Q-2}{2}-(\alpha+j)\right)^{2}.$$

If $C_{k,Q}\neq 0$, this can be written as

$$\left\|\frac{f}{|x|^{k+\alpha}}\right\|^{2}_{L^{2}(\mathbb{G})}\leq\frac{1}{C_{k,Q}}\left\|\frac{1}{|x|^{\alpha}}\mathcal{R}^{k}f\right\|^{2}_{L^{2}(\mathbb{G})},$$

which gives (3.33).

Now it remains to show the sharpness of the constant and the equality in (3.33). To do this, we rewrite the equality

$$\frac{\mathcal{R}^{k-l}f}{|x|^{l+\alpha}}+\frac{Q-2(l+1+\alpha)}{2|x|^{l+1+\alpha}}\mathcal{R}^{k-l-1}f=0$$

as

$$|x|\mathcal{R}(\mathcal{R}^{k-l-1}f)+\frac{Q-2(l+1+\alpha)}{2}(\mathcal{R}^{k-l-1}f)=0,$$

and by Proposition 1.3.1, Part (i), this means that $\mathcal{R}^{k-l-1}f$ is positively homogeneous of degree $-\frac{Q}{2}+l+1+\alpha$. So, all the remainder terms vanish if f is positively homogeneous of degree $k-\frac{Q}{2}+\alpha$. As this can be approximated by functions in $C_0^{\infty}(\mathbb{G}\backslash\{0\})$, the constant $C_{k,Q}$ is sharp.

If this constant was attained, it would be on functions f which are positively homogeneous of degree $k-\frac{Q}{2}+\alpha$, in which case $\frac{f}{|x|^{k+\alpha}}$ would be positively homogeneous of degree $-\frac{Q}{2}$. These are in L^2 if and only if they are zero. □

Theorem 3.1.4 and Theorem 3.1.10 were proved in [RS17b]. They can be extended further to derivatives of higher order. Such results have been recently obtained in [Ngu17] and in the rest of this subsection we describe such extensions.

Theorem 3.1.12 (Further higher-order Hardy–Rellich identities and inequalities). *Let $\mathbb{G}$ be a homogeneous group of homogeneous dimension Q and let $|\cdot|$ be a homogeneous quasi-norm on $\mathbb{G}$. We denote*

$$\tilde{\mathcal{R}}f:=\mathcal{R}^{2}f+\frac{Q-1}{|x|}\mathcal{R}f, \quad C_{\beta}:=\frac{(Q+2\beta)(Q-4-2\beta)}{4}, \tag{3.37}$$

for $\beta \in \mathbb{R}$, for the constants appearing below. Let $k \in \mathbb{N}$ be a positive integer and let $\alpha \in \mathbb{R}$. Then for all complex-valued functions $f \in C_0^\infty(\mathbb{G}\backslash\{0\})$ we have the following identities:

(1) *If $Q \geq 4k+1$, then we have*

$$\begin{aligned}
&\left(\prod_{i=0}^{k-1} C_{2i+\alpha}\right)^2 \left\|\frac{f}{|x|^{2k+\alpha}}\right\|^2_{L^2(\mathbb{G})} \\
&= \left\|\frac{\tilde{\mathcal{R}}^k f}{|x|^\alpha}\right\|^2_{L^2(\mathbb{G})} - \left\|\frac{1}{|x|^\alpha}\left|\tilde{\mathcal{R}}^k f + C_\alpha \frac{\tilde{\mathcal{R}}^{k-1} f}{|x|^2}\right|\right\|^2_{L^2(\mathbb{G})} \\
&\quad - \sum_{j=1}^{k-1}\left(\prod_{i=0}^{k-1} C_{2i+\alpha}\right)^2 \left\|\frac{1}{|x|^{2j+\alpha}}\left|\tilde{\mathcal{R}}^{k-j} f + C_{2j+\alpha}\frac{\tilde{\mathcal{R}}^{k-j-1} f}{|x|^2}\right|\right\|^2_{L^2(\mathbb{G})} \\
&\quad - 2C_\alpha \left\|\frac{1}{|x|^{1+\alpha}}\left|\mathcal{R}(\tilde{\mathcal{R}}^{k-1} f) + \frac{Q-4-2\alpha}{2|x|}\tilde{\mathcal{R}}^{k-1} f\right|\right\|^2_{L^2(\mathbb{G})} \\
&\quad - 2\sum_{j=1}^{k-1}\left(\prod_{i=0}^{k-1} C_{2i+\alpha}\right)^2 C_{2j+\alpha} \qquad (3.38) \\
&\quad \times \left\|\frac{1}{|x|^{1+\alpha+2j}}\left|\mathcal{R}(\tilde{\mathcal{R}}^{k-j-1} f) + \frac{Q-4-2\alpha-4j}{2|x|}\tilde{\mathcal{R}}^{k-j-1} f\right|\right\|^2_{L^2(\mathbb{G})}.
\end{aligned}$$

(2) *If $Q \geq 4k+3$, then we have*

$$\begin{aligned}
&\left(\frac{Q-2-2\alpha}{2}\prod_{i=0}^{k-1} C_{2i+1+\alpha}\right)^2 \left\|\frac{f}{|x|^{2k+1+\alpha}}\right\|^2_{L^2(\mathbb{G})} \\
&= \left\|\frac{\mathcal{R}(\tilde{\mathcal{R}}^k f)}{|x|^\alpha}\right\|^2_{L^2(\mathbb{G})} - \left\|\frac{1}{|x|^\alpha}\left|\mathcal{R}(\tilde{\mathcal{R}}^k f) + \frac{Q-2-2\alpha}{2|x|}\tilde{\mathcal{R}}^k f\right|\right\|^2_{L^2(\mathbb{G})} \\
&\quad - \frac{(Q-2-2\alpha)^2}{4}\left\|\frac{1}{|x|^\alpha}\left|\tilde{\mathcal{R}}^k f + C_{1+\alpha}\frac{\tilde{\mathcal{R}}^{k-1} f}{|x|^2}\right|\right\|^2_{L^2(\mathbb{G})} \\
&\quad - \frac{(Q-2-2\alpha)^2}{4}\sum_{j=1}^{k-1}\left(\prod_{i=0}^{k-1} C_{2i+\alpha}\right)^2 \\
&\quad \times \left\|\frac{1}{|x|^{1+2j+\alpha}}\left|\tilde{\mathcal{R}}^{k-j} f + C_{2j+1+\alpha}\frac{\tilde{\mathcal{R}}^{k-j-1} f}{|x|^2}\right|\right\|^2_{L^2(\mathbb{G})} \\
&\quad - \frac{(Q-2-2\alpha)^2}{4} C_{\alpha+1}\left\|\frac{1}{|x|^{2+\alpha}}\left|\mathcal{R}(\tilde{\mathcal{R}}^{k-1} f) + \frac{Q-6-2\alpha}{2|x|}\tilde{\mathcal{R}}^{k-1} f\right|\right\|^2_{L^2(\mathbb{G})}
\end{aligned}$$

$$- \frac{(Q-2-2\alpha)^2}{4} \sum_{j=1}^{k-1} \left(\prod_{i=0}^{k-1} C_{2i+1+\alpha} \right)^2 C_{2j+1+\alpha} \tag{3.39}$$
$$\times \left\| \frac{1}{|x|^{2+\alpha+2j}} \left| \mathcal{R}(\tilde{\mathcal{R}}^{k-j-1} f) + \frac{Q-6-2\alpha-4j}{2|x|} \tilde{\mathcal{R}}^{k-j-1} f \right| \right\|^2_{L^2(\mathbb{G})}.$$

As consequences, for all complex-valued function $f \in C_0^\infty(\mathbb{G}\backslash\{0\})$ we have the following estimates:

(3) *If $Q \geq 4k+1$ and $\alpha \in (-Q/2, (Q-4k)/2)$, then we have*

$$\left(\prod_{i=0}^{k-1} C_{2i+\alpha} \right) \left\| \frac{f}{|x|^{2k+\alpha}} \right\|_{L^2(\mathbb{G})} \leq \left\| \frac{\tilde{\mathcal{R}}^k f}{|x|^\alpha} \right\|_{L^2(\mathbb{G})}. \tag{3.40}$$

(4) *If $Q \geq 4k+3$ and $\alpha \in (-(Q+2)/2, (Q-4k-2)/2)$, then we have*

$$\left(\frac{Q-2-2\alpha}{2} \prod_{i=0}^{k-1} C_{2i+1+\alpha} \right) \left\| \frac{f}{|x|^{2k+1+\alpha}} \right\|_{L^2(\mathbb{G})} \leq \left\| \frac{\mathcal{R}(\tilde{\mathcal{R}}^k f)}{|x|^\alpha} \right\|_{L^2(\mathbb{G})}. \tag{3.41}$$

Moreover, these inequalities in Parts (3) *and* (4) *are sharp and the equalities hold if only if $f = 0$.*

Proof of Theorem 3.1.12. The equality (3.38) follows by induction using Theorem 3.1.4, which gives the case of $k = 1$. Furthermore, we have

$$\int_{\mathbb{G}} \frac{|\mathcal{R}(\tilde{\mathcal{R}}^k f)|^2}{|x|^{2\alpha}} dx = \frac{(Q-2-2\alpha)^2}{4} \int_{\mathbb{G}} \frac{|\tilde{\mathcal{R}}^k f|^2}{|x|^{2+2\alpha}} dx$$
$$+ \int_{\mathbb{G}} \frac{1}{|x|^{2\alpha}} \left| \mathcal{R}(\tilde{\mathcal{R}}^k f) + \frac{Q-2-2\alpha}{2|x|} \tilde{\mathcal{R}}^k f \right|^2 dx. \tag{3.42}$$

Thus, (3.39) follows from (3.38) and (3.42). The inequalities (3.40) and (3.41) follow directly from these equalities since $C_{\alpha+2i} > 0, C_{\alpha+2i+1} > 0$ for any $i = 0, \ldots, k-1$ corresponding to each case. To check the sharpness of constants, we use the arguments similar to those in the proof of Theorem 3.1.4, considering approximations of the function $r^{-(Q-4k-2\alpha)/2}$. Indeed, one has

$$\int_{\mathbb{G}} \frac{|f_\epsilon|^2}{|x|^{4k+2\alpha}} dx = (-\ln \epsilon)|\wp| + O(1),$$

and

$$\int_{\mathbb{G}} \frac{|\tilde{\mathcal{R}}^k f|^2}{|x|^{2\alpha}} dx = \left(\prod_{i=0}^{k-1} C_{2i+\alpha} \right)^2 (-\ln \epsilon)|\wp| + O(1).$$

This proves the sharpness of (3.40). For (3.41), we use the approximation of the function $r^{-(Q-2-4k-2\alpha)/2}$ and similar calculations.

Finally, suppose that for some function f we have an equality in (3.40). Then by (3.38), we must have

$$\mathcal{R}f + \frac{Q-4k-2\alpha}{2|x|}f = 0.$$

This is equivalent to

$$\mathbb{E}f = -\frac{Q-4k-2\alpha}{2}f.$$

By Proposition 1.3.1, Part (i), it follows that f is positively homogeneous of order $-(Q-2\alpha-4k)/2$. Since $f/|x|^{\alpha+2k} \in L^2(\mathbb{G})$, this forces to have $f = 0$. The sharpness of (3.41) is proved in a similar way. □

Theorem 3.1.12 can be used to derive further weighted Rellich type inequalities. First we start with one iteration, continuing using the notation $\tilde{\mathcal{R}}$ as in (3.37).

Theorem 3.1.13 (Iterated weighted Rellich inequality). *Let $\mathbb{G}$ be a homogeneous group of homogeneous dimension $Q \geq 5$, and let $|\cdot|$ be a homogeneous quasi-norm on $\mathbb{G}$. Then for any $\alpha \in \mathbb{R}$ and for all complex-valued functions $f \in C_0^\infty(\mathbb{G}\backslash\{0\})$ we have the identity*

$$\left\|\frac{\tilde{\mathcal{R}}f}{|x|^\alpha}\right\|^2_{L^2(\mathbb{G})} = \frac{(Q+2\alpha)^2}{4}\left\|\frac{\mathcal{R}f}{|x|^{1+\alpha}}\right\|^2_{L^2(\mathbb{G})} + \left\|\frac{1}{|x|^\alpha}\left|\mathcal{R}^2 f + \frac{Q-2-2\alpha}{2|x|}\mathcal{R}f\right|\right\|^2_{L^2(\mathbb{G})}. \tag{3.43}$$

As a consequence, for any $\alpha \in \mathbb{R}$ we get the inequality

$$\frac{|Q+2\alpha|}{2}\left\|\frac{\mathcal{R}f}{|x|^{1+\alpha}}\right\|_{L^2(\mathbb{G})} \leq \left\|\frac{\tilde{\mathcal{R}}f}{|x|^\alpha}\right\|_{L^2(\mathbb{G})}, \tag{3.44}$$

for all complex-valued functions $f \in C_0^\infty(\mathbb{G}\backslash\{0\})$. For $Q+2\alpha \neq 0$ the inequality (3.44) is sharp and the equality holds if and only if $f = 0$.

Proof of Theorem 3.1.13. By definition (3.37) we have

$$\int_{\mathbb{G}}\frac{|\tilde{\mathcal{R}}f|^2}{|x|^{2\alpha}}dx = \int_{\mathbb{G}}\frac{|\mathcal{R}(\mathcal{R}f)|^2}{|x|^{2\alpha}}dx + 2(Q-1)\mathrm{Re}\int_{\mathbb{G}}\frac{\mathcal{R}(\mathcal{R}f)\overline{\mathcal{R}f}}{|x|^{2\alpha+1}}dx + (Q-1)^2\int_{\mathbb{G}}\frac{|\mathcal{R}f|^2}{|x|^{2+2\alpha}}dx.$$

Moreover, we have

$$2\mathrm{Re}\int_{\mathbb{G}}\frac{\mathcal{R}(\mathcal{R}f)\overline{\mathcal{R}f}}{|x|^{2\alpha+1}}dx = -(Q-2-2\alpha)\int_{\mathbb{G}}\frac{|\mathcal{R}f|^2}{|x|^{2+2\alpha}}dx.$$

On the other hand, by Theorem 3.1.1 we have the identity

$$\int_{\mathbb{G}} \frac{|\mathcal{R}(\mathcal{R}f)|^2}{|x|^{2\alpha}} dx = \frac{(Q-2-2\alpha)^2}{4} \int_{\mathbb{G}} \frac{|\mathcal{R}f|^2}{|x|^{2+2\alpha}} dx + \int_{\mathbb{G}} \frac{1}{|x|^{2\alpha}} \left| \mathcal{R}^2 f + \frac{Q-2-2\alpha}{2|x|} \mathcal{R}f \right|^2 dx.$$

Combining these inequalities we obtain (3.43).

Identity (3.43) then implies inequality (3.44). The sharpness of inequality (3.44) can be verified by using approximations of the function $r^{-(Q-2\alpha-4k)/2}$. If the equality in (3.44) holds for some function f, then we must have

$$\mathcal{R}^2 f + \frac{Q-2-2\alpha}{2|x|} \mathcal{R}f = 0.$$

This is equivalent to

$$\mathbb{E}(\mathcal{R}f) = -\frac{Q-2-2\alpha}{2} \mathcal{R}f$$

which implies, by Proposition 1.3.1, Part (i), that $\mathcal{R}f$ is positively homogeneous of degree $-(Q-2\alpha-2k)/2$. Since $\mathcal{R}f/|x|^{1+\alpha}$ is in $L^2(\mathbb{G})$, this implies that $\mathcal{R}f = 0$. Consequently, we must also have $f = 0$. □

Theorem 3.1.13 implies further identities and inequalities.

Theorem 3.1.14 (Further higher-order Rellich type identities and inequalities). *Let $\mathbb{G}$ be a homogeneous group of homogeneous dimension Q with a homogeneous quasi-norm $|\cdot|$. Let $k, l \in \mathbb{N}$ be such that $Q \geq 4k+1$ and $k \geq l+1$. Then for all complex-valued functions $f \in C_0^\infty(\mathbb{G}\backslash\{0\})$ we have*

$$\left\| \frac{\tilde{\mathcal{R}}^k f}{|x|^\alpha} \right\|^2_{L^2(\mathbb{G})} = \frac{4}{(Q-2\alpha)^2} \left(\prod_{i=0}^{k-l-1} \frac{Q^2 - 4(2i+\alpha)^2}{4} \right)^2 \left\| \frac{\mathcal{R}\tilde{\mathcal{R}}^l f}{|x|^{2(k-l)-1+\alpha}} \right\|^2_{L^2(\mathbb{G})}$$
$$+ \left(\prod_{i=0}^{k-l-2} C_{2i+\alpha} \right)^2 \left\| \frac{\left| \mathcal{R}^2 \tilde{\mathcal{R}}^l f + \frac{Q+2-4(k-l)-2\alpha}{2|x|} \mathcal{R}\tilde{\mathcal{R}}^l f \right|}{|x|^{2(k-l-1)+\alpha}} \right\|^2_{L^2(\mathbb{G})}$$
$$+ \left\| \frac{\left| \tilde{\mathcal{R}}^k f + C_\alpha \frac{\tilde{\mathcal{R}}^{k-1} f}{|x|^2} \right|}{|x|^\alpha} \right\|^2_{L^2(\mathbb{G})} + \sum_{j=1}^{k-l-2} \left(\prod_{i=0}^{j-1} C_{2i+\alpha} \right)^2 \left\| \frac{\left| \tilde{\mathcal{R}}^{k-j} f + C_{2j+\alpha} \frac{\tilde{\mathcal{R}}^{k-j-1} f}{|x|^2} \right|}{|x|^{2j+\alpha}} \right\|^2_{L^2(\mathbb{G})}$$
$$+ 2C_\alpha \left\| \frac{\left| \mathcal{R}(\tilde{\mathcal{R}}^{k-1} f) + \frac{Q-4-2\alpha}{2|x|} \tilde{\mathcal{R}}^{k-1} f \right|}{|x|^{1+\alpha}} \right\|^2_{L^2(\mathbb{G})}$$

$$+2\sum_{j=1}^{k-l-2}\left(\prod_{i=0}^{j-1}C_{2i+\alpha}\right)^2 C_{2j+\alpha}\left\|\frac{\left|\mathcal{R}(\tilde{\mathcal{R}}^{k-j-1}f)+\frac{Q-4-2\alpha-4j}{2|x|}\tilde{\mathcal{R}}^{k-j-1}f\right|}{|x|^{1+2j+\alpha}}\right\|^2_{L^2(\mathbb{G})} \tag{3.45}$$

and

$$\left\|\frac{\mathcal{R}\tilde{\mathcal{R}}^k f}{|x|^{\alpha}}\right\|^2_{L^2(\mathbb{G})}=\left(\prod_{i=0}^{k-l-1}\frac{Q^2-4(1+2i+\alpha)^2}{4}\right)^2\left\|\frac{\mathcal{R}\tilde{\mathcal{R}}^l f}{|x|^{2(k-l)+\alpha}}\right\|^2_{L^2(\mathbb{G})}$$
$$+\frac{(Q-2-2\alpha)^2}{4}\left(\prod_{i=0}^{k-l-2}C_{2i+1+\alpha}\right)^2\left\|\frac{\left|\mathcal{R}^2\tilde{\mathcal{R}}^l f+\frac{Q+2-4(k-l)-2\alpha}{2|x|}\mathcal{R}\tilde{\mathcal{R}}^l f\right|}{|x|^{2(k-l-1)+1+\alpha}}\right\|^2_{L^2(\mathbb{G})}$$
$$+\frac{(Q-2-2\alpha)^2}{4}\left\|\frac{\left|\tilde{\mathcal{R}}^k f+C_\alpha\frac{\tilde{\mathcal{R}}^{k-1}f}{|x|^2}\right|}{|x|^{\alpha}}\right\|^2_{L^2(\mathbb{G})}$$
$$+\frac{(Q-2-2\alpha)^2}{4}\sum_{j=1}^{k-l-2}\left(\prod_{i=0}^{j-1}C_{2i+1+\alpha}\right)^2\left\|\frac{\left|\tilde{\mathcal{R}}^{k-j}f+C_{2j+1+\alpha}\frac{\tilde{\mathcal{R}}^{k-j-1}f}{|x|^2}\right|}{|x|^{2j+\alpha+1}}\right\|^2_{L^2(\mathbb{G})}$$
$$+2C_{1+\alpha}\frac{(Q-2-2\alpha)^2}{4}\left\|\frac{\left|\mathcal{R}(\tilde{\mathcal{R}}^{k-1}f)+\frac{Q-6-2\alpha}{2|x|}\tilde{\mathcal{R}}^{k-1}f\right|}{|x|^{2+\alpha}}\right\|^2_{L^2(\mathbb{G})}$$
$$+2\frac{(Q-2-2\alpha)^2}{4}\sum_{j=1}^{k-l-2}\left(\prod_{i=0}^{j-1}C_{2i+1+\alpha}\right)^2 C_{2j+1+\alpha}$$
$$\times\left\|\frac{\left|\mathcal{R}(\tilde{\mathcal{R}}^{k-j-1}f)+\frac{Q-6-2\alpha-4j}{2|x|}\tilde{\mathcal{R}}^{k-j-1}f\right|}{|x|^{1+2j+\alpha}}\right\|^2_{L^2(\mathbb{G})}$$
$$+\left\|\frac{\left|\mathcal{R}\tilde{\mathcal{R}}^k f+\frac{Q-2-2\alpha}{2}\frac{\tilde{\mathcal{R}}^k f}{|x|}\right|}{|x|^{\alpha}}\right\|^2_{L^2(\mathbb{G})}. \tag{3.46}$$

As consequences, we have the following weighted Rellich type inequalities for all complex-valued functions $f\in C_0^\infty(\mathbb{G}\backslash\{0\})$*:*

(1) *For any* $\alpha\in(-Q/2,(Q-4(k-l-1))/2)$ *if* $k\geq l+2$ *and* $\alpha\in\mathbb{R}$ *if* $k=l+1$, *we have*

$$\frac{4}{(Q-2\alpha)^2}\left(\prod_{i=0}^{k-l-1}\frac{Q^2-4(2i+\alpha)^2}{4}\right)^2\left\|\frac{\mathcal{R}\tilde{\mathcal{R}}^l f}{|x|^{2(k-l)-1+\alpha}}\right\|^2_{L^2(\mathbb{G})}\leq\left\|\frac{\tilde{\mathcal{R}}^k f}{|x|^{\alpha}}\right\|^2_{L^2(\mathbb{G})}. \tag{3.47}$$

(2) *For any* $\alpha \in (-(Q+2)/2, (Q-4(k-l)+2)/2)$ *if* $k \geq l+2$ *and* $\alpha \in \mathbb{R}$ *if* $k = l+1$, *we have*

$$\left(\prod_{i=0}^{k-l-1} \frac{Q^2-4(1+2i+\alpha)^2}{4}\right)^2 \left\| \frac{\mathcal{R}\tilde{\mathcal{R}}^l f}{|x|^{2(k-l)+\alpha}} \right\|^2_{L^2(\mathbb{G})} \leq \left\| \frac{\mathcal{R}\tilde{\mathcal{R}}^k f}{|x|^{\alpha}} \right\|^2_{L^2(\mathbb{G})}. \tag{3.48}$$

Moreover, inequalities (3.47) *and* (3.48) *are sharp and there is an equality in each of them if and only if* $f = 0$.

Proof of Theorem 3.1.14. Let $g := \tilde{\mathcal{R}}^{k+1} f$. Applying (3.38) to $\tilde{\mathcal{R}}^{k-l-1} g$ and then Theorem 3.1.13 to $\mathcal{R}\tilde{\mathcal{R}}^l f$, one obtains equality (3.45). Note that

$$\frac{Q+4(k-l-1)+2\alpha}{2} \prod_{i=0}^{k-l-2} C_{2i+\alpha} = \frac{2}{Q-2\alpha} \prod_{i=0}^{k-l-1} \frac{Q^2-4(2i+\alpha)^2}{4}.$$

By applying Theorem 3.1.1 to $\mathcal{R}\tilde{\mathcal{R}}^k f$, and then using (3.45) for $\tilde{\mathcal{R}}^k f$ with weights $|x|^{2(1+\alpha)}$, together with the equality

$$\frac{Q-2-2\alpha}{2} \frac{(Q+4(k-l-1)+2+2\alpha)^2}{4} \prod_{i=0}^{k-l-2} C_{2i+\alpha} = \prod_{i=0}^{k-l-1} \frac{Q^2-4(1+\alpha+2i)^2}{4},$$

we also obtain (3.46).

Consequently, inequalities (3.47) and (3.48) follow from respective identities (3.45) and (3.46) by dropping non-negative remainder terms on the right-hand side. The sharpness of (3.47) and (3.48) follows by considering, respectively, as in the previous theorems, approximations of the function $r^{-(Q-2\alpha-4k)/2}$ and of the function $r^{-(Q-2\alpha-2-4k)/2}$.

If for some function f we have the equality in (3.47), then (3.45) implies that we have

$$\mathcal{R}(\tilde{\mathcal{R}}^{k-1} f) + \frac{Q-4-2\alpha}{2|x|} \tilde{\mathcal{R}}^{k-1} f = 0.$$

This means that

$$\mathbb{E}(\tilde{\mathcal{R}}^{k-1} f) = -\frac{Q-4-2\alpha}{2} \tilde{\mathcal{R}}^{k-1} f.$$

By Proposition 1.3.1, Part (i), it follows that $\tilde{\mathcal{R}}^{k-1} f$ is positively homogeneous of order $-(Q-4-2\alpha)/2$. Since $\tilde{\mathcal{R}}^{k-1} f/|x|^{2\alpha} \in L^2(\mathbb{G})$, we then must have $\tilde{\mathcal{R}}^{k-1} f = 0$, which in turn implies $f = 0$. The same arguments also work for (3.48). □

Using the established theorems, as a corollary one gets the following weighted L^p-Hardy–Rellich type identities which will, in turn, imply the corresponding inequalities.

Theorem 3.1.15 (Higher-order weighted L^p-Hardy–Rellich type identities). *Let $\mathbb{G}$ be a homogeneous group of homogeneous dimension Q, with a homogeneous quasi-norm $|\cdot|$ on $\mathbb{G}$. Let $k \in \mathbb{N}$ be a positive integer, let $1 < p < Q/k$, and let $\alpha \in \mathbb{R}$. Then for all complex-valued functions $f \in C_0^\infty(\mathbb{G}\backslash\{0\})$ we have the identities:*

(1) *If $k = 2l$, $l \geq 2$, then*

$$
\begin{aligned}
&\left\| \frac{\tilde{\mathcal{R}}^l f}{|x|^\alpha} \right\|^p_{L^p(\mathbb{G})} \\
&= \left| \prod_{i=0}^{l-1} C_{p,2i+\alpha} \right|^p \int_{\mathbb{G}} \frac{|f|^p}{|x|^{p(k+\alpha)}} dx + p \int_{\mathbb{G}} \frac{1}{|x|^{p\alpha}} R_p \left(C_{p,\alpha} \frac{\tilde{\mathcal{R}}^{l-1} f}{|x|^2}, -\tilde{\mathcal{R}}^l f \right) dx \\
&+ p \sum_{j=1}^{l-1} \left| \prod_{i=0}^{j-1} C_{p,2i+\alpha} \right|^p \int_{\mathbb{G}} \frac{1}{|x|^{p(2j+\alpha)}} R_p \left(C_{p,2j+\alpha} \frac{\tilde{\mathcal{R}}^{l-j-1} f}{|x|^2}, -\tilde{\mathcal{R}}^{l-j} f \right) dx \\
&+ B_{p,\alpha}(p-1) \int_{\mathbb{G}} \frac{|\tilde{\mathcal{R}}^{l-1} f|^{p-2}}{|x|^{p(2j+\alpha)-2}} \left| \mathcal{R} |\tilde{\mathcal{R}}^{l-1} f| + \frac{Q - p(2+\alpha)}{p|x|} |\tilde{\mathcal{R}}^{l-1} f| \right|^2 dx \\
&+ B_{p,\alpha} \int_{\mathbb{G}} \frac{|\tilde{\mathcal{R}}^{l-1} f|^{p-4} (\operatorname{Im}(\tilde{\mathcal{R}}^{l-1} f \overline{\mathcal{R}\tilde{\mathcal{R}}^{l-1} f}))^2}{|x|^{p(2+\alpha)-2}} dx \qquad (3.49) \\
&+ \sum_{j=1}^{l-1} \left| \prod_{i=0}^{j-1} C_{p,2i+\alpha} \right|^p B_{p,\alpha+2j} \int_{\mathbb{G}} \frac{|\tilde{\mathcal{R}}^{l-j-1} f|^{p-4} (\operatorname{Im}(\tilde{\mathcal{R}}^{l-j-1} f \overline{\mathcal{R}\tilde{\mathcal{R}}^{l-j-1} f}))^2}{|x|^{p(2(j+1)+\alpha)-2}} dx \\
&+ (p-1) \int_{\mathbb{G}} \frac{|\tilde{\mathcal{R}}^{l-j-1} f|^{p-2}}{|x|^{p(2(j+1)+\alpha)-2}} \left| \mathcal{R} |\tilde{\mathcal{R}}^{l-j-1} f| + \frac{Q - p(2(j+1)+\alpha)}{p|x|} |\tilde{\mathcal{R}}^{l-j-1} f| \right|^2 dx,
\end{aligned}
$$

where R_p is as in (2.22), i.e.,

$$
R_p(\xi, \eta) := \frac{1}{p} |\eta|^p + \frac{p-1}{p} |\xi|^p - \operatorname{Re}(|\xi|^{p-2} \xi \overline{\eta}),
$$

as well as

$$
B_{p,\alpha+2j} = p |C_{p,\alpha+2j}|^{p-2} C_{p,\alpha+2j},
$$

and

$$
C_{p,\alpha} = \frac{(Q - 2p - p\alpha)(Q + p'\alpha)}{pp'}, \quad p' = \frac{p}{p-1}.
$$

(2) *If $k = 2l+1$, $l \geq 1$, then*

$$
\begin{aligned}
\left\| \frac{\mathcal{R}(\tilde{\mathcal{R}}^l f)}{|x|^\alpha} \right\|^p_{L^p(\mathbb{G})} &= \left| \frac{Q - p(1+\alpha)}{p} \right|^p \left| \prod_{i=0}^{l-1} C_{p,2i+1+\alpha} \right|^p \int_{\mathbb{G}} \frac{|f|^p}{|x|^{p(k+\alpha)}} dx \\
&+ p \int_{\mathbb{G}} \frac{1}{|x|^{p\alpha}} R_p \left(-\frac{Q - p(1+\alpha)}{p} \frac{\tilde{\mathcal{R}}^l f}{|x|}, \mathcal{R}\tilde{\mathcal{R}}^l f \right) dx \\
&+ A_{p,\alpha} \int_{\mathbb{G}} \frac{1}{|x|^{p(1+\alpha)}} R_p \left(C_{p,1+\alpha} \frac{\tilde{\mathcal{R}}^{l-1} f}{|x|^2}, -\tilde{\mathcal{R}}^l f \right) dx
\end{aligned}
$$

$$
\begin{aligned}
&+ A_{p,\alpha}\sum_{j=1}^{l-1}\left|\prod_{i=0}^{j-1} C_{p,2i+1+\alpha}\right|^p \int_{\mathbb{G}} \frac{1}{|x|^{p(2j+1+\alpha)}} R_p\left(C_{p,2j+1+\alpha}\frac{\tilde{\mathcal{R}}^{l-j-1}f}{|x|^2}, -\tilde{\mathcal{R}}^{l-j}f\right)dx \\
&+ A_{p,\alpha}\frac{B_{p,\alpha+1}}{p}\left[(p-1)\int_{\mathbb{G}} \frac{|\tilde{\mathcal{R}}^{l-1}f|^{p-2}}{|x|^{p(3+\alpha)-2}}\left|\mathcal{R}|\tilde{\mathcal{R}}^{l-1}f| + \frac{Q-p(3+\alpha)}{p|x|}|\tilde{\mathcal{R}}^{l-1}f|\right|^2 dx\right] \\
&+ A_{p,\alpha}\frac{B_{p,\alpha+1}}{p}\left[\int_{\mathbb{G}} \frac{|\tilde{\mathcal{R}}^{l-1}f|^{p-4}(\mathrm{Im}(\tilde{\mathcal{R}}^{l-1}f\overline{\mathcal{R}\tilde{\mathcal{R}}^{l-1}f}))^2}{|x|^{p(3+\alpha)-2}}dx\right] \qquad (3.50) \\
&+ A_{p,\alpha}\sum_{j=1}^{l-1}\left|\prod_{i=0}^{j-1} C_{p,2i+1+\alpha}\right|^p \frac{B_{p,\alpha+2j+1}}{p}\left[\int_{\mathbb{G}} \frac{|\tilde{\mathcal{R}}^{l-j-1}f|^{p-4}(\mathrm{Im}(\tilde{\mathcal{R}}^{l-j-1}f\overline{\mathcal{R}\tilde{\mathcal{R}}^{l-j-1}f}))^2}{|x|^{p(2j+3+\alpha)-2}}dx\right. \\
&\left.+ (p-1)\int_{\mathbb{G}} \frac{|\tilde{\mathcal{R}}^{l-j-1}f|^{p-2}}{|x|^{p(2j+3+\alpha)-2}}\left|\mathcal{R}|\tilde{\mathcal{R}}^{l-j-1}f| + \frac{Q-p(2j+3+\alpha)}{p|x|}|\tilde{\mathcal{R}}^{l-j-1}f|\right|^2 dx\right],
\end{aligned}
$$

where $A_{p,\alpha} = p\left|\frac{Q-p(1+\alpha)}{p}\right|^p$.

Proof of Theorem 3.1.15. The equality (3.49) follows from (3.25). The equality (3.50) is consequence of (3.49) and (3.32). Note that in (3.50), if $k = 2l+1, l \geq 1$, then the terms concerning the sum from 1 to $l-1$ do not appear if $l = 1$. □

By dropping the non-negative remainder terms in (3.49) and (3.50), we obtain the following higher-order weighted L^p-Hardy–Rellich type inequalities.

Corollary 3.1.16 (Higher-order weighted L^p-Hardy–Rellich type inequalities). *Let $\mathbb{G}$ be a homogeneous group of homogeneous dimension Q and let $|\cdot|$ be a homogeneous quasi-norm on $\mathbb{G}$. Then for any $\alpha \in \mathbb{R}$ and all complex-valued functions $f \in C_0^\infty(\mathbb{G}\backslash\{0\})$ we have:*

(1) *if $1 < p < Q/2k$ and $\alpha \in (-Q(p-1))/p, (Q-2pk)/p)$, then*

$$
\left(\prod_{i=0}^{k-1} C_{p,2i+\alpha}\right)\left\|\frac{f}{|x|^{(2k+\alpha)}}\right\|_{L^p(\mathbb{G})} \leq \left\|\frac{\tilde{\mathcal{R}}^k f}{|x|^\alpha}\right\|_{L^p(\mathbb{G})}. \qquad (3.51)
$$

(2) *if $1 < p < Q/(2k+1)$ and $\alpha \in (-(Q+p')/p', (Q-p(2k+1))/p)$, then*

$$
\frac{Q-p(1+\alpha)}{p}\left(\prod_{i=0}^{k-1} C_{p,2i+1+\alpha}\right)\left\|\frac{f}{|x|^{(2k+1+\alpha)}}\right\|_{L^p(\mathbb{G})} \leq \left\|\frac{\mathcal{R}\tilde{\mathcal{R}}^k f}{|x|^\alpha}\right\|_{L^p(\mathbb{G})}. \qquad (3.52)
$$

In the above inequalities

$$
C_{p,\alpha} = \frac{(Q-2p-p\alpha)(Q+p'\alpha)}{pp'}, \quad p' = \frac{p}{p-1}.
$$

Inequalities (3.51) *and* (3.52) *are sharp and equalities hold in each of them if and only if $f = 0$.*

Proof of Corollary 3.1.16. Inequalities (3.51) and (3.52) are immediate consequences of equalities (3.49) and (3.50), respectively, since $C_{p,2i+\alpha+1}\geq 0$ for $0\leq i\leq k-1$ under the respective assumptions on indices. The sharpness of constants in (3.51) and (3.52) can be checked by approximating the function $r^{-(Q-p(2k+\alpha))/p}$ and the function $r^{-(Q-p(2k+1+\alpha))/p}$, respectively, by smooth compactly supported functions. Moreover, if for some function f an equality holds in any of these inequalities, then it follows from Theorem 3.1.15 that $|f|$ is positively homogeneous of degree $-(Q-p(2k+\alpha))/p$ and $-(Q-p(2k+1+\alpha))/p$, respectively, which implies that $f=0$ in view of the condition of L^p integrability. □

As usual, Hardy inequalities imply the corresponding uncertainty principles.

Corollary 3.1.17 (Higher-order Hardy–Rellich uncertainty principles). *Let $\mathbb{G}$ be a homogeneous group of homogeneous dimension Q and let $|\cdot|$ be a homogeneous quasi-norm on $\mathbb{G}$. Let $k\in\mathbb{N}$ be a positive integer and let $p>1$. Then for all complex-valued functions $f\in C_0^\infty(\mathbb{G}\backslash\{0\})$ we have:*

(1) *if $k=2l$, $l\geq 1$, and $\alpha\in(-Q(p-1)/p),(Q-2pl)/p)$, then*

$$\left(\prod_{i=0}^{l-1}C_{p,2i+\alpha}\right)\int_{\mathbb{G}}|f|^2dx\leq\left(\int_{\mathbb{G}}\frac{|\tilde{\mathcal{R}}^l f|^p}{|x|^{p\alpha}}dx\right)^{1/p}\left(\int_{\mathbb{G}}|f|^{p'}|x|^{p'(2l+\alpha)}dx\right)^{1/p'}. \tag{3.53}$$

(2) *if $k=2l+1$, $l\geq 0$, and $\alpha\in(-(Q+p')/p',(Q-p(2l+1))/p)$ if $l\geq 1$ and $\alpha\in\mathbb{R}$ if $l=0$, then*

$$\begin{aligned}&\left|\frac{\frac{Q-p(1+\alpha)}{p}}{C_{p,2l+1+\alpha}}\prod_{i=0}^{l}C_{p,2i+1+\alpha}\right|\int_{\mathbb{G}}|f|^2dx\\&\leq\left(\int_{\mathbb{G}}\frac{|\mathcal{R}\tilde{\mathcal{R}}^l f|^p}{|x|^{p\alpha}}dx\right)^{1/p}\left(\int_{\mathbb{G}}|f|^{p'}|x|^{p'(2l+1+\alpha)}dx\right)^{1/p'}.\end{aligned} \tag{3.54}$$

Proof of Corollary 3.1.17. By the Hölder inequality, we have

$$\int_{\mathbb{G}}|f|^2dx=\int_{\mathbb{G}}\frac{|f|}{|x|^{2l+\alpha}}|f||x|^{2l+\alpha}dx\leq\left(\int_{\mathbb{G}}\frac{|f|^p}{|x|^{p(2l+\alpha)}}dx\right)^{\frac{1}{p}}\left(\int_{\mathbb{G}}|f|^{p'}|x|^{p'(2l+\alpha)}dx\right)^{\frac{1}{p'}}.$$

Further, applying inequality (3.51) to this, we get (3.53) with $C_{2i+\alpha}>0$ for $0\leq i\leq l-1$ and $\alpha\in(-Q(p-1)/p,(Q-2pl)/p)$. Inequality (3.54) is proved in the same way. □

Combining Corollary 3.28 and Theorem 2.24, we obtain the weighted L^p-Rellich type inequality below which is an L^p-analogue of Theorem 3.1.14.

Theorem 3.1.18 (Higher-order L^p-weighted Rellich identities and inequalities). *Let $\mathbb{G}$ be a homogeneous group of homogeneous dimension Q and let $|\cdot|$ be any* *homogeneous quasi-norm on* $\mathbb{G}$. *Let* $k, l \in \mathbb{N}$ *be non-negative integers such that* $k \geq l+1$ *and let* $p > 1$. *Then for any* $\alpha \in \mathbb{R}$ *and for all complex-valued functions* $f \in C_0^\infty(\mathbb{G}\backslash\{0\})$ *we have*

$$
\begin{aligned}
&\left\|\frac{\tilde{\mathcal{R}}^k f}{|x|^\alpha}\right\|^p_{L^p(\mathbb{G})} \\
&= \frac{p^p}{|Q-p\alpha|^p}\left|\prod_{i=0}^{k-l-1}\frac{(Q-p(2i+\alpha))(Q+p'(2i+\alpha))}{pp'}\right|^p \int_{\mathbb{G}} \frac{|\mathcal{R}\tilde{\mathcal{R}}^l f|^p}{|x|^{p(2(k-l)-1+\alpha)}}dx \\
&\quad + p\left|\prod_{i=0}^{k-l-2} C_{p,2i+\alpha}\right|^p \int_{\mathbb{G}} \frac{R_p\left(\frac{Q+p'(2(k-l-1)+\alpha)}{p'}\frac{\mathcal{R}\tilde{\mathcal{R}}^l f}{|x|}, \tilde{\mathcal{R}}^{l+1} f\right)}{|x|^{p(2(k-l-1)+\alpha)}}dx \\
&\quad + p\int_{\mathbb{G}} \frac{R_p\left(-C_{p,\alpha}\frac{\tilde{\mathcal{R}}^{k-1} f}{|x|^2}, \tilde{\mathcal{R}}^k f\right)}{|x|^{p\alpha}}dx \\
&\quad + p\sum_{j=1}^{k-l-2}\left|\prod_{i=0}^{j-1} C_{p,2i+\alpha}\right|^p \int_{\mathbb{G}} \frac{R_p\left(-C_{p,2j+\alpha}\frac{\tilde{\mathcal{R}}^{k-j-1} f}{|x|^2}, \tilde{\mathcal{R}}^{k-j} f\right)}{|x|^{p(2j+\alpha)}}dx \\
&\quad + B_{p,\alpha}(p-1)\int_{\mathbb{G}} \frac{|\tilde{\mathcal{R}}^{k-1} f|^{p-2}}{|x|^{p(2+\alpha)-2}}\left|\mathcal{R}|\tilde{\mathcal{R}}^{k-1} f| + \frac{Q-p(2+\alpha)}{p|x|}|\tilde{\mathcal{R}}^{k-1} f|\right|^2 dx \\
&\quad + B_{p,\alpha}\int_{\mathbb{G}} \frac{|\tilde{\mathcal{R}}^{k-1} f|^{p-4}(\mathrm{Im}(\tilde{\mathcal{R}}^{l-1} f\overline{\mathcal{R}\tilde{\mathcal{R}}^{k-1} f}))^2}{|x|^{p(2+\alpha)-2}}dx \\
&\quad + \sum_{j=1}^{k-l-2}\left|\prod_{i=0}^{j-1} C_{p,2i+\alpha}\right|^p B_{p,\alpha+2j}\int_{\mathbb{G}} \frac{|\tilde{\mathcal{R}}^{k-j-1} f|^{p-4}(\mathrm{Im}(\tilde{\mathcal{R}}^{k-j-1} f\overline{\mathcal{R}\tilde{\mathcal{R}}^{k-j-1} f}))^2}{|x|^{p(2(j+1)+\alpha)-2}}dx \\
&\quad + (p-1)\int_{\mathbb{G}} \frac{|\tilde{\mathcal{R}}^{k-j-1} f|^{p-2}}{|x|^{p(2(j+1)+\alpha)-2}}\left|\mathcal{R}|\tilde{\mathcal{R}}^{k-j-1} f|\right. \\
&\quad \left.+\frac{Q-p(2(j+1)+\alpha)}{p|x|}|\tilde{\mathcal{R}}^{k-j-1} f|\right|^2 dx, \qquad (3.55)
\end{aligned}
$$

where R_p *is as in* (2.22):

$$
R_p(\xi,\eta) := \frac{1}{p}|\eta|^p + \frac{p-1}{p}|\xi|^p - \mathrm{Re}(|\xi|^{p-2}\xi\overline{\eta}),
$$

$$
B_{p,\alpha+2j} = p|C_{p,\alpha+2j}|^{p-2}C_{p,\alpha+2j},
$$

and

$$
C_{p,\alpha} = \frac{(Q-2p-p\alpha)(Q+p'\alpha)}{pp'}, \quad p' = \frac{p}{p-1}.
$$

We also have

$$\begin{aligned}
&\left\|\frac{\mathcal{R}\tilde{\mathcal{R}}^k f}{|x|^\alpha}\right\|^p_{L^p(\mathbb{G})}\\
&= \left|\prod_{i=0}^{k-l-1}\frac{(Q-p(2i+1+\alpha))(Q+p'(2i+1+\alpha))}{pp'}\right|^p \int_{\mathbb{G}}\frac{|\mathcal{R}\tilde{\mathcal{R}}^l f|^p}{|x|^{p(2(k-l)-1+\alpha)}}dx\\
&+ A_{p,\alpha}\left|\prod_{i=0}^{k-l-2} C_{p,2i+1+\alpha}\right|^p \int_{\mathbb{G}}\frac{R_p\left(\frac{Q+p'(2(k-l)-1+\alpha)}{p'}\frac{\mathcal{R}\tilde{\mathcal{R}}^l f}{|x|},\tilde{\mathcal{R}}^{l+1}f\right)}{|x|^{p(2(k-l)-1+\alpha)}}dx\\
&+ A_{p,\alpha}\int_{\mathbb{G}}\frac{R_p\left(-C_{p,1+\alpha}\frac{\tilde{\mathcal{R}}^{k-j-1}f}{|x|^2},\tilde{\mathcal{R}}^{k-j}f\right)}{|x|^{p(1+\alpha)}}dx\\
&+ A_{p,\alpha}\sum_{j=1}^{k-l-2}\left|\prod_{i=0}^{j-1} C_{p,2i+1+\alpha}\right|^p \int_{\mathbb{G}}\frac{R_p\left(-C_{p,2j+1+\alpha}\frac{\tilde{\mathcal{R}}^{k-j-1}f}{|x|^2},\tilde{\mathcal{R}}^{k-j}f\right)}{|x|^{p(2j+1+\alpha)}}dx\\
&+ A_{p,\alpha}|\frac{B_{p,\alpha+1}}{p}(p-1)\int_{\mathbb{G}}\frac{|\tilde{\mathcal{R}}^{k-1}f|^{p-2}}{|x|^{p(3+\alpha)-2}}\left|\mathcal{R}|\tilde{\mathcal{R}}^{k-1}f|+\frac{Q-p(3+\alpha)}{p|x|}|\tilde{\mathcal{R}}^{k-1}f|\right|^2dx\\
&+ A_{p,\alpha}\frac{B_{p,\alpha+1}}{p}\int_{\mathbb{G}}\frac{|\tilde{\mathcal{R}}^{k-1}f|^{p-4}(\mathrm{Im}(\tilde{\mathcal{R}}^{k-1}f\overline{\mathcal{R}\tilde{\mathcal{R}}^{k-1}f}))^2}{|x|^{p(3+\alpha)-2}}dx\\
&+ A_{p,\alpha}\sum_{j=1}^{k-l-2}\left|\prod_{i=0}^{j-1} C_{p,2i+1+\alpha}\right|^p \frac{B_{p,2j+\alpha+1}}{p}\\
&\times\int_{\mathbb{G}}\frac{|\tilde{\mathcal{R}}^{k-j-1}f|^{p-4}(\mathrm{Im}(\tilde{\mathcal{R}}^{k-j-1}f\overline{\mathcal{R}\tilde{\mathcal{R}}^{k-j-1}f}))^2}{|x|^{p(2(j+1)+1+\alpha)-2}}dx\\
&+(p-1)\int_{\mathbb{G}}\frac{|\tilde{\mathcal{R}}^{k-j-1}f|^{p-2}}{|x|^{p(2(j+1)+1+\alpha)-2}}\\
&\times\left|\mathcal{R}|\tilde{\mathcal{R}}^{k-j-1}f|+\frac{Q-p(2(j+1)+\alpha)}{p|x|}|\tilde{\mathcal{R}}^{k-j-1}f|\right|^2dx\\
&+p\int_{\mathbb{G}}\frac{1}{|x|^{p\alpha}}R_p\left(-\frac{Q-p-p\alpha}{p}\frac{\tilde{\mathcal{R}}^k f}{|x|},\mathcal{R}\tilde{\mathcal{R}}^k f\right)dx,
\end{aligned}\tag{3.56}$$

where

$$A_{p,\alpha}=p\left|\frac{Q-p(1+\alpha)}{p}\right|^p.$$

Consequently, we obtain the following inequalities for all functions $f\in C_0^\infty(\mathbb{G}\backslash\{0\})$: for any $\alpha\in(-Q(p-1)/p,(Q-2p(k-l-1))/2)$ if $k-l\geq 2$ and for any $\alpha\in\mathbb{R}$

if $k = l+1$, we have

$$\frac{p}{|Q-p\alpha|}\left|\prod_{i=0}^{k-l-1}\frac{(Q-p(2i+\alpha))(Q+p'(2i+\alpha))}{pp'}\right|\left\|\frac{\mathcal{R}\tilde{\mathcal{R}}^l f}{|x|^{(2(k-l)-1+\alpha)}}\right\|_{L^p(\mathbb{G})} \leq \left\|\frac{\tilde{\mathcal{R}}^k f}{|x|^{\alpha}}\right\|_{L^p(\mathbb{G})}; \tag{3.57}$$

and for any $\alpha \in (-(Q+p')/p', (Q-p(2(k-l-1)+1))/p)$ if $k-l \geq 2$ and for any $\alpha \in \mathbb{R}$ if $k = l+1$, we have

$$\left|\prod_{i=0}^{k-l-1}\frac{(Q-p(2i+1+\alpha))(Q+p'(2i+1+\alpha))}{pp'}\right|\left\|\frac{\mathcal{R}\tilde{\mathcal{R}}^l f}{|x|^{(2(k-l)-1+\alpha)}}\right\|_{L^p(\mathbb{G})} \leq \left\|\frac{\mathcal{R}\tilde{\mathcal{R}}^k f}{|x|^{\alpha}}\right\|_{L^p(\mathbb{G})}. \tag{3.58}$$

Moreover, these inequalities (3.57) and (3.58) are sharp and equality in any of them holds if and only if $f = 0$.

Now let us present an extension of the critical Hardy inequality to the higher-order derivatives. To do this, still following [Ngu17] until the end of this section, we present a critical Rellich inequality for $\tilde{\mathcal{R}}$ as follows.

Theorem 3.1.19 (Critical Rellich identity and inequality for $\tilde{\mathcal{R}}$). *Let $\mathbb{G}$ be a homogeneous group of homogeneous dimension $Q \geq 3$ and let $|\cdot|$ be a homogeneous quasi-norm on $\mathbb{G}$. Let $f \in C_0^\infty(\mathbb{G}\backslash\{0\})$ be any complex-valued function and denote*

$$f_R(x) := f(Rx/|x|)$$

for any $x \in \mathbb{G}$ and $R > 0$. Then we have

$$\begin{aligned}\left\|\frac{\tilde{\mathcal{R}}f}{|x|^{\frac{Q}{p}-2}}\right\|^p_{L^p(\mathbb{G})} &= \left(\frac{(p-1)(Q-2)}{p}\right)^p \int_{\mathbb{G}} \frac{|f-f_R|^p}{|x|^Q\left|\ln\frac{R}{|x|}\right|^p}dx \\ &+ p\int_{\mathbb{G}}\frac{1}{|x|^{Q-2p}}R_p\left((Q-2)\frac{\mathcal{R}f}{|x|}, \tilde{\mathcal{R}}f\right)dx \\ &+ p(Q-2)^p\int_{\mathbb{G}}\frac{1}{|x|^{Q-p}}R_p\left(-\frac{p-1}{p}\frac{f-f_R}{|x|\ln\frac{R}{|x|}}, \mathcal{R}f\right)dx,\end{aligned} \tag{3.59}$$

for any $1 < p < \infty$ and any $R > 0$. Here R_p is as in (2.22). As a consequence, for all complex-valued functions $f \in C_0^\infty(\mathbb{G}\backslash\{0\})$ we have

$$\frac{(p-1)(Q-2)}{p}\sup_{R>0}\left\|\frac{f-f_R}{|x|^{\frac{Q}{p}}\left|\ln\frac{R}{|x|}\right|}\right\|_{L^p(\mathbb{G})} \leq \left\|\frac{\tilde{\mathcal{R}}f}{|x|^{\frac{Q}{p}-2}}\right\|_{L^p(\mathbb{G})}, \quad 1<p<\infty, \tag{3.60}$$

with the constant sharp $(p-1)(Q-2)/p$.

Proof of Theorem 3.1.19. Let $1 < p < \infty$. Let us first restate equality (2.48) in the form

$$\begin{aligned}\int_{\mathbb{G}} \frac{|\mathcal{R}f|^p}{|x|^{Q-p}}dx = \left(\frac{p-1}{p}\right)^p \int_{\mathbb{G}} \frac{|f-f_R|^p}{|x|^Q|\ln\frac{R}{|x|}|^p}dx \\ + p\int_{\mathbb{G}} \frac{1}{|x|^{Q-p}}R_p\left(-\frac{p-1}{p}\frac{f-f_R}{|x|\ln\frac{R}{|x|}}, \mathcal{R}f\right)dx,\end{aligned} \tag{3.61}$$

for any $R > 0$. It follows from Theorem 3.1.7 for $\alpha = (Q-2p)/p$ that

$$\begin{aligned}\int_{\mathbb{G}} \frac{|\tilde{\mathcal{R}}f|^p}{|x|^{Q-2p}}dx = (Q-2)^p \int_{\mathbb{G}} \frac{|\mathcal{R}f|^p}{|x|^{Q-p}}dx \\ + p\int_{\mathbb{G}} \frac{1}{|x|^{Q-2p}}R_p\left((Q-2)\frac{\mathcal{R}f}{|x|}, \tilde{\mathcal{R}}f\right)dx.\end{aligned} \tag{3.62}$$

The combination of (3.61) and (3.62) gives (3.59), which in turn implies (3.60).

Let us now check the sharpness of (3.60). For small enough $\epsilon, \delta > 0$ and for $R > 2$, let us define the function

$$f_\delta(x) := \left(\ln\frac{R}{|x|}\right)^{1-\frac{1}{p}-\delta} g(|x|),$$

where g is the function as in the proof of Theorem 3.1.4. A straightforward calculation gives

$$\mathcal{R}f_\delta(r) = -\left(1-\frac{1}{p}-\delta\right)\frac{1}{r}\left(\ln\frac{R}{r}\right)^{-\frac{1}{p}-\delta} g(r) + \left(\ln\frac{R}{r}\right)^{1-\frac{1}{p}-\delta} g'(r)$$

and

$$\begin{aligned}\mathcal{R}^2 f_\delta(r) = & -2\left(1-\frac{1}{p}-\delta\right)\frac{1}{r}\left(\ln\frac{R}{r}\right)^{-\frac{1}{p}-\delta} g'(r) \\ & -\left(1-\frac{1}{p}-\delta\right)\left(\frac{1}{p}+\delta\right)\frac{1}{r^2}\left(\ln\frac{R}{r}\right)^{-1-\frac{1}{p}-\delta} g(r) \\ & +\left(1-\frac{1}{p}-\delta\right)\frac{1}{r^2}\left(\ln\frac{R}{r}\right)^{-\frac{1}{p}-\delta} g(r) + \left(\ln\frac{R}{r}\right)^{1-\frac{1}{p}-\delta} g''(r).\end{aligned}$$

Therefore, we have

$$\begin{aligned}\tilde{\mathcal{R}}f_\delta(r) = & -2\left(1-\frac{1}{p}-\delta\right)\frac{1}{r}\left(\ln\frac{R}{r}\right)^{-\frac{1}{p}-\delta} g'(r) \\ & -(Q-2)\left(1-\frac{1}{p}-\delta\right)\frac{1}{r^2}\left(\ln\frac{R}{r}\right)^{-\frac{1}{p}-\delta} g(r) \\ & -\left(1-\frac{1}{p}-\delta\right)\left(\frac{1}{p}+\delta\right)\frac{1}{r^2}\left(\ln\frac{R}{r}\right)^{-1-\frac{1}{p}-\delta} g(r) + \left(\ln\frac{R}{r}\right)^{1-\frac{1}{p}-\delta} \tilde{\mathcal{R}}g(r).\end{aligned}$$

Since $(f_\delta)_R = 0$ we obtain

$$\begin{aligned}\int_{\mathbb{G}} \frac{|f_\delta - (f_\delta)_R|^p}{|x|^Q |\ln \frac{R}{|x|}|^p} dx &= |\wp| \int_0^2 \frac{1}{r} (\ln R - \ln r)^{-1-\delta p} g(r)^p dr \\ &\geq |\wp| \int_0^1 \frac{1}{r} (\ln R - \ln r)^{-1-\delta p} dr \\ &= \frac{1}{\delta p} (\ln R)^{p\sigma} |\wp|.\end{aligned}$$

Thus,

$$\lim_{\delta \to 0} \int_{\mathbb{G}} \frac{|f_\delta - (f_\delta)_R|^p}{|x|^Q |\ln \frac{R}{|x|}|^p} dx = \infty.$$

At the same time, direct calculations also give

$$\int_{\mathbb{G}} \frac{1}{|x|^{Q-2p}} \left| \frac{1}{|x|} \left(\ln \frac{R}{|x|} \right)^{-\frac{1}{p}-\delta} g'(|x|) \right|^p dx = O(1),$$

$$\int_{\mathbb{G}} \frac{1}{|x|^{Q-2p}} \left| \frac{1}{|x|^2} \left(\ln \frac{R}{|x|} \right)^{-1-\frac{1}{p}-\delta} g(|x|) \right|^p dx = O(1),$$

$$\int_{\mathbb{G}} \frac{1}{|x|^{Q-2p}} \left| \left(\ln \frac{R}{|x|} \right)^{1-\frac{1}{p}-\delta} \tilde{\mathcal{R}} g(|x|) \right|^p dx = O(1),$$

and

$$\begin{aligned}\int_{\mathbb{G}} \frac{1}{|x|^{Q-2p}} \left| \frac{1}{|x|} \left(\ln \frac{R}{|x|} \right)^{-\frac{1}{p}-\delta} g'(|x|) \right|^p dx &= |\wp| \int_0^2 \frac{1}{r} (\ln R - \ln r)^{-1-\delta p} g(r)^p dr \\ &= \int_{\mathbb{G}} \frac{|f_\delta - (f_\delta)_R|^p}{|x|^Q |\ln \frac{R}{|x|}|^p} dx\end{aligned}$$

Consequently, we obtain

$$\lim_{\delta \to 0} \frac{\int_{\mathbb{G}} \frac{|\tilde{\mathcal{R}} f_\delta|^p}{|x|^{Q-2p}} dx}{\int_{\mathbb{G}} \frac{|f_\delta - (f_\delta)_R|^p}{|x|^Q |\ln \frac{R}{|x|}|^p} dx} = \left(\frac{(p-1)(Q-2)}{p} \right)^p.$$

This proves the sharpness of (3.60). □

Consequently, the following identities hold true.

Theorem 3.1.20 (Higher-order critical Rellich identities for $\tilde{\mathcal{R}}$). *Let $\mathbb{G}$ be a homogeneous group of homogeneous dimension $Q \geq 3$ and let $|\cdot|$ be a homogeneous quasi-norm on $\mathbb{G}$. Let $f \in C_0^\infty(\mathbb{G}\backslash\{0\})$ be any complex-valued function and denote*

$$f_R(x) := f(Rx/|x|)$$

for $x \in \mathbb{G}$ and $R > 0$.

Then for $2 \leq k < Q/2$ we have

$$\left\|\frac{\tilde{\mathcal{R}}^k f}{|x|^{\frac{Q}{p}-2k}}\right\|^p_{L^p(\mathbb{G})} = \left(\frac{Q-2}{p'}\right)^p \left(\prod_{i=1}^{k-1} a_{i,Q}\right)^p \int_{\mathbb{G}} \frac{|f-f_R|^p}{|x|^Q \left|\ln\frac{R}{|x|}\right|^p} dx \tag{3.63}$$

$$+ p\left(\prod_{i=1}^{k-1} a_{i,Q}\right)^p \int_{\mathbb{G}} \frac{1}{|x|^{Q-2p}} R_p\left((Q-2)\frac{\mathcal{R}f}{|x|}, \tilde{\mathcal{R}}f\right) dx$$

$$+ p(Q-2)^p \left(\prod_{i=1}^{k-1} a_{i,Q}\right)^p \int_{\mathbb{G}} \frac{1}{|x|^{Q-p}} R_p\left(-\frac{p-1}{p}\frac{f-f_R}{|x|\ln\frac{R}{|x|}}, \mathcal{R}f\right) dx$$

$$+ p\int_{\mathbb{G}} \frac{1}{|x|^{Q-2kp}} R_p\left(-a_{k-1,Q}\frac{\tilde{\mathcal{R}}^{k-1}f}{|x|^2}, \tilde{\mathcal{R}}^k f\right) dx$$

$$+ p\sum_{j=1}^{k-2} \left(\prod_{i=1}^{k-1} a_{i,Q}\right)^p \int_{\mathbb{G}} \frac{1}{|x|^{Q-2(k-j)p}} R_p\left(-a_{k-j-1,Q}\frac{\tilde{\mathcal{R}}^{k-j-1}f}{|x|^2}, \tilde{\mathcal{R}}^{k-j} f\right) dx$$

$$+ pa^{p-1}_{k-1,Q}(p-1)\int_{\mathbb{G}} \frac{|\tilde{\mathcal{R}}^{k-1}f|^{p-2}}{|x|^{Q-2(k-1)p-2}} \left|\mathcal{R}|\tilde{\mathcal{R}}^{k-1}f| + \frac{2(k-1)}{|x|}|\tilde{\mathcal{R}}^{k-1}f|\right|^2 dx$$

$$+ pa^{p-1}_{k-1,Q}\int_{\mathbb{G}} \frac{|\tilde{\mathcal{R}}^{k-1}f|^{p-4}(\mathrm{Im}(\tilde{\mathcal{R}}^{k-1}f\overline{\mathcal{R}\tilde{\mathcal{R}}^{k-1}f}))^2}{|x|^{Q-2(k-1)p-2}} dx$$

$$+ p\sum_{j=1}^{l-1} \left(\prod_{i=k-j}^{k-1} a_{k-i-1,Q}\right)^p a^{p-1}_{k-i-1,Q}$$

$$\times \int_{\mathbb{G}} \frac{|\tilde{\mathcal{R}}^{k-j-1}f|^{p-4}(\mathrm{Im}(\tilde{\mathcal{R}}^{k-j-1}f\overline{\mathcal{R}\tilde{\mathcal{R}}^{k-j-1}f}))^2}{|x|^{Q-2(k-j-1)p-2}}$$

$$+ p\sum_{j=1}^{l-1} \left(\prod_{i=k-j}^{k-1} a_{k-i-1,Q}\right)^p a^{p-1}_{k-i-1,Q}(p-1)$$

$$\times \int_{\mathbb{G}} \frac{|\tilde{\mathcal{R}}^{k-j-1}f|^{p-2}}{|x|^{Q-2(k-j-1)p-2}} \left|\mathcal{R}|\tilde{\mathcal{R}}^{k-j-1}f| + \frac{Q-2p(2(j+1)+\alpha)}{p|x|}|\tilde{\mathcal{R}}^{k-j-1}f|\right|^2 dx,$$

where

$$a_{j,Q} = 2j(Q-2j-2)$$

and R_p is as in (2.22). *For $1 \leq k < (Q-1)/2$ we also have*

$$\left\|\frac{\mathcal{R}\tilde{\mathcal{R}}^k f}{|x|^{\frac{Q}{p}-(2k+1)}}\right\|^p_{L^p(\mathbb{G})} = (2k)^p\left(\frac{(p-1)(Q-2)}{p}\right)^p \left(\prod_{i=1}^{k-1} a_{i,Q}\right)^p \int_{\mathbb{G}} \frac{|f-f_R|^p}{|x|^Q \left|\ln\frac{R}{|x|}\right|^p} dx$$

$$+ p(2k)^p \left(\prod_{i=1}^{k-1} a_{i,Q}\right)^p \int_{\mathbb{G}} \frac{1}{|x|^{Q-2p}} R_p\left((Q-2)\frac{\mathcal{R}f}{|x|}, \tilde{\mathcal{R}}f\right) dx$$

$$+p(2k)^p(Q-2)^p\left(\prod_{i=1}^{k-1}a_{i,Q}\right)^p\int_{\mathbb{G}}\frac{1}{|x|^{Q-p}}R_p\left(-\frac{p-1}{p}\frac{f-f_R}{|x|\ln\frac{R}{|x|}},\mathcal{R}f\right)dx$$

$$+p(2k)^p\int_{\mathbb{G}}\frac{1}{|x|^{Q-2kp}}R_p\left(-a_{k-1,Q}\frac{\tilde{\mathcal{R}}^{k-1}f}{|x|^2},\tilde{\mathcal{R}}^k f\right)dx$$

$$+p(2k)^p\sum_{j=1}^{k-2}\left(\prod_{i=k-j}^{k-1}a_{i,Q}\right)^p\int_{\mathbb{G}}\frac{1}{|x|^{Q-2(k-j)p}}R_p\left(-a_{k-j-1,Q}\frac{\tilde{\mathcal{R}}^{k-j-1}f}{|x|^2},\tilde{\mathcal{R}}^{k-j}f\right)dx$$

$$+p(2k)^p a_{k-1,Q}^{p-1}(p-1)\int_{\mathbb{G}}\frac{|\tilde{\mathcal{R}}^{k-1}f|^{p-2}}{|x|^{Q-2(k-1)p-2}}\left|\mathcal{R}|\tilde{\mathcal{R}}^{k-1}f|+\frac{2(k-1)}{|x|}|\tilde{\mathcal{R}}^{k-1}f|\right|^2dx$$

$$+p(2k)^p a_{k-1,Q}^{p-1}\int_{\mathbb{G}}\frac{|\tilde{\mathcal{R}}^{k-1}f|^{p-4}(\mathrm{Im}(\tilde{\mathcal{R}}^{k-1}f\overline{\mathcal{R}\tilde{\mathcal{R}}^{k-1}f}))^2}{|x|^{Q-2(k-1)p-2}}dx$$

$$+p(2k)^p\sum_{j=1}^{k-2}\left(\prod_{i=1}^{k-1}a_{k-i-1,Q}\right)^p a_{k-i-1,Q}^{p-1}$$

$$\times\int_{\mathbb{G}}\frac{|\tilde{\mathcal{R}}^{k-j-1}f|^{p-4}(\mathrm{Im}(\tilde{\mathcal{R}}^{k-j-1}f\overline{\mathcal{R}\tilde{\mathcal{R}}^{k-j-1}f}))^2}{|x|^{Q-2(k-j-1)p-2}}dx$$

$$+p(2k)^p\sum_{j=1}^{k-2}\left(\prod_{i=1}^{k-1}a_{k-i-1,Q}\right)^p a_{k-i-1,Q}^{p-1}(p-1)$$

$$\times\int_{\mathbb{G}}\frac{|\tilde{\mathcal{R}}^{k-j-1}f|^{p-2}}{|x|^{Q-2(k-j-1)p-2}}\left|\mathcal{R}|\tilde{\mathcal{R}}^{k-j-1}f|+\frac{Q-2p(2(j+1)+\alpha)}{p|x|}|\tilde{\mathcal{R}}^{k-j-1}f|\right|^2dx$$

$$+p\int_{\mathbb{G}}\frac{1}{|x|^{Q-(2k+1)p}}R_p\left(-2k\frac{\tilde{\mathcal{R}}^k f}{|x|},\mathcal{R}\tilde{\mathcal{R}}^k f\right)dx. \tag{3.64}$$

Proof of Theorem 3.1.20. Denoting $g=\tilde{\mathcal{R}}f$ and applying (3.49) to the function g with $l=k-1$, and then using (3.59), we arrive at (3.63). To prove (3.64) we use Theorem 2.1.8 with $\alpha=(Q-2(k+1)p)/p$ which gives

$$\begin{aligned}\left\|\frac{\mathcal{R}\tilde{\mathcal{R}}^k f}{|x|^{\frac{Q}{p}-(2k+1)}}\right\|_{L^p(\mathbb{G})}^p&=(2k)^p\int_{\mathbb{G}}\frac{|\tilde{\mathcal{R}}^k f|^p}{|x|^{Q-2kp}}dx\\&\quad+p\int_{\mathbb{G}}\frac{1}{|x|^{Q-(2k+1)p}}R_p\left(-2k\frac{\tilde{\mathcal{R}}^k f}{|x|},\mathcal{R}\tilde{\mathcal{R}}^k f\right)dx.\end{aligned} \tag{3.65}$$

Now, the combination of (3.65) and (3.63) implies (3.64). □

As a consequence of Theorem 3.1.20 we have the corresponding inequalities.

Corollary 3.1.21 (Higher-order critical Rellich inequalities for $\tilde{\mathcal{R}}$)**.** *Let $\mathbb{G}$ be a homogeneous group of homogeneous dimension $Q\geq 3$ and let $|\cdot|$ be a homogeneous* *quasi-norm on* $\mathbb{G}$. *Let* $f \in C_0^\infty(\mathbb{G}\backslash\{0\})$ *be any complex-valued function and denote*

$$f_R(x) := f(Rx/|x|)$$

with $x \in \mathbb{G}$ *and* $R > 0$. *Let* $1 < p < \infty$. *Then for any* $2 \le k < Q/2$ *we have*

$$\frac{2^{k-1}(k-1)!(p-1)}{p}\prod_{i=0}^{k-1}(Q-2i-2)\sup_{R>0}\left\|\frac{f-f_R}{|x|^{\frac{Q}{p}}\left|\ln\frac{R}{|x|}\right|}\right\|_{L^p(\mathbb{G})} \le \left\|\frac{\tilde{\mathcal{R}}f}{|x|^{\frac{Q}{p}-2k}}\right\|_{L^p(\mathbb{G})}. \tag{3.66}$$

Furthermore, for any $1 \le k \le (Q-1)/2$ *we have*

$$\frac{2^k k!(p-1)}{p}\prod_{i=0}^{k-1}(Q-2i-2)\sup_{R>0}\left\|\frac{f-f_R}{|x|^{\frac{Q}{p}}\left|\ln\frac{R}{|x|}\right|}\right\|_{L^p(\mathbb{G})} \le \left\|\frac{\mathcal{R}\tilde{\mathcal{R}}^k f}{|x|^{\frac{Q}{p}-2(k+1)}}\right\|_{L^p(\mathbb{G})}. \tag{3.67}$$

Moreover, the constants in inequalities (3.66) *and* (3.67) *are sharp.*

Proof of Corollary 3.1.21. Inequalities (3.66) and (3.67) are immediate consequences of identities (3.63) and (3.64), respectively, since we have $a_{i,Q} \ge 0$ for $1 \le i \le k-1$, as well as the equalities

$$\frac{(Q-2)(p-1)}{p}\prod_{i=1}^{k-1}a_{i,Q} = \frac{2^{k-1}(k-1)!}{p'}\prod_{i=0}^{k-1}(Q-2i-2),$$

and

$$\frac{2k(Q-2)(p-1)}{p}\prod_{i=1}^{k-1}a_{i,Q} = \frac{2^k k!(p-1)}{p}\prod_{i=0}^{k-1}(Q-2i-2).$$

To show the sharpness of the constants in (3.66) and (3.67) we can use the same argument as in the proof of Theorem 3.1.19 with the test function

$$f_\delta(x) := \left(\ln\frac{R}{|x|}\right)^{1-\frac{1}{p}-\delta} g(|x|).$$

This completes the proof. □

3.2 Sobolev type inequalities

In this section we restate some of Hardy inequalities from the previous chapter in terms of the Euler operator $\mathbb{E}$ from Section 1.3.2, relating them to Sobolev type inequalities. We also briefly recast some of their proofs for the convenience of the reader since fixing the notation in terms of the Euler and radial operators will be useful in further arguments, especially for the analysis in Chapter 10.

The classical *Sobolev inequality* in $\mathbb{R}^n$ has the form

$$\|g\|_{L^p(\mathbb{R}^n)} \leq C(p)\|\nabla g\|_{L^{p^*}(\mathbb{R}^n)}, \quad 1 < p, p^* < \infty, \tag{3.68}$$

where ∇ is the standard gradient in $\mathbb{R}^n$ and

$$\frac{1}{p} = \frac{1}{p^*} - \frac{1}{n}.$$

The aim of this section is to discuss another version of the Sobolev inequality (we call it *Sobolev type inequality*) with respect to the operator $x \cdot \nabla$ instead of ∇, that is, the inequality

$$\|g\|_{L^p(\mathbb{R}^n)} \leq C'(p)\|x \cdot \nabla g\|_{L^q(\mathbb{R}^n)}. \tag{3.69}$$

For any $\lambda > 0$, by setting $g(x) = h(\lambda x)$ in (3.69), it is straightforward to see that $p = q$ is a necessary condition to have inequality (3.69). So, one may concentrate on the case $p = q$. In the case of $\mathbb{R}^n$ this inequality was analysed by Ozawa and Sasaki [OS09], now we concentrate on the setting of general homogeneous groups.

We also note that the classical Sobolev inequality (3.68) can be extended to nilpotent Lie groups: for stratified groups see Folland [Fol75], for graded groups see [FR17], with a general summary presented also in [FR16], and for another version on general homogeneous groups see Section 4.3.

3.2.1 Hardy and Sobolev type inequalities

The inequality (3.70) below is such an extension of (3.69) formulated in terms of the Euler operator $\mathbb{E}$. Moreover, it turns out to be possible to derive a formula for the remainder in this inequality. For indices $1 < p < Q$ this Sobolev type inequality implies the Hardy inequality and for $p = 2$ they are equivalent. All this is the subject of the following statement, an analogue of Theorem 2.1.1.

In this section, unless stated otherwise, $\mathbb{G}$ is a homogeneous group of homogeneous dimension $Q \geq 1$ and $|\cdot|$ is a homogeneous quasi-norm on $\mathbb{G}$.

Proposition 3.2.1 (Sobolev type and Hardy inequalities). *We have the following properties.*

(i) *For all complex-valued functions $f \in C_0^\infty(\mathbb{G}\backslash\{0\})$,*

$$\|f\|_{L^p(\mathbb{G})} \leq \frac{p}{Q}\|\mathbb{E}f\|_{L^p(\mathbb{G})}, \quad 1 < p < \infty, \tag{3.70}$$

where the constant $\frac{p}{Q}$ is sharp and the equality is attained if and only if $f = 0$.

(ii) *Let*

$$u := u(x) = -\frac{p}{Q}\mathbb{E}f(x), \; v := v(x) = f(x).$$

Then we have the following expression for the remainder:

$$\|u\|^p_{L^p(\mathbb{G})} - \|v\|^p_{L^p(\mathbb{G})} = p\int_{\mathbb{G}} I_p(v,u)|v-u|^2 dx, \quad 1<p<\infty, \tag{3.71}$$

for all real-valued functions $f \in C_0^\infty(\mathbb{G}\backslash\{0\})$, where

$$I_p(h,g) = (p-1)\int_0^1 |\xi h + (1-\xi)g|^{p-2}\xi d\xi.$$

(iii) *For all complex-valued functions $f \in C_0^\infty(\mathbb{G}\backslash\{0\})$ the identity (3.71) with $p=2$ holds and can be written in the form*

$$\|\mathbb{E}f\|^2_{L^2(\mathbb{G})} = \left(\frac{Q}{2}\right)^2 \|f\|^2_{L^2(\mathbb{G})} + \left\|\mathbb{E}f + \frac{Q}{2}f\right\|^2_{L^2(\mathbb{G})}, \quad Q \geq 1. \tag{3.72}$$

(iv) *In the case $p=2$ and $Q \geq 3$ the inequality (3.70) is equivalent to Hardy's inequality, i.e., for any $g \in C_0^\infty(\mathbb{G}\backslash\{0\})$ the inequality*

$$\left\|\frac{g}{|x|}\right\|_{L^2(\mathbb{G})} \leq \frac{2}{Q-2}\|\mathcal{R}g\|_{L^2(\mathbb{G})} \equiv \frac{2}{Q-2}\left\|\frac{1}{|x|}\mathbb{E}g\right\|_{L^2(\mathbb{G})}. \tag{3.73}$$

(v) *In the case $1<p<Q$ the inequality (3.70) yields Hardy's inequality for any $f \in C_0^\infty(\mathbb{G}\backslash\{0\})$, i.e., the inequality*

$$\left\|\frac{f}{|x|}\right\|_{L^p(\mathbb{G})} \leq \frac{p}{Q-p}\|\mathcal{R}f\|_{L^p(\mathbb{G})}. \tag{3.74}$$

Remark 3.2.2.

1. As mentioned above in the Euclidian case, for any $\lambda > 0$, substituting $g(x) = h(\lambda x)$ into the Sobolev type inequality

$$\|g\|_{L^p(\mathbb{G})} \leq C(p)\|\mathbb{E}g\|_{L^q(\mathbb{G})}, \quad 1<p,q<\infty, \tag{3.75}$$

and using the fact that the Euler operator is a homogeneous operator of order zero, we obtain that $p=q$ is a necessary condition for having inequality (3.75).

2. In the Euclidean case $\mathbb{G}=\mathbb{R}^n$ the inequality (3.70) was observed in [BEHL08]: for any $n \geq 1$ and $1 \leq p < \infty$, for all $f \in C_0^\infty(\mathbb{R}^n)$ we have

$$\|f\|_{L^p(\mathbb{R}^n)} \leq \frac{p}{n}\|x\cdot\nabla f\|_{L^p(\mathbb{R}^n)}.$$

Indeed, this is a consequence of a simple integration by parts:

$$\begin{aligned} n\int_{\mathbb{R}^n}|f(x)|^p dx &= \int_{\mathbb{R}^n}\operatorname{div}(x)|f(x)|^p dx \\ &= -p\operatorname{Re}\int_{\mathbb{R}^n} x\cdot\nabla f(x)|f(x)|^{p-2}\overline{f(x)}dx \\ &\le p\left(\int_{\mathbb{R}^n}|x\cdot\nabla f(x)|^p dx\right)^{1/p}\left(\int_{\mathbb{R}^n}|f(x)|^p dx\right)^{\frac{p-1}{p}}, \end{aligned}$$

using Hölder's inequality in the last line.

There is a weighted version of the above inequality given in Theorem 6.6.1, see also Remark 6.6.2, Part 3.

3. The analysis in this section is based on [RSY18d].

Proof of Proposition 3.2.1. Let us first show that Part (ii) implies Part (i). By dropping non-negative term in the right-hand side of (3.71), we get

$$\|f\|_{L^p(\mathbb{G})} \le \frac{p}{Q}\|\mathbb{E}f\|_{L^p(\mathbb{G})}, \quad 1<p<\infty,\ \ Q\ge 1, \tag{3.76}$$

for all real-valued functions $f\in C_0^\infty(\mathbb{G}\backslash\{0\})$. Consequently, this inequality is valid for all complex-valued functions if we use the identity

$$\forall z\in\mathbb{C}:\ |z|^p=\left(\int_{-\pi}^{\pi}|\cos\theta|^p d\theta\right)^{-1}\int_{-\pi}^{\pi}|\operatorname{Re}(z)\cos\theta+\operatorname{Im}(z)\sin\theta|^p\,d\theta, \tag{3.77}$$

see (2.8).

So, the inequality (3.70) holds true and the expression for the remainder implies that the constant $\frac{p}{Q}$ is sharp.

Let us show that this constant is attained only for $f=0$. In view of the identity (3.77), it is enough to check this only for real-valued functions f. If the right-hand side of (3.71) is zero, then we must have the equality

$$-\frac{p}{Q}\mathbb{E}f(x)=f(x),$$

which yields that

$$\mathbb{E}f=-\frac{Q}{p}f.$$

By the property of the Euler operator in Lemma 1.3.1 this means that f is positively homogeneous function of order $-\frac{Q}{p}$, i.e., there exists a function $h:\wp\to\mathbb{C}$, where $\wp$ is defined by (1.12), such that

$$f(x)=|x|^{-\frac{Q}{p}}h\left(\frac{x}{|x|}\right). \tag{3.78}$$

In particular, (3.78) means that f cannot be compactly supported unless it is identically equal to zero. Therefore, we have shown that Part (ii), namely (3.71) implies Part (i) of Theorem 3.2.1.

Nevertheless, let us also give another direct proof of (3.70) for complex-valued functions without using formula (3.77) and without using the remainder formula in Part (ii).

Proof of Part (i). Introducing polar coordinates $(r, y) = (|x|, \frac{x}{|x|}) \in (0, \infty) \times \wp$ on $\mathbb{G}$, where the sphere $\wp$ is defined in (1.12), we now apply the polar decomposition formula (1.13). This and integrating by parts yield

$$\begin{aligned}\int_{\mathbb{G}} |f(x)|^p dx &= \int_0^\infty \int_\wp |f(ry)|^p r^{Q-1} d\sigma(y) dr \\ &= -\frac{p}{Q} \int_0^\infty r^Q \operatorname{Re} \int_\wp |f(ry)|^{p-2} f(ry) \overline{\frac{df(ry)}{dr}} d\sigma(y) dr \qquad (3.79) \\ &= -\frac{p}{Q} \operatorname{Re} \int_{\mathbb{G}} |f(x)|^{p-2} f(x) \overline{\mathbb{E} f(x)} dx.\end{aligned}$$

By using the Hölder inequality with an index q such that $\frac{1}{p} + \frac{1}{q} = 1$ we obtain

$$\begin{aligned}\int_{\mathbb{G}} |f(x)|^p dx &= -\frac{p}{Q} \operatorname{Re} \int_{\mathbb{G}} |f(x)|^{p-2} f(x) \overline{\mathbb{E} f(x)} dx \\ &\leq \frac{p}{Q} \left(\int_{\mathbb{G}} \left| |f(x)|^{p-2} f(x) \right|^q dx \right)^{\frac{1}{q}} \left(\int_{\mathbb{G}} |\mathbb{E} f(x)|^p dx \right)^{\frac{1}{p}} \\ &= \frac{p}{Q} \left(\int_{\mathbb{G}} |f(x)|^p dx \right)^{1-\frac{1}{p}} \|\mathbb{E} f\|_{L^p(\mathbb{G})},\end{aligned}$$

which gives inequality (3.70) in Part (i).

Proof of Part (ii). With the notation

$$u := u(x) = -\frac{p}{Q} \mathbb{E} f \quad \text{and} \quad v := v(x) = f(x),$$

the formula (3.79) can be reformulated as

$$\|v\|^p_{L^p(\mathbb{G})} = \operatorname{Re} \int_{\mathbb{G}} |v|^{p-2} v \bar{u} dx. \qquad (3.80)$$

For any real-valued function f formula (3.79) becomes

$$\int_{\mathbb{G}} |f(x)|^p dx = -\frac{p}{Q} \int_{\mathbb{G}} |f(x)|^{p-2} f(x) \mathbb{E} f(x) dx,$$

and (3.80) becomes

$$\|v\|^p_{L^p(\mathbb{G})} = \int_{\mathbb{G}} |v|^{p-2} v u dx. \qquad (3.81)$$

Moreover, for all L^p-integrable real-valued functions u and v the following equalities hold

$$\begin{aligned}
&\|u\|^p_{L^p(\mathbb{G})} - \|v\|^p_{L^p(\mathbb{G})} + p\int_{\mathbb{G}}(|v|^p - |v|^{p-2}vu)dx \\
&= \int_{\mathbb{G}}(|u|^p + (p-1)|v|^p - p|v|^{p-2}vu)dx \\
&= p\int_{\mathbb{G}} I_p(v,u)|v-u|^2dx, \quad 1<p<\infty,
\end{aligned}$$

where

$$I_p(v,u) = (p-1)\int_0^1 |\xi v + (1-\xi)u|^{p-2}\xi d\xi.$$

Combining this with (3.81) we arrive at

$$\|u\|^p_{L^p(\mathbb{G})} - \|v\|^p_{L^p(\mathbb{G})} = p\int_{\mathbb{G}} I_p(v,u)|v-u|^2dx,$$

which proves the equality (3.71).

Now we prove Part (iii). If $p=2$, then the identity (3.80) can be rewritten as

$$\|v\|^2_{L^2(\mathbb{G})} = \mathrm{Re}\int_{\mathbb{G}} v\overline{u}dx.$$

Thus, we have

$$\begin{aligned}
\|u\|^2_{L^2(\mathbb{G})} - \|v\|^2_{L^2(\mathbb{G})} &= \|u\|^2_{L^2(\mathbb{G})} - \|v\|^2_{L^2(\mathbb{G})} + 2\int_{\mathbb{G}}(|v|^2 - \mathrm{Re}\, v\overline{u})dx \\
&= \int_{\mathbb{G}}(|u|^2 + |v|^2 - 2\mathrm{Re}\, v\overline{u})dx \\
&= \int_{\mathbb{G}} |u-v|^2dx,
\end{aligned}$$

which gives (3.72).

To show Part (iv) first we verify that the inequality (3.70) implies (3.73). A direct calculation shows

$$\begin{aligned}
\|\mathbb{E}f\|^2_{L^2(\mathbb{G})} &= \left\|\mathbb{E}\frac{g}{|x|}\right\|^2_{L^2(\mathbb{G})} \\
&= \int_0^\infty \int_{\wp} \left|\left(r\frac{d}{dr}\right)\frac{g(ry)}{r}\right|^2 r^{Q-1}d\sigma(y)dr \\
&= \left\|-\frac{g}{|x|} + \frac{d}{d|x|}g\right\|^2_{L^2(\mathbb{G})} \\
&= \left\|\frac{g}{|x|}\right\|^2_{L^2(\mathbb{G})} - 2\mathrm{Re}\int_{\mathbb{G}} \frac{g}{|x|}\overline{\frac{d}{d|x|}g}dx + \left\|\frac{d}{d|x|}g\right\|^2_{L^2(\mathbb{G})},
\end{aligned}$$

where $g = |x|f$. From

$$
\begin{aligned}
-2\mathrm{Re}\int_{\mathbb{G}} \frac{g}{|x|}\overline{\frac{d}{d|x|}g}dx &= -2\mathrm{Re}\int_0^{\infty}\int_{\wp} \frac{g(ry)}{r}\overline{\frac{d}{dr}g(ry)}r^{Q-1}d\sigma(y)dr \\
&= -\mathrm{Re}\int_0^{\infty}\int_{\wp} \frac{d}{dr}(|\overline{g}|^2)r^{Q-2}d\sigma(y)dr \\
&= (Q-2)\mathrm{Re}\int_0^{\infty}\int_{\wp} |\overline{g}|^2 r^{Q-3}d\sigma(y)dr \\
&= (Q-2)\left\|\frac{g}{|x|}\right\|^2_{L^2(\mathbb{G})},
\end{aligned}
$$

it follows that

$$
\|\mathbb{E}f\|^2_{L^2(\mathbb{G})} = (Q-1)\left\|\frac{g}{|x|}\right\|^2_{L^2(\mathbb{G})} + \left\|\frac{d}{d|x|}g\right\|^2_{L^2(\mathbb{G})}. \tag{3.82}
$$

Combining (3.70) and (3.82) we obtain

$$
\left\|\frac{g}{|x|}\right\|^2_{L^2(\mathbb{G})} \leq \frac{4}{Q^2}\left((Q-1)\left\|\frac{g}{|x|}\right\|^2_{L^2(\mathbb{G})} + \left\|\frac{d}{d|x|}g\right\|^2_{L^2(\mathbb{G})}\right),
$$

which gives (3.73).

Conversely, let us assume that (3.73) is valid. Then with the notation $f = g/|x|$ we get

$$
\begin{aligned}
\left\|\frac{d}{d|x|}(|x|f)\right\|^2_{L^2(\mathbb{G})} &= \|f + \mathbb{E}f\|^2_{L^2(\mathbb{G})} \\
&= \|f\|^2_{L^2(\mathbb{G})} + 2\mathrm{Re}\int_{\mathbb{G}} f(x)\overline{\mathbb{E}f(x)}dx + \|\mathbb{E}f\|^2_{L^2(\mathbb{G})}.
\end{aligned}
$$

Hence by (1.42) and (3.73) it follows that

$$
\|f\|^2_{L^2(\mathbb{G})} \leq \frac{4}{(Q-2)^2}\left(\|\mathbb{E}f\|^2_{L^2(\mathbb{G})} - (Q-1)\|f\|^2_{L^2(\mathbb{G})}\right),
$$

which gives

$$
\|f\|_{L^2(\mathbb{G})} \leq \frac{2}{Q}\|\mathbb{E}f\|_{L^2(\mathbb{G})}.
$$

Now it remains to prove Part (v). We will show that the inequality (3.70) gives (3.74). We have

$$
\|\mathcal{R}(|x|f)\|_{L^p(\mathbb{G})} = \|\mathbb{E}f + f\|_{L^p(\mathbb{G})} \geq \|\mathbb{E}f\|_{L^p(\mathbb{G})} - \|f\|_{L^p(\mathbb{G})}.
$$

Finally, by using the inequality (3.70) we establish

$$
\|\mathcal{R}(|x|f)\|_{L^p(\mathbb{G})} \geq \frac{Q-p}{p}\|f\|_{L^p(\mathbb{G})},
$$

which implies the Hardy inequality (3.74). □

3.2.2 Weighted L^p-Sobolev type inequalities

Now we establish weighted L^p-Sobolev type inequalities on the homogeneous group $\mathbb{G}$ of homogeneous dimension $Q \geq 1$.

Theorem 3.2.3 (Weighted L^p-Sobolev type inequalities). *For all complex-valued functions $f \in C_0^\infty(\mathbb{G}\backslash\{0\})$, $1 < p < \infty$, and any homogeneous quasi-norm $|\cdot|$ on $\mathbb{G}$ for $\alpha p \neq Q$ we have*

$$\left\| \frac{f}{|x|^\alpha} \right\|_{L^p(\mathbb{G})} \leq \left| \frac{p}{Q-\alpha p} \right| \left\| \frac{1}{|x|^\alpha} \mathbb{E} f \right\|_{L^p(\mathbb{G})} \quad \text{for all } \alpha \in \mathbb{R}. \tag{3.83}$$

If $\alpha p \neq Q$ then the constant $\left|\frac{p}{Q-\alpha p}\right|$ is sharp.

For $\alpha p = Q$ we have

$$\left\| \frac{f}{|x|^{\frac{Q}{p}}} \right\|_{L^p(\mathbb{G})} \leq p \left\| \frac{\log |x|}{|x|^{\frac{Q}{p}}} \mathbb{E} f \right\|_{L^p(\mathbb{G})}, \tag{3.84}$$

where the constant p is sharp.

Proof of Theorem 3.2.3. Using the integration by parts formula from Proposition 1.2.10, for $\alpha p \neq Q$ we obtain

$$\begin{aligned}
\int_{\mathbb{G}} \frac{|f(x)|^p}{|x|^{\alpha p}} dx &= \int_0^\infty \int_\wp |f(ry)|^p r^{Q-1-\alpha p} d\sigma(y) dr \\
&= -\frac{p}{Q-\alpha p} \int_0^\infty r^{Q-\alpha p} \operatorname{Re} \int_\wp |f(ry)|^{p-2} f(ry) \overline{\frac{df(ry)}{dr}} d\sigma(y) dr \\
&\leq \left| \frac{p}{Q-\alpha p} \right| \int_{\mathbb{G}} \frac{|\mathbb{E} f(x)| |f(x)|^{p-1}}{|x|^{\alpha p}} dx = \left| \frac{p}{Q-\alpha p} \right| \int_{\mathbb{G}} \frac{|\mathbb{E} f(x)| |f(x)|^{p-1}}{|x|^{\alpha+\alpha(p-1)}} dx.
\end{aligned}$$

By Hölder's inequality, it follows that

$$\int_{\mathbb{G}} \frac{|f(x)|^p}{|x|^{\alpha p}} dx \leq \left| \frac{p}{Q-\alpha p} \right| \left(\int_{\mathbb{G}} \frac{|\mathbb{E} f(x)|^p}{|x|^{\alpha p}} dx \right)^{\frac{1}{p}} \left(\int_{\mathbb{G}} \frac{|f(x)|^p}{|x|^{\alpha p}} dx \right)^{\frac{p-1}{p}},$$

which gives (3.83).

Now we show the sharpness of the constant. We need to check the equality condition in the above Hölder inequality. Let us consider the function

$$g(x) = \frac{1}{|x|^C},$$

where $C \in \mathbb{R}, C \neq 0$ and $\alpha p \neq Q$. Then by a direct calculation we obtain

$$\left| \frac{1}{C} \right|^p \left(\frac{|\mathbb{E} g(x)|}{|x|^\alpha} \right)^p = \left(\frac{|g(x)|^{p-1}}{|x|^{\alpha(p-1)}} \right)^{\frac{p}{p-1}},$$

which satisfies the equality condition in Hölder's inequality. This gives the sharpness of the constant $\left|\frac{p}{Q-\alpha p}\right|$ in (3.83).

Now let us prove (3.84). Using integration by parts, we have

$$\int_{\mathbb{G}} \frac{|f(x)|^p}{|x|^Q} dx = \int_0^\infty \int_{\wp} |f(ry)|^p r^{Q-1-Q} d\sigma(y) dr$$

$$= -p \int_0^\infty \log r \operatorname{Re} \int_{\wp} |f(ry)|^{p-2} f(ry) \overline{\frac{df(ry)}{dr}} d\sigma(y) dr$$

$$\leq p \int_{\mathbb{G}} \frac{|\mathbb{E}f(x)||f(x)|^{p-1}}{|x|^Q} |\log |x|| dx = p \int_{\mathbb{G}} \frac{|\mathbb{E}f(x)| \log |x||| }{|x|^{\frac{Q}{p}}} \frac{|f(x)|^{p-1}}{|x|^{\frac{Q(p-1)}{p}}} dx.$$

By Hölder's inequality, it follows that

$$\int_{\mathbb{G}} \frac{|f(x)|^p}{|x|^Q} dx \leq p \left(\int_{\mathbb{G}} \frac{|\mathbb{E}f(x)|^p |\log |x||^p}{|x|^Q} dx \right)^{1/p} \left(\int_{\mathbb{G}} \frac{|f(x)|^p}{|x|^Q} dx \right)^{(p-1)/p},$$

which gives (3.84).

Now we show the sharpness of the constant. We need to check the equality condition in the above Hölder inequality. Let us consider the function

$$h(x) = (\log |x|)^C,$$

where $C \in \mathbb{R}$ and $C \neq 0$. Then by a direct calculation we obtain

$$\left|\frac{1}{C}\right|^p \left(\frac{|\mathbb{E}h(x)|| \log |x||}{|x|^{\frac{Q}{p}}} \right)^p = \left(\frac{|h(x)|^{p-1}}{|x|^{\frac{Q(p-1)}{p}}} \right)^{p/(p-1)},$$

which satisfies the equality condition in Hölder's inequality. This gives the sharpness of the constant p in (3.84). □

Let us consider separately the case $p = 2$, that is, let us restate Theorem 2.1.5 in terms of the operator $\mathbb{E}$:

Proposition 3.2.4 (An identity for Euler operator). *For every complex-valued function $f \in C_0^\infty(\mathbb{G}\backslash\{0\})$ we have*

$$\left\| \frac{1}{|x|^\alpha} \mathbb{E}f \right\|^2_{L^2(\mathbb{G})} = \left(\frac{Q}{2} - \alpha \right)^2 \left\| \frac{f}{|x|^\alpha} \right\|^2_{L^2(\mathbb{G})} + \left\| \frac{1}{|x|^\alpha} \mathbb{E}f + \frac{Q-2\alpha}{2|x|^\alpha} f \right\|^2_{L^2(\mathbb{G})}, \quad (3.85)$$

for any $\alpha \in \mathbb{R}$.

Remark 3.2.5.

1. By dropping the non-negative last term in (3.85) we immediately get the following inequality for $\alpha \in \mathbb{R}$ with $Q - 2\alpha \neq 0$:

$$\left\| \frac{f}{|x|^\alpha} \right\|_{L^2(\mathbb{G})} \leq \frac{2}{|Q-2\alpha|} \left\| \frac{1}{|x|^\alpha} \mathbb{E}f \right\|_{L^2(\mathbb{G})}, \quad (3.86)$$

for all complex-valued functions $f \in C_0^\infty(\mathbb{G}\backslash\{0\})$, where the constant in (3.86) is sharp and the equality is attained if and only if $f = 0$. This statement on the constant and the equality follows by the same argument as that in Remark 2.1.7.

2. By iterating the established weighted Sobolev inequality (3.83) one obtains inequalities of higher order. Thus, for $1 < p < \infty$, $k \in \mathbb{N}$ and $\alpha \in \mathbb{R}$ with $Q \neq \alpha p$ we have

$$\left\| \frac{f}{|x|^\alpha} \right\|_{L^p(\mathbb{G})} \leq \left| \frac{p}{Q - \alpha p} \right|^k \left\| \frac{\mathbb{E}^k f}{|x|^\alpha} \right\|_{L^p(\mathbb{G})} \tag{3.87}$$

for any complex-valued function $f \in C_0^\infty(\mathbb{G}\backslash\{0\})$.

3. For $k = 1$ (3.87) implies the weighted Sobolev inequality and for $k = 1$ and $\alpha = 0$ this gives the Sobolev inequality. In the case $k = 2$ this can be thought of as a (weighted) Sobolev–Rellich type inequality.

3.2.3 Stubbe type remainder estimates

The remainders in Hardy inequalities may be described in different ways: there may be equalities or estimates of different forms. These are discussed in some detail at various spaces of this book. Here, we give a remainder estimate in the most basic case of L^2. Such a type of inequalities have been analysed on $\mathbb{R}^n$ by Stubbe [Stu90], and here we give its general version on homogeneous groups.

In the proof of the following statement we will use the useful feature that some estimates involving radial derivatives of the Euler operator can be proved first for radial functions, and then extended to non-radial ones by a more abstract argument, see Section 1.3.3.

Theorem 3.2.6 (Stubbe type remainder estimate). *Let $\mathbb{G}$ be a homogeneous group of homogeneous dimension $Q \geq 3$. Let $|\cdot|$ be a homogeneous quasi-norm. Then we have for all $f \in C_0^\infty(\mathbb{G})$ and $0 \leq \delta < \frac{Q^2}{4}$, the inequality*

$$\int_{\mathbb{G}} |\mathbb{E}f(x)|^2 dx - \delta \int_{\mathbb{G}} |f(x)|^2 dx \geq \frac{\left(\frac{Q^2}{4} - \delta\right)^{\frac{Q-1}{Q}}}{\left(\frac{(Q-2)^2}{4}\right)^{\frac{Q-1}{Q}}} S_Q \left(\int_{\mathbb{G}} |x|^{2^*} g(|x|)|^{2^*} dx \right)^{2/2^*} \tag{3.88}$$

with sharp constant, where

$$g(|x|) = \mathcal{M}(f)(|x|) := \frac{1}{|\wp|} \int_{\wp} f(|x|, y) d\sigma(y)$$

and

$$S_Q := |\wp|^{\frac{2}{Q}} Q^{\frac{Q-2}{Q}} (Q-2) \left(\frac{\Gamma(Q/2)\Gamma(1+Q/2)}{\Gamma(Q)} \right)^{2/Q}. \tag{3.89}$$

Remark 3.2.7.

1. In the Abelian case $\mathbb{G} = (\mathbb{R}^n, +)$, we have $Q = n$, and inequality (3.88) becomes

$$\int_{\mathbb{R}^n} |(x\cdot\nabla)f(x)|^2dx - \delta\int_{\mathbb{R}^n} |f(x)|^2dx \geq \frac{\left(\frac{n^2}{4}-\delta\right)^{\frac{n-1}{n}}}{\left(\frac{(n-2)^2}{4}\right)^{\frac{n-1}{n}}} S_n\left(\int_{\mathbb{G}} |x|^{2^*}|g(|x|)|^{2^*}dx\right)^{\frac{2}{2^*}}. \tag{3.90}$$

2. An interesting observation is that the constant in the above inequality on $\mathbb{R}^n$ is sharp for any quasi-norm $|\cdot|$, that is, it does not depend on the quasi-norm $|\cdot|$. Therefore, this inequality is new already in the Euclidean setting of $\mathbb{R}^n$. When $|x| = \sqrt{x_1^2+x_2^2+\cdots+x_n^2}$ is the Euclidean distance, the inequality (3.90) was investigated in $\mathbb{R}^n$ in [BEHL08, Corollary 4.4] and in [Xia11, Theorem 1.1].
3. The following result, proved in [Bli30], will be useful in the proof: Let f be a non-negative function. Then for $s \geq 0$ and $q > p > 1$ we have

$$\left(\int_0^\infty \left|\int_0^s f(r)dr\right|^q r^{q/p-q-1}dr\right)^{p/q} \leq C_{p,q}\int_0^\infty |f(r)|^p dr, \tag{3.91}$$

where

$$C_{p,q} = (q-q/p)^{-p/q}\left(\frac{(q/p-1)\Gamma\left(\frac{pq}{q-p}\right)}{\Gamma\left(\frac{p}{q-p}\right)\Gamma\left(\frac{p(q-1)}{q-p}\right)}\right)^{(q-p)/q}$$

is sharp. Moreover, the equality in (3.91) is attained for functions of the form

$$f(r) = c_1(c_2r^{q/p-1}+1)^{q/(p-q)},\ c_1>0,\ c_2>0. \tag{3.92}$$

Proof of Theorem 3.2.6. First we prove inequality (3.88) for $|\cdot|$-radial functions $f(x) = \widetilde{f}(|x|)$. Then we have $g(r) = \widetilde{f}(r)$ since

$$g(|x|) = \mathcal{M}(f)(|x|) := \frac{1}{|\wp|}\int_{\wp} f(|x|, y)d\sigma(y),$$

and we also have $\mathbb{E}f(x) = |x|\widetilde{f}'(|x|)$. We calculate

$$\int_{\mathbb{G}} |\mathbb{E}\widetilde{f}(|x|)|^2dx - \beta(Q-\beta)\int_{\mathbb{G}} |\widetilde{f}(|x|)|^2dx = |\wp|\left(\int_0^\infty |\widetilde{f}'(r)|^2r^{Q+1}dr - \beta(Q-\beta)\int_0^\infty |\widetilde{f}(r)|^2r^{Q-1}dr\right), \tag{3.93}$$

where $0 \leq \beta < Q/2$. Using the notation $h(|x|) := |x|^{\beta}\widetilde{f}(|x|)$, and integrating by parts we obtain

$$\begin{aligned}
&\int_0^\infty |h'(r)|^2 r^{Q+1-2\beta} dr \\
&= \int_0^\infty |\beta r^{\beta-1}\widetilde{f}(r) + r^{\beta}\widetilde{f}'(r)|^2 r^{Q+1-2\beta} dr \\
&= \beta^2 \int_0^\infty |\widetilde{f}(r)|^2 r^{Q-1} dr + \int_0^\infty |\widetilde{f}'(r)|^2 r^{Q+1} dr + \beta \int_0^\infty \frac{d}{dr}|\widetilde{f}(r)|^2 r^{Q} dr \\
&= \int_0^\infty |\widetilde{f}'(r)|^2 r^{Q+1} dr - \beta(Q-\beta) \int_0^\infty |\widetilde{f}(r)|^2 r^{Q-1} dr.
\end{aligned} \tag{3.94}$$

Moreover, by changing the variables $s = r^{Q-2\beta}$ we get

$$\begin{aligned}
\int_0^\infty |h'(r)|^2 r^{Q+1-2\beta} dr &= \int_0^\infty |(Q-2\beta) s^{\frac{Q-2\beta-1}{Q-2\beta}} h'(s)|^2 s^{\frac{Q-2\beta+1}{Q-2\beta}} \frac{s^{\frac{1}{Q-2\beta}-1} ds}{Q-2\beta} \\
&= (Q-2\beta) \int_0^\infty s^2 |h'(s)|^2 ds.
\end{aligned} \tag{3.95}$$

Combining (3.94) and (3.95), we restate (3.93) as

$$\int_{\mathbb{G}} |\mathbb{E}\widetilde{f}(|x|)|^2 dx - \beta(Q-\beta) \int_{\mathbb{G}} |\widetilde{f}(|x|)|^2 dx = |\wp|(Q-2\beta) \left(\int_0^\infty s^2 |h'(s)|^2 ds \right). \tag{3.96}$$

Now setting $\phi(s) := h'(s)$ and $\psi(s) := s^{-2}\phi(s^{-1})$, and using (3.91) with $p = 2$ and $q = 2^*$, we obtain

$$\begin{aligned}
&\int_0^\infty s^2 |h'(s)|^2 ds = \int_0^\infty s^2 |\phi(s)|^2 ds = \int_0^\infty |\psi(s)|^2 ds \\
&\geq \left(\frac{Q}{Q-2}\right)^{\frac{Q-2}{Q}} \left(\frac{\Gamma(Q/2)\Gamma(1+Q/2)}{\Gamma(Q)}\right)^{\frac{2}{Q}} \left(\int_0^\infty \left| \int_0^s |\psi(t)| dt \right|^{2^*} s^{\frac{2-2Q}{Q-2}} ds \right)^{\frac{2}{2^*}} \\
&= \left(\frac{Q}{Q-2}\right)^{\frac{Q-2}{Q}} \left(\frac{\Gamma(Q/2)\Gamma(1+Q/2)}{\Gamma(Q)}\right)^{\frac{2}{Q}} \left(\int_0^\infty \left| \int_{s^{-1}}^\infty |\phi(t)| dt \right|^{2^*} s^{\frac{2-2Q}{Q-2}} ds \right)^{\frac{2}{2^*}} \\
&\geq \left(\frac{Q}{Q-2}\right)^{\frac{Q-2}{Q}} \left(\frac{\Gamma(Q/2)\Gamma(1+Q/2)}{\Gamma(Q)}\right)^{\frac{2}{Q}} \left(\int_0^\infty |h(s^{-1})|^{2^*} s^{\frac{2-2Q}{Q-2}} ds \right)^{\frac{2}{2^*}} \\
&= \left(\frac{Q}{Q-2}\right)^{\frac{Q-2}{Q}} \left(\frac{\Gamma(Q/2)\Gamma(1+Q/2)}{\Gamma(Q)}\right)^{\frac{2}{Q}} \left(\int_0^\infty |h(s)|^{2^*} s^{\frac{2}{Q-2}} ds \right)^{\frac{2}{2^*}} \\
&= (Q-2\beta)^{\frac{2}{2^*}} \left(\frac{Q}{Q-2}\right)^{\frac{Q-2}{Q}} \left(\frac{\Gamma(Q/2)\Gamma(1+Q/2)}{\Gamma(Q)}\right)^{\frac{2}{Q}} \left(\int_0^\infty |r\widetilde{f}(r)|^{2^*} r^{Q-1} dr \right)^{\frac{2}{2^*}},
\end{aligned}$$

where we have used $s = r^{Q-2\beta}$ and $h(r) = r^{\beta}\widetilde{f}(r)$ in the last line. Thus, now (3.96) implies that

$$\begin{aligned}
&\int_{\mathbb{G}}|\mathbb{E}\widetilde{f}(|x|)|^2dx-\beta(Q-\beta)\int_{\mathbb{G}}|\widetilde{f}(|x|)|^2dx\\
&\geq|\wp|(Q-2\beta)^{\frac{2Q-2}{Q}}\left(\frac{Q}{Q-2}\right)^{\frac{Q-2}{Q}}\left(\frac{\Gamma(Q/2)\Gamma(1+Q/2)}{\Gamma(Q)}\right)^{\frac{2}{Q}}\left(\int_0^{\infty}|r\widetilde{f}(r)|^{2^*}r^{Q-1}dr\right)^{\frac{2}{2^*}}\\
&=|\wp|^{\frac{2}{Q}}(Q-2\beta)^{\frac{2Q-2}{Q}}\left(\frac{Q}{Q-2}\right)^{\frac{Q-2}{Q}}\left(\frac{\Gamma(Q/2)\Gamma(1+Q/2)}{\Gamma(Q)}\right)^{\frac{2}{Q}}\left(\int_{\mathbb{G}}|x|^{2^*}|\widetilde{f}(|x|)|^{2^*}dx\right)^{\frac{2}{2^*}}.
\end{aligned} \tag{3.97}$$

Here, denoting $\beta = (Q-\sqrt{Q^2-4\delta})/2$ for $0 \leq \delta < Q^2/4$ and recalling that $g(|x|) = \widetilde{f}(|x|)$, we see that (3.97) (with (3.89)) yields

$$\begin{aligned}
&\int_{\mathbb{G}}|\mathbb{E}\widetilde{f}(|x|)|^2dx-\delta\int_{\mathbb{G}}|\widetilde{f}(|x|)|^2dx\\
&\geq|\wp|^{\frac{2}{Q}}(Q^2-4\delta)^{\frac{Q-1}{Q}}\left(\frac{Q}{Q-2}\right)^{\frac{Q-2}{Q}}\left(\frac{\Gamma(Q/2)\Gamma(1+Q/2)}{\Gamma(Q)}\right)^{\frac{2}{Q}}\left(\int_{\mathbb{G}}|x|^{2^*}|g(|x|)|^{2^*}dx\right)^{\frac{2}{2^*}}\\
&=\left(\frac{Q^2-4\delta}{(Q-2)^2}\right)^{\frac{Q-1}{Q}}S_Q\left(\int_{\mathbb{G}}|x|^{2^*}|g(|x|)|^{2^*}dx\right)^{\frac{2}{2^*}}.
\end{aligned} \tag{3.98}$$

That is, we obtain (3.88) with sharp constant for all $|\cdot|$-radial functions $f \in C_0^{\infty}(\mathbb{G})$.

Finally, using Proposition 1.3.3, and in (3.98) with $g(|x|) = \widetilde{f}(|x|)$, we get (3.88) for non-radial functions. Clearly, the constant in (3.88) is sharp, since this constant is sharp for radial functions by Remark 3.2.7, Part (3). □

3.3 Caffarelli–Kohn–Nirenberg inequalities

This section is devoted to deriving the Caffarelli–Kohn–Nirenberg inequalities in the setting of homogeneous groups. Here we will be working with the radial operators and general quasi-norms. The case of stratified groups with the horizontal gradient and weights will be discussed in Section 6.7.

First, we recall the classical Caffarelli–Kohn–Nirenberg inequalities on $\mathbb{R}^n$ due to Caffarelli, Kohn and Nirenberg [CKN84], with $|\cdot|$ denoting the usual Euclidean distance:

Theorem 3.3.1 (Classical Caffarelli–Kohn–Nirenberg inequality). *Let $n \in \mathbb{N}$ and let p, q, r, a, b, d, $\delta \in \mathbb{R}$ be such that $p, q \geq 1$, $r > 0$, $0 \leq \delta \leq 1$, and*

$$\frac{1}{p}+\frac{a}{n},\ \frac{1}{q}+\frac{b}{n},\ \frac{1}{r}+\frac{c}{n}>0, \tag{3.99}$$

where

$$c = \delta d + (1-\delta)b.$$

Then there exists a positive constant C such that

$$\| |x|^c f \|_{L^r(\mathbb{R}^n)} \leq C \| |x|^a |\nabla f| \|^{\delta}_{L^p(\mathbb{R}^n)} \| |x|^b f \|^{1-\delta}_{L^q(\mathbb{R}^n)} \tag{3.100}$$

holds for all $f \in C_0^\infty(\mathbb{R}^n)$, if and only if the following conditions hold:

$$\frac{1}{r} + \frac{c}{n} = \delta\left(\frac{1}{p} + \frac{a-1}{n}\right) + (1-\delta)\left(\frac{1}{q} + \frac{b}{n}\right), \tag{3.101}$$

$$a - d \geq 0 \quad \textit{if} \quad \delta > 0, \tag{3.102}$$

$$a - d \leq 1 \quad \text{if} \quad \delta > 0 \quad \textit{and} \quad \frac{1}{r} + \frac{c}{n} = \frac{1}{p} + \frac{a-1}{n}. \tag{3.103}$$

Thus, in this section we are interested in inequalities of this type, and we show that some Caffarelli–Kohn–Nirenberg inequalities continue to hold in the setting of homogeneous groups, in particular, including the cases of anisotropic structures on $\mathbb{R}^n$, i.e., the quasi-norm $|\cdot|$ does not need to be the Euclidean norm $|\cdot|_E$ on $\mathbb{R}^n$.

Some inequalities will be obtained as a consequence of the weighted Hardy inequalities. As a particular case of such weighted inequalities we can think of the inequalities

$$\left(\int_{\mathbb{R}^n} |x|_E^{-p\beta} |f|^p dx\right)^{\frac{2}{p}} \leq C_{\alpha,\beta} \int_{\mathbb{R}^n} |x|_E^{-2\alpha} |\nabla f|^2 dx, \tag{3.104}$$

for $f \in C_0^\infty(\mathbb{R}^n)$, where for $n \geq 3$:

$$-\infty < \alpha < \frac{n-2}{2}, \ \alpha \leq \beta \leq \alpha + 1, \text{ and } p = \frac{2n}{n-2+2(\beta-\alpha)},$$

and for $n = 2$:

$$-\infty < \alpha < 0, \ \alpha < \beta \leq \alpha + 1, \text{ and } p = \frac{2}{\beta - \alpha}.$$

Here

$$|x|_E = \sqrt{x_1^2 + \cdots + x_n^2}$$

is the standard Euclidean norm. Moreover, we are interested in replacing the Euclidean norm by a general quasi-norm as well as extending such inequalities to general homogeneous groups.

As a special case we can highlight the case of $p = 2$ that was also studied by [WW03] in the Euclidean setting. Here, for all $f \in C_0^\infty(\mathbb{R}^n)$ we have

$$\int_{\mathbb{R}^n} |x|_E^{-2(\alpha+1)} |f|^2 dx \leq \widetilde{C}_\alpha \int_{\mathbb{R}^n} |x|_E^{-2\alpha} |\nabla f|^2 dx,$$

with any $n \geq 2$ and $-\infty < \alpha < 0$, which in turn can be written for any $f \in C_0^\infty(\mathbb{R}^n \backslash \{0\})$ as

$$\left\| \frac{1}{|x|_E^{\alpha+1}} |f| \right\|_{L^2(\mathbb{R}^n)} \leq C_\alpha \left\| \frac{1}{|x|_E^{\alpha}} |\nabla f| \right\|_{L^2(\mathbb{R}^n)}, \tag{3.105}$$

for all $\alpha \in \mathbb{R}$.

A homogeneous group version of the inequality (3.105) was obtained in Corollary 2.1.6, that is, it was proved that if $\mathbb{G}$ is a homogeneous group of homogeneous dimension Q, then for any homogeneous quasi-norm $|\cdot|$ on $\mathbb{G}$ and for every $f \in C_0^\infty(\mathbb{G}\backslash\{0\})$ we have

$$\frac{|Q-2-2\alpha|}{2} \left\| \frac{f}{|x|^{\alpha+1}} \right\|_{L^2(\mathbb{G})} \leq \left\| \frac{1}{|x|^\alpha} \mathcal{R} f \right\|_{L^2(\mathbb{G})} \quad \text{for all } \alpha \in \mathbb{R}, \tag{3.106}$$

where $\mathcal{R}$ is the radial derivative operator with respect to the norm $|\cdot|$. Note that if $\alpha \neq \frac{Q-2}{2}$, then the constant in (3.106) was shown to be sharp for any homogeneous quasi-norm $|\cdot|$ on $\mathbb{G}$.

Remark 3.3.2.

1. An alternative formulation of Theorem 3.3.1 emphasizing the appearing indices was given by D'Ancona and Luca [DL12].
2. The improved versions of the Caffarelli–Kohn–Nirenberg inequality for radially symmetric functions with respect to the range of parameters were investigated in [NDD12]. In [ZHD15] and [HZ11], weighted Hardy type inequalities were obtained for the generalized Baouendi–Grushin vector fields: for $\gamma = 0$ it gives the standard gradient in $\mathbb{R}^n$. We also refer to [HNZ11], [Han15] for weighted Hardy inequalities on the Heisenberg group, to [HZD11] and [ZHD14] on the H-type groups, and a recent paper [Yac18] on Lie groups of polynomial growth.
3. The analysis in this section is based on [ORS18] as well as on [RSY17b] and [RSY18b].

3.3.1 L^p-Caffarelli–Kohn–Nirenberg inequalities

In this section we generalize inequality (3.106) to L^p-cases for all $1 < p < \infty$. Since all the inequalities are of similar type we will keep calling them the Caffarelli–Kohn–Nirenberg inequalities.

Theorem 3.3.3 (Caffarelli–Kohn–Nirenberg inequality for L^p-norms)**.** *Let $\mathbb{G}$ be a homogeneous group of homogeneous dimension $Q \geq 2$ and let $|\cdot|$ be a homogeneous quasi-norm on $\mathbb{G}$. Then we have*

$$\frac{|Q-\gamma|}{p} \left\| \frac{f}{|x|^{\frac{\gamma}{p}}} \right\|_{L^p(\mathbb{G})}^p \leq \left\| \frac{1}{|x|^\alpha} \mathcal{R} f \right\|_{L^p(\mathbb{G})} \left\| \frac{f}{|x|^{\frac{\beta}{p-1}}} \right\|_{L^p(\mathbb{G})}^{p-1}, \tag{3.107}$$

for all complex-valued functions $f \in C_0^\infty(\mathbb{G}\backslash\{0\})$, $1 < p < \infty$, *and all* $\alpha, \beta \in \mathbb{R}$ *with*

$$\gamma = \alpha + \beta + 1.$$

If $\gamma \neq Q$ *then the constant* $\frac{|Q-\gamma|}{p}$ *is sharp.*

Before proving this theorem let us point out some of its implications.

Remark 3.3.4.

1. In the Euclidean case $\mathbb{G} = (\mathbb{R}^n, +)$, Theorem 3.3.3 gives the inequality

$$\frac{|n-\gamma|}{p}\left\|\frac{f}{|x|^{\frac{\gamma}{p}}}\right\|_{L^p(\mathbb{R}^n)}^p \leq \left\|\frac{1}{|x|^\alpha}\frac{df}{d|x|}\right\|_{L^p(\mathbb{R}^n)}\left\|\frac{f}{|x|^{\frac{\beta}{p-1}}}\right\|_{L^p(\mathbb{R}^n)}^{p-1},$$

with the optimal constant. In particular, for the standard Euclidean distance $|x|_E = \sqrt{x_1^2 + \cdots + x_n^2}$, by using the Cauchy–Schwarz inequality, it follows that

$$\frac{|n-\gamma|}{p}\left\|\frac{f}{|x|_E^{\frac{\gamma}{p}}}\right\|_{L^p(\mathbb{R}^n)}^p \leq \left\|\frac{1}{|x|_E^\alpha}\nabla f\right\|_{L^p(\mathbb{R}^n)}\left\|\frac{f}{|x|_E^{\frac{\beta}{p-1}}}\right\|_{L^p(\mathbb{R}^n)}^{p-1},$$

for all $f \in C_0^\infty(\mathbb{R}^n\backslash\{0\})$, with the sharp constant.

2. In the case $\alpha = 0$, $\beta = p-1$, and $1 < p < Q$, the inequality (3.107) implies the homogeneous group version of the L^p-Hardy inequality

$$\left\|\frac{1}{|x|}f\right\|_{L^p(\mathbb{G})} \leq \frac{p}{Q-p}\|\mathcal{R}f\|_{L^p(\mathbb{G})}, \tag{3.108}$$

again with $\frac{p}{Q-p}$ being the best constant, see Section 2.1.1.

3. For $\mathbb{G} = (\mathbb{R}^n, +)$, $n \geq 3$, inequality (3.108) gives

$$\left\|\frac{f}{|x|}\right\|_{L^p(\mathbb{R}^n)} \leq \frac{p}{n-p}\left\|\frac{df}{d|x|}\right\|_{L^p(\mathbb{R}^n)}. \tag{3.109}$$

For the Euclidean distance, by the Cauchy–Schwarz inequality, it implies the classical Hardy inequality:

$$\left\|\frac{f}{|x|_E}\right\|_{L^p(\mathbb{R}^n)} \leq \frac{p}{n-p}\|\nabla f\|_{L^p(\mathbb{R}^n)},$$

for all $f \in C_0^\infty(\mathbb{R}^n\backslash\{0\})$.

4. For the Euclidean distance, the exact formulae of the difference between the right-hand side and the left-hand side of inequality (3.109) were investigated by Ioku, Ishiwata and Ozawa [IIO16b], see also Machihara, Ozawa and Wadade [MOW17a] as well as [IIO16a].

5. One can obtain a number of Heisenberg–Pauli–Weyl type uncertainty inequities directly from the inequality (3.107) which have various consequences and applications. For instance, if $\alpha p = \alpha + \beta + 1$, inequality (3.107) gives

$$\frac{|Q-\alpha p|}{p}\left\|\frac{f}{|x|^{\alpha}}\right\|^{p}_{L^p(\mathbb{G})} \leq \left\|\frac{\mathcal{R}f}{|x|^{\alpha}}\right\|_{L^p(\mathbb{G})}\left\||x|^{\frac{1}{p-1}}\frac{f}{|x|^{\alpha}}\right\|^{p-1}_{L^p(\mathbb{G})}. \tag{3.110}$$

For $\alpha+\beta+1=0$ and $\alpha=-p$, inequality (3.107) gives

$$\frac{Q}{p}\|f\|^{p}_{L^p(\mathbb{G})} \leq \||x|^p\mathcal{R}f\|_{L^p(\mathbb{G})}\left\|\frac{f}{|x|}\right\|^{p-1}_{L^p(\mathbb{G})}, \tag{3.111}$$

all with sharp constants.

The Hardy inequality immediately implies a version of the Heisenberg–Pauli–Weyl uncertainty principle.

Corollary 3.3.5 (Heisenberg–Pauli–Weyl type uncertainty principle). *Let $\mathbb{G}$ be a homogeneous group of homogeneous dimension $Q \geq 2$ and let $|\cdot|$ be a homogeneous quasi-norm on $\mathbb{G}$. Then we have*

$$\|f\|^2_{L^2(\mathbb{G})} \leq \frac{p}{Q-p}\|\mathcal{R}f\|_{L^p(\mathbb{G})}\||x|f\|_{L^{\frac{p}{p-1}}(\mathbb{G})}, \quad 1<p<Q, \tag{3.112}$$

for all $f \in C_0^\infty(\mathbb{G}\backslash\{0\})$.

Proof of Corollary 3.3.5. By applying the Hölder inequality to (3.108) with $1 < p < Q$, we get

$$\|f\|^2_{L^2(\mathbb{G})} \leq \left\|\frac{1}{|x|}f\right\|_{L^p(\mathbb{G})}\||x|f\|_{L^{\frac{p}{p-1}}(\mathbb{G})} \leq \frac{p}{Q-p}\|\mathcal{R}f\|_{L^p(\mathbb{G})}\||x|f\|_{L^{\frac{p}{p-1}}(\mathbb{G})}, \tag{3.113}$$

giving (3.112). □

Remark 3.3.6.

1. For the Euclidean space $\mathbb{G}=(\mathbb{R}^n,+)$ and $p=2$, inequality (3.112) implies the uncertainty principle for any homogeneous quasi-norm $|x|$ on $\mathbb{R}^n$:

$$\left(\int_{\mathbb{R}^n}|f(x)|^2dx\right)^2 \leq \left(\frac{2}{n-2}\right)^2\int_{\mathbb{R}^n}\left|\frac{df(x)}{d|x|}\right|^2dx\int_{\mathbb{R}^n}|x|^2|f(x)|^2dx. \tag{3.114}$$

In turn this gives the classical uncertainty principle on $\mathbb{R}^n$ with the standard Euclidean distance:

$$\left(\int_{\mathbb{R}^n}|f(x)|^2dx\right)^2 \leq \left(\frac{2}{n-2}\right)^2\int_{\mathbb{R}^n}|\nabla f(x)|^2dx\int_{\mathbb{R}^n}|x|_E^2|f(x)|^2dx,$$

which is the Heisenberg–Pauli–Weyl uncertainty principle on the Euclidean spaces $\mathbb{R}^n$, $n>2$.

2. The stratified groups version of Corollary 3.3.5 will be discussed in Section 4.7.

Proof of Theorem 3.3.3. In the case $\gamma = Q$ the inequality is trivial, so we may assume that $\gamma \neq Q$. To use the polar decomposition in Proposition 1.2.10 we denote $(r, y) = (|x|, \frac{x}{|x|}) \in (0, \infty) \times \wp$ on $\mathbb{G}$, where $\wp$ is the unit quasi-sphere. Then a direct calculation gives that

$$\begin{aligned}
\int_{\mathbb{G}} \frac{|f(x)|^p}{|x|^\gamma} dx &= \int_0^\infty \int_\wp \frac{|f(ry)|^p}{r^\gamma} r^{Q-1} d\sigma(y) dr \\
&= \frac{1}{Q-\gamma} \int_0^\infty \int_\wp |f(ry)|^p \frac{d\, r^{Q-\gamma}}{dr} d\sigma(y) dr \\
&= -\frac{1}{Q-\gamma} \mathrm{Re} \int_0^\infty \int_\wp p f(ry) |f(ry)|^{p-2} \left(\overline{\frac{df(ry)}{dr}} \right) \frac{1}{r^{\gamma-1}} r^{Q-1} d\sigma(y) dr \\
&= -\frac{p}{Q-\gamma} \mathrm{Re} \int_{\mathbb{G}} f(x) \frac{|f(x)|^{p-2}}{|x|^{\gamma-1}} \left(\overline{\frac{d}{d|x|} f(x)} \right) dx \\
&\leq \left| \frac{p}{Q-\gamma} \right| \int_{\mathbb{G}} \frac{|f(x)|^{p-1}}{|x|^{\gamma-1}} \left| \frac{d}{d|x|} f(x) \right| dx \\
&= \left| \frac{p}{Q-\gamma} \right| \int_{\mathbb{G}} \frac{|f(x)|^{p-1}}{|x|^{\alpha+\beta}} |\mathcal{R} f(x)| \, dx \\
&\leq \left| \frac{p}{Q-\gamma} \right| \left(\int_{\mathbb{G}} \frac{|\mathcal{R} f(x)|^p}{|x|^{\alpha p}} dx \right)^{1/p} \left(\int_{\mathbb{G}} \frac{|f(x)|^p}{|x|^{\frac{\beta p}{p-1}}} dx \right)^{(p-1)/p},
\end{aligned}$$

using the Hölder inequality in the last line. Thus, we obtain

$$\left| \frac{Q-\gamma}{p} \right| \int_{\mathbb{G}} \frac{|f(x)|^p}{|x|^\gamma} dx \leq \left(\int_{\mathbb{G}} \frac{|\mathcal{R} f(x)|^p}{|x|^{\alpha p}} dx \right)^{1/p} \left(\int_{\mathbb{G}} \frac{|f(x)|^p}{|x|^{\frac{\beta p}{p-1}}} dx \right)^{(p-1)/p},$$

yielding (3.107).

Now it remains to show the sharpness of the constant. To do it we have to examine the equality condition in the Hölder inequality. Let us consider the following function

$$g(x) = \begin{cases} e^{-\frac{C}{\lambda}|x|^\lambda}, & \lambda := \alpha - \frac{\beta}{p-1} + 1 \neq 0, \\ \frac{1}{|x|^C}, & \alpha - \frac{\beta}{p-1} + 1 = 0, \end{cases}$$

where $C = \left| \frac{Q-\gamma}{p} \right|$ and $\gamma \neq Q$. Then one can readily check that

$$\left| \frac{p}{Q-\gamma} \right|^p \frac{|\mathcal{R} g(x)|^p}{|x|^{\alpha p}} = \frac{|g(x)|^p}{|x|^{\frac{\beta p}{p-1}}},$$

satisfying the equality condition in the Hölder inequality. This means that the constant $|(Q-\gamma)/p|$ is sharp. □

3.3.2 Higher-order L^p-Caffarelli–Kohn–Nirenberg inequalities

By iterating the L^p-Caffarelli–Kohn–Nirenberg type inequalities from Theorem 3.3.3 we can establish higher-order inequalities.

Theorem 3.3.7 (Higher-order L^p-Caffarelli–Kohn–Nirenberg inequalities). *Let $\mathbb{G}$ be a homogeneous group of homogeneous dimension $Q \geq 2$ and let $|\cdot|$ be a homogeneous quasi-norm on $\mathbb{G}$. Then for all $k, m \in \mathbb{N}$ and all $1 < p < \infty$ we have*

$$\frac{|Q-\gamma|}{p}\left\|\frac{f}{|x|^{\frac{\gamma}{p}}}\right\|^p_{L^p(\mathbb{G})} \leq \widetilde{A}_{\alpha,m}\widetilde{A}_{\beta,k}\left\|\frac{1}{|x|^{\alpha-m}}\mathcal{R}^{m+1}f\right\|_{L^p(\mathbb{G})}\left\|\frac{1}{|x|^{\frac{\beta}{p-1}-k}}\mathcal{R}^k f\right\|^{p-1}_{L^p(\mathbb{G})}, \tag{3.115}$$

for all complex-valued function $f \in C_0^\infty(\mathbb{G}\backslash\{0\})$,

$$\gamma = \alpha + \beta + 1,$$

and $\alpha \in \mathbb{R}$ such that $\prod_{j=0}^{m-1}|Q - p(\alpha - j)| \neq 0$, and

$$\widetilde{A}_{\alpha,m} := p^m\left[\prod_{j=0}^{m-1}|Q-p(\alpha-j)|\right]^{-1},$$

as well as $\beta \in \mathbb{R}$ such that $\prod_{j=0}^{k-1}\left|Q - p(\frac{\beta}{p-1} - j)\right| \neq 0$, and

$$\widetilde{A}_{\beta,k} := p^{k(p-1)}\left[\prod_{j=0}^{k-1}\left|Q-p\left(\frac{\beta}{p-1}-j\right)\right|\right]^{-(p-1)}.$$

For $p = 2$ the above constants are sharp.

Proof of Theorem 3.3.7. First, let us consider in (3.107) the case

$$\beta = \gamma\left(1 - \frac{1}{p}\right).$$

In this case we have $\beta = (\alpha+1)(p-1)$ and $\gamma = p(\alpha+1)$, so that inequality (3.107) becomes

$$\left\|\frac{f}{|x|^{\alpha+1}}\right\|_{L^p(\mathbb{G})} \leq \frac{p}{|Q-p(\alpha+1)|}\left\|\frac{1}{|x|^\alpha}\mathcal{R}f\right\|_{L^p(\mathbb{G})}, \quad 1 < p < \infty, \tag{3.116}$$

for all $f \in C_0^\infty(\mathbb{G}\backslash\{0\})$ and every $\alpha \in \mathbb{R}$ with $\alpha \neq \frac{Q}{p} - 1$.

Now taking $\mathcal{R}f$ instead of f and $\alpha-1$ instead of α in (3.116) we consequently obtain

$$\left\|\frac{\mathcal{R}f}{|x|^{\alpha}}\right\|_{L^p(\mathbb{G})} \leq \frac{p}{|Q-p\alpha|}\left\|\frac{1}{|x|^{\alpha-1}}\mathcal{R}^2 f\right\|_{L^p(\mathbb{G})},$$

for $\alpha \neq \frac{Q}{p}$. Combining it with (3.116) we get

$$\left\|\frac{f}{|x|^{\alpha+1}}\right\|_{L^p(\mathbb{G})} \leq \frac{p}{|Q-p(\alpha+1)|}\frac{p}{|Q-p\alpha|}\left\|\frac{1}{|x|^{\alpha-1}}\mathcal{R}^2 f\right\|_{L^p(\mathbb{G})},$$

for every $\alpha \in \mathbb{R}$ such that $\alpha \neq \frac{Q}{p}-1$ and $\alpha \neq \frac{Q}{p}$. This iteration process yields

$$\left\|\frac{f}{|x|^{\theta+1}}\right\|_{L^p(\mathbb{G})} \leq A_{\theta,k}\left\|\frac{1}{|x|^{\theta+1-k}}\mathcal{R}^k f\right\|_{L^p(\mathbb{G})}, \quad 1<p<\infty, \tag{3.117}$$

for all $f \in C_0^\infty(\mathbb{G}\backslash\{0\})$ and all $\theta \in \mathbb{R}$ such that $\prod_{j=0}^{k-1}|Q-p(\vartheta+1-j)| \neq 0$, and

$$A_{\theta,k} := p^k\left[\prod_{j=0}^{k-1}|Q-p(\theta+1-j)|\right]^{-1}.$$

Similarly, we get

$$\left\|\frac{\mathcal{R}f}{|x|^{\vartheta+1}}\right\|_{L^p(\mathbb{G})} \leq A_{\vartheta,m}\left\|\frac{1}{|x|^{\vartheta+1-m}}\mathcal{R}^{m+1} f\right\|_{L^p(\mathbb{G})}, \quad 1<p<\infty, \tag{3.118}$$

for all $f \in C_0^\infty(\mathbb{G}\backslash\{0\})$ and all $\vartheta \in \mathbb{R}$ such that $\prod_{j=0}^{m-1}|Q-p(\vartheta+1-j)| \neq 0$, and

$$A_{\vartheta,m} := p^m\left[\prod_{j=0}^{m-1}|Q-p(\vartheta+1-j)|\right]^{-1}.$$

Now putting $\vartheta+1=\alpha$ and $\theta+1=\frac{\beta}{p-1}$ into (3.118) and (3.117), respectively, we arrive at (3.115).

Let us now show the sharpness of the constants in the case $p=2$. This will follow from having an exact form of the remainder in these inequalities. Recall Theorem 3.1.10 saying that if $Q \geq 3$, $\alpha \in \mathbb{R}$, $k \in \mathbb{N}$ and $\prod_{j=0}^{k-1}\left|\frac{Q-2}{2}-(\alpha+j)\right| \neq 0$, then for all complex-valued functions $f \in C_0^\infty(\mathbb{G}\backslash\{0\})$ we have

$$\left\|\frac{f}{|x|^{k+\alpha}}\right\|_{L^2(\mathbb{G})} \leq \left[\prod_{j=0}^{k-1}\left|\frac{Q-2}{2}-(\alpha+j)\right|\right]^{-1}\left\|\frac{1}{|x|^{\alpha}}\mathcal{R}^k f\right\|_{L^2(\mathbb{G})}, \tag{3.119}$$

where the constant is sharp. In addition, from (3.34) we have the identity

$$\begin{aligned}\left\|\frac{1}{|x|^{\alpha}}\mathcal{R}^{k}f\right\|^{2}_{L^{2}(\mathbb{G})} &= \left[\prod_{j=0}^{k-1}\left(\frac{Q-2}{2}-(\alpha+j)\right)^{2}\right]\left\|\frac{f}{|x|^{k+\alpha}}\right\|^{2}_{L^{2}(\mathbb{G})} \\ &+\sum_{l=1}^{k-1}\left[\prod_{j=0}^{l-1}\left(\frac{Q-2}{2}-(\alpha+j)\right)^{2}\right] \\ &\times\left\|\frac{1}{|x|^{l+\alpha}}\mathcal{R}^{k-l}f+\frac{Q-2(l+1+\alpha)}{2|x|^{l+1+\alpha}}\mathcal{R}^{k-l-1}f\right\|^{2}_{L^{2}(\mathbb{G})} \\ &+\left\|\frac{1}{|x|^{\alpha}}\mathcal{R}^{k}f+\frac{Q-2-2\alpha}{2|x|^{1+\alpha}}\mathcal{R}^{k-1}f\right\|^{2}_{L^{2}(\mathbb{G})},\end{aligned} \tag{3.120}$$

for all $k\in\mathbb{N}$ and $\alpha\in\mathbb{R}$. When $p=2$, Theorem 3.3.3 can be restated that for each $f\in C_0^{\infty}(\mathbb{G}\backslash\{0\})$ we have

$$\frac{|Q-\gamma|}{2}\left\|\frac{f}{|x|^{\frac{\gamma}{2}}}\right\|^{2}_{L^{2}(\mathbb{G})}\leq\left\|\frac{1}{|x|^{\alpha}}\mathcal{R}f\right\|_{L^{2}(\mathbb{G})}\left\|\frac{f}{|x|^{\beta}}\right\|_{L^{2}(\mathbb{G})},\quad\forall\alpha,\,\beta\in\mathbb{R}, \tag{3.121}$$

where $\gamma=\alpha+\beta+1$. The sharpness then follows from the following remark. □

Remark 3.3.8. Combining (3.121) with (3.119) (or with (3.120)), one can obtain a number of inequalities with sharp constants, for example:

$$\frac{|Q-\gamma|}{2}\left\|\frac{f}{|x|^{\frac{\gamma}{2}}}\right\|^{2}_{L^{2}(\mathbb{G})}\leq C_{j}(\beta,k)\left\|\frac{1}{|x|^{\alpha}}\mathcal{R}f\right\|_{L^{2}(\mathbb{G})}\left\|\frac{1}{|x|^{\beta-k}}\mathcal{R}^{k}f\right\|_{L^{2}(\mathbb{G})}, \tag{3.122}$$

for $\gamma=\alpha+\beta+1$ and all $\alpha,\,\beta\in\mathbb{R}$ and $k\in\mathbb{N}$, such that,

$$C_{j}(\beta,k):=\left[\prod_{j=0}^{k-1}\left|\frac{Q-2}{2}-(\beta-k+j)\right|\right]^{-1}\neq0,$$

as well as

$$\frac{|Q-\gamma|}{2}\left\|\frac{f}{|x|^{\frac{\gamma}{2}}}\right\|^{2}_{L^{2}(\mathbb{G})}\leq C_{j}(\alpha,k)\left\|\frac{1}{|x|^{\alpha-k}}\mathcal{R}^{k+1}f\right\|_{L^{2}(\mathbb{G})}\left\|\frac{f}{|x|^{\beta}}\right\|_{L^{2}(\mathbb{G})}, \tag{3.123}$$

for $\gamma=\alpha+\beta+1$ and all $\alpha,\,\beta\in\mathbb{R}$ and $k\in\mathbb{N}$, such that,

$$C_{j}(\alpha,k):=\left[\prod_{j=0}^{k-1}\left|\frac{Q-2}{2}-(\alpha+k+j)\right|\right]^{-1}\neq0.$$

It follows from (3.120) that these constants $C_j(\beta,k)$ and $C_j(\alpha,k)$ in (3.122) and (3.123) are sharp.

3.3.3 New type of L^p-Caffarelli–Kohn–Nirenberg inequalities

In this section, we introduce new Caffarelli–Kohn–Nirenberg type inequalities on homogeneous groups.

Theorem 3.3.9 (New types of L^p-Caffarelli–Kohn–Nirenberg inequalities). *Let $\mathbb{G}$ be a homogeneous group of homogeneous dimension $Q \geq 2$. Let $|\cdot|$ be a homogeneous quasi-norm on $\mathbb{G}$. Let $0 < \delta < 1$. Then for all $f \in C_0^\infty(\mathbb{G}\backslash\{0\})$ we have*

$$\left\| \frac{f}{|x|^{\frac{Q-p}{p}+\delta}} \right\|_{L^p(\mathbb{G})} \leq p^\delta \left\| \frac{\log|x|}{|x|^{\frac{Q-p}{p}}} \mathcal{R} f \right\|^{\delta}_{L^p(\mathbb{G})} \left\| \frac{f}{|x|^{\frac{Q-p}{p}}} \right\|^{1-\delta}_{L^p(\mathbb{G})}, \quad 1 < p < \infty. \tag{3.124}$$

Moreover, we have

$$\left\| \frac{f}{|x|^{\frac{Q-p}{p}+\delta}} \right\|_{L^p(\mathbb{G})} \leq \left\| \frac{\mathcal{R} f}{|x|^{\frac{Q-2p}{p}}} \right\|^{1-\delta}_{L^p(\mathbb{G})} \left\| \frac{f}{|x|^{\frac{Q}{p}}} \right\|^{\delta}_{L^p(\mathbb{G})}, \quad 1 < p < \infty. \tag{3.125}$$

Remark 3.3.10. In the Abelian case $\mathbb{G} = (\mathbb{R}^n, +)$ and $Q = n$, (3.124) implies a new type of the Caffarelli–Kohn–Nirenberg inequality for any quasi-norm on $\mathbb{R}^n$: For any function $f \in C_0^\infty(\mathbb{R}^n\backslash\{0\})$ and any $1 < p < \infty$ we have

$$\left\| \frac{f}{|x|^{\frac{n-p}{p}+\delta}} \right\|_{L^p(\mathbb{R}^n)} \leq p^\delta \left\| \frac{\log|x|}{|x|^{\frac{n-p}{p}}} \left(\frac{x}{|x|} \cdot \nabla f \right) \right\|^{\delta}_{L^p(\mathbb{R}^n)} \left\| \frac{f}{|x|^{\frac{n-p}{p}}} \right\|^{1-\delta}_{L^p(\mathbb{R}^n)}. \tag{3.126}$$

By the Schwarz inequality with the standard Euclidean distance given by $|x|_E = \sqrt{x_1^2 + x_2^2 + \cdots + x_n^2}$, we obtain the Euclidean form of the Caffarelli–Kohn–Nirenberg type inequality for any quasi-norm on $\mathbb{R}^n$, for $1 < p < \infty$, and for all functions $f \in C_0^\infty(\mathbb{R}^n\backslash\{0\})$:

$$\left\| \frac{f}{|x|_E^{\frac{n-p}{p}+\delta}} \right\|_{L^p(\mathbb{R}^n)} \leq p^\delta \left\| \frac{\log|x|_E}{|x|_E^{\frac{n-p}{p}}} \nabla f \right\|^{\delta}_{L^p(\mathbb{R}^n)} \left\| \frac{f}{|x|_E^{\frac{n-p}{p}}} \right\|^{1-\delta}_{L^p(\mathbb{R}^n)}, \tag{3.127}$$

where ∇ is the standard gradient in $\mathbb{R}^n$. Similarly, we can write the inequality (3.125) in the Euclidean case as

$$\left\| \frac{f}{|x|_E^{\frac{n-p}{p}+\delta}} \right\|_{L^p(\mathbb{R}^n)} \leq \left\| \frac{\nabla f}{|x|_E^{\frac{n-2p}{p}}} \right\|^{1-\delta}_{L^p(\mathbb{R}^n)} \left\| \frac{f}{|x|_E^{\frac{n}{p}}} \right\|^{\delta}_{L^p(\mathbb{R}^n)}, \quad 1 < p < \infty. \tag{3.128}$$

Note that since $\frac{1}{p} + (-\frac{n}{p})\frac{1}{n} = 0$, the inequality (3.128) does not follow from the Caffarelli–Kohn–Nirenberg inequality in Theorem 3.3.3, thus providing an extension of (3.107) in terms of indices but also in terms of a possibility of choosing any homogeneous quasi-norm on $\mathbb{R}^n$.

Proof of Theorem 3.3.9. A direct calculation shows that we have

$$\left\|\frac{f}{|x|^{\frac{Q-p}{p}+\delta}}\right\|_{L^p(\mathbb{G})}^p=\int_{\mathbb{G}}\frac{|f(x)|^p}{|x|^{Q-p+\delta p}}dx=\int_{\mathbb{G}}\frac{|f(x)|^{\delta p}}{|x|^{\delta Q}}\cdot\frac{|f(x)|^{p(1-\delta)}}{|x|^{(Q-p)(1-\delta)}}dx.$$

Using Hölder's inequality it follows that

$$\left\|\frac{f}{|x|^{\frac{Q-p}{p}+\delta}}\right\|_{L^p(\mathbb{G})}^p\le\left(\int_{\mathbb{G}}\frac{|f(x)|^p}{|x|^Q}dx\right)^\delta\left(\int_{\mathbb{G}}\frac{|f(x)|^p}{|x|^{Q-p}}dx\right)^{1-\delta}. \tag{3.129}$$

By Theorem 3.2.3, we have

$$\int_{\mathbb{G}}\frac{|f(x)|^p}{|x|^Q}dx\le p^p\int_{\mathbb{G}}\frac{(\log|x|)^p}{|x|^Q}|\mathbb{E}f(x)|^p dx,\ \ 1<p<\infty,$$

where $\mathbb{E}=|x|\mathcal{R}$ is the Euler operator. It implies that

$$\int_{\mathbb{G}}\frac{|f(x)|^p}{|x|^Q}dx\le p^p\int_{\mathbb{G}}\frac{(\log|x|)^p}{|x|^{Q-p}}|\mathcal{R}f(x)|^p dx,\ \ 1<p<\infty.$$

Using this in (3.129), one obtains

$$\left\|\frac{f}{|x|^{\frac{Q-p}{p}+\delta}}\right\|_{L^p(\mathbb{G})}^p\le p^{p\delta}\left(\int_{\mathbb{G}}\frac{(\log|x|)^p}{|x|^{Q-p}}|\mathcal{R}f(x)|^p dx\right)^\delta\left(\int_{\mathbb{G}}\frac{|f(x)|^p}{|x|^{Q-p}}dx\right)^{1-\delta},$$

which implies (3.124).

Now let us prove (3.125). Using Theorem 3.2.3, one has

$$\int_{\mathbb{G}}\frac{|f(x)|^p}{|x|^{Q-p}}dx\le\int_{\mathbb{G}}\frac{|\mathbb{E}f(x)|^p}{|x|^{Q-p}}dx,\ \ 1<p<\infty.$$

Then, using this in (3.130), we obtain

$$\begin{aligned}\left\|\frac{f}{|x|^{\frac{Q-p}{p}+\delta}}\right\|_{L^p(\mathbb{G})}^p&\le\left(\int_{\mathbb{G}}\frac{|f(x)|^p}{|x|^Q}dx\right)^\delta\left(\int_{\mathbb{G}}\frac{|\mathbb{E}f(x)|^p}{|x|^{Q-p}}dx\right)^{1-\delta}\\&=\left(\int_{\mathbb{G}}\frac{|f(x)|^p}{|x|^Q}dx\right)^\delta\left(\int_{\mathbb{G}}\frac{|\mathcal{R}f(x)|^p}{|x|^{Q-2p}}dx\right)^{1-\delta},\end{aligned}$$

which gives (3.125), completing the proof. □

3.3.4 Extended Caffarelli–Kohn–Nirenberg inequalities

In this section, we extend the range of indices for Theorem 3.3.1. Again, we work in the setting of general homogeneous groups: $\mathbb{G}$ is a homogeneous group of homogeneous dimension $Q\ge 1$ and $|\cdot|$ is a homogeneous quasi-norm on $\mathbb{G}$.

Theorem 3.3.11 (Extended Caffarelli–Kohn–Nirenberg inequalities). *Let $1 < p, q < \infty$, $0 < r < \infty$, with $p + q \geq r$. Let $\delta \in [0,1] \cap \left[\frac{r-q}{r}, \frac{p}{r}\right]$ and a, b, $c \in \mathbb{R}$. Assume that*

$$\frac{\delta r}{p} + \frac{(1-\delta)r}{q} = 1 \quad \text{and} \quad c = \delta(a-1) + b(1-\delta).$$

Then for all $f \in C_0^\infty(\mathbb{G}\backslash\{0\})$ we have the following inequalities:

(i) *If $Q \neq p(1-a)$, then we have*

$$\||x|^c f\|_{L^r(\mathbb{G})} \leq \left|\frac{p}{Q - p(1-a)}\right|^\delta \||x|^a \mathcal{R} f\|^\delta_{L^p(\mathbb{G})} \||x|^b f\|^{1-\delta}_{L^q(\mathbb{G})}. \tag{3.130}$$

(ii) *If $Q = p(1-a)$, then we have*

$$\||x|^c f\|_{L^r(\mathbb{G})} \leq p^\delta \||x|^a \log|x| \mathcal{R} f\|^\delta_{L^p(\mathbb{G})} \||x|^b f\|^{1-\delta}_{L^q(\mathbb{G})}. \tag{3.131}$$

The constant in the inequality (3.130) is sharp for $p = q$ with $a - b = 1$ or $p \neq q$ with $p(1-a) + bq \neq 0$. Moreover, the constants in (3.130) and (3.131) are sharp for $\delta = 0$ or $\delta = 1$.

To compare these inequalities with those in Theorem 3.3.1 let us first formulate the isotropic version of Theorem 3.3.11 in the usual setting of $\mathbb{R}^n$, and its further implication in the case of the Euclidean norm.

Corollary 3.3.12. *Let $|\cdot|$ be a homogeneous quasi-norm on $\mathbb{R}^n$, $n \in \mathbb{N}$. Let $1 < p, q < \infty$, $0 < r < \infty$, with $p+q \geq r$, $\delta \in [0,1] \cap \left[\frac{r-q}{r}, \frac{p}{r}\right]$ and a, b, $c \in \mathbb{R}$. Assume that*

$$\frac{\delta r}{p} + \frac{(1-\delta)r}{q} = 1 \quad \text{and} \quad c = \delta(a-1) + b(1-\delta).$$

Then we have the following estimates:

(i) *If $n \neq p(1-a)$, then for any function $f \in C_0^\infty(\mathbb{R}^n\backslash\{0\})$ we have*

$$\||x|^c f\|_{L^r(\mathbb{R}^n)} \leq \left|\frac{p}{n - p(1-a)}\right|^\delta \left\||x|^a \left(\frac{x}{|x|} \cdot \nabla f\right)\right\|^\delta_{L^p(\mathbb{R}^n)} \||x|^b f\|^{1-\delta}_{L^q(\mathbb{R}^n)} \tag{3.132}$$

(ii) *In the critical case $n = p(1-a)$ for any function $f \in C_0^\infty(\mathbb{R}^n\backslash\{0\})$ we have*

$$\||x|^c f\|_{L^r(\mathbb{R}^n)} \leq p^\delta \left\||x|^a \log|x| \left(\frac{x}{|x|} \cdot \nabla f\right)\right\|^\delta_{L^p(\mathbb{R}^n)} \||x|^b f\|^{1-\delta}_{L^q(\mathbb{R}^n)}. \tag{3.133}$$

(iii) *If $|\cdot|_E$ is the Euclidean norm on $\mathbb{R}^n$, inequalities (3.132) and (3.133) imply, respectively,*

$$\||x|_E^c f\|_{L^r(\mathbb{R}^n)} \leq \left|\frac{p}{n - p(1-a)}\right|^\delta \||x|_E^a \nabla f\|^\delta_{L^p(\mathbb{R}^n)} \||x|_E^b f\|^{1-\delta}_{L^q(\mathbb{R}^n)} \tag{3.134}$$

for $n \neq p(1-a)$, *and*

$$\left\||x|_E^c f\right\|_{L^r(\mathbb{R}^n)} \le p^\delta \left\||x|_E^a \log|x| \nabla f\right\|_{L^p(\mathbb{R}^n)}^\delta \left\||x|_E^b f\right\|_{L^q(\mathbb{R}^n)}^{1-\delta}, \tag{3.135}$$

for $n = p(1-a)$.

The inequality (3.132) *holds for any homogeneous quasi-norm* $|\cdot|$, *and the constant* $\left|\frac{p}{n-p(1-a)}\right|^\delta$ *is sharp for* $p = q$ *with* $a - b = 1$, *or for* $p \neq q$ *with* $p(1-a) + bq \neq 0$. *Furthermore, the constants* $\left|\frac{p}{n-p(1-a)}\right|^\delta$ *and* p^δ *are sharp for* $\delta = 0, 1$.

Remark 3.3.13.

1. If the conditions (3.99) on the parameters hold, then the inequality (3.134) is contained in the inequalities in Theorem 3.3.1. However, already in this case, if we require $p = q$ with $a - b = 1$ or $p \neq q$ with $p(1-a) + bq \neq 0$, then (3.134) yields the inequality (3.100) with sharp constant. Moreover, the constants $\left|\frac{p}{n-p(1-a)}\right|^\delta$ and p^δ are sharp for $\delta = 0$ or $\delta = 1$. The conditions

   $$\frac{\delta r}{p} + \frac{(1-\delta)r}{q} = 1 \text{ and } c = \delta(a-1) + b(1-\delta)$$

   imply the condition (3.101) of Theorem 3.3.1, as well as conditions (3.102)–(3.103) which are all necessary for having estimates of this type, at least under the conditions (3.99).
2. If the conditions (3.99) are not satisfied, then the inequality (3.134) is not covered by Theorem 3.3.1. So, this gives an extension of Theorem 3.3.1 with respect to the range of parameters. Indeed, let us take, for example,

   $$1 < p = q = r < \infty,\ a = -\frac{n-2p}{p},\ b = -\frac{n}{p},\ c = -\frac{n-\delta p}{p}.$$

   Then by (3.134), for all $f \in C_0^\infty(\mathbb{R}^n \backslash \{0\})$ we have the inequalities

   $$\left\|\frac{f}{|x|_E^{\frac{n-\delta p}{p}}}\right\|_{L^p(\mathbb{R}^n)} \le \left\|\frac{\nabla f}{|x|_E^{\frac{n-2p}{p}}}\right\|_{L^p(\mathbb{R}^n)}^\delta \left\|\frac{f}{|x|_E^{\frac{n}{p}}}\right\|_{L^p(\mathbb{R}^n)}^{1-\delta}, \tag{3.136}$$

   for all $1 < p < \infty$ and $0 \le \delta \le 1$, where ∇ is the standard gradient in $\mathbb{R}^n$. Since we have

   $$\frac{1}{q} + \frac{b}{n} = \frac{1}{p} + \frac{1}{n}\left(-\frac{n}{p}\right) = 0,$$

   we see that conditions (3.99) fail, so that the inequality (3.136) is not covered by Theorem 3.3.1.

Proof of Theorem 3.3.11. Case $\delta = 0$. In this case we have $q = r$ and $b = c$ by $\frac{\delta r}{p} + \frac{(1-\delta)r}{q} = 1$ and $c = \delta(a-1) + b(1-\delta)$, respectively. Then, the inequalities (3.130) and (3.131) are equivalent to the trivial estimate

$$\||x|^b f\|_{L^q(\mathbb{G})} \leq \||x|^b f\|_{L^q(\mathbb{G})},$$

with clearly a sharp constant.

Case $\delta = 1$. In this case we have $p = r$ and $a - 1 = c$. By Theorem 3.2.3, we have for $Q + pc = Q + p(a-1) \neq 0$ the inequality

$$\||x|^c f\|_{L^r(\mathbb{G})} \leq \left|\frac{p}{Q+pc}\right| \||x|^c \mathbb{E} f\|_{L^r(\mathbb{G})},$$

where $\mathbb{E} = |x|\mathcal{R}$ is the Euler operator. Using this estimate we get

$$\||x|^c f\|_{L^r(\mathbb{G})} \leq \left|\frac{p}{Q+pc}\right| \||x|^{c+1}\mathcal{R} f\|_{L^r(\mathbb{G})} = \left|\frac{p}{Q-p(1-a)}\right| \||x|^a \mathcal{R} f\|_{L^p(\mathbb{G})},$$

which implies (3.130). For $Q + pc = Q + p(a-1) = 0$ by Theorem 3.2.3 we obtain

$$\begin{aligned}\||x|^c f\|_{L^r(\mathbb{G})} &\leq p\||x|^c \log|x| \mathbb{E} f\|_{L^r(\mathbb{G})} \\ &= p\||x|^{c+1}\log|x|\mathcal{R} f\|_{L^r(\mathbb{G})} \\ &= p\||x|^a \log|x|\mathcal{R} f\|_{L^p(\mathbb{G})},\end{aligned}$$

which gives (3.131). In this case, the constants in (3.130) and (3.131) are sharp, since the constants in Theorem 3.2.3 are sharp.

Case $\delta \in (0,1) \cap \left[\frac{r-q}{r}, \frac{p}{r}\right]$. Using $c = \delta(a-1) + b(1-\delta)$, a direct calculation gives

$$\||x|^c f\|_{L^r(\mathbb{G})} = \left(\int_{\mathbb{G}} |x|^{cr}|f(x)|^r dx\right)^{1/r} = \left(\int_{\mathbb{G}} \frac{|f(x)|^{\delta r}}{|x|^{\delta r(1-a)}} \frac{|f(x)|^{(1-\delta)r}}{|x|^{-br(1-\delta)}} dx\right)^{1/r}.$$

Since we have $\delta \in (0,1) \cap \left[\frac{r-q}{r}, \frac{p}{r}\right]$ and $p + q \geq r$, then by using Hölder's inequality for $\frac{\delta r}{p} + \frac{(1-\delta)r}{q} = 1$, we obtain

$$\begin{aligned}\||x|^c f\|_{L^r(\mathbb{G})} &\leq \left(\int_{\mathbb{G}} \frac{|f(x)|^p}{|x|^{p(1-a)}} dx\right)^{\delta/p} \left(\int_{\mathbb{G}} \frac{|f(x)|^q}{|x|^{-bq}} dx\right)^{(1-\delta)/q} \\ &= \left\|\frac{f}{|x|^{1-a}}\right\|^{\delta}_{L^p(\mathbb{G})} \left\|\frac{f}{|x|^{-b}}\right\|^{1-\delta}_{L^q(\mathbb{G})}.\end{aligned} \tag{3.137}$$

Here we note that when $p = q$ and $a - b = 1$, the equality in Hölder's inequality holds for any function. We also note that in the case $p \neq q$ the function

$$h(x) = |x|^{\frac{1}{(p-q)}(p(1-a)+bq)} \tag{3.138}$$

satisfies Hölder's equality condition

$$\frac{|h|^p}{|x|^{p(1-a)}} = \frac{|h|^q}{|x|^{-bq}}.$$

If $Q \neq p(1-a)$, then by Theorem 3.2.3 we have

$$\begin{aligned}\left\|\frac{f}{|x|^{1-a}}\right\|^{\delta}_{L^p(\mathbb{G})} &\leq \left|\frac{p}{Q-p(1-a)}\right|^{\delta}\left\|\frac{\mathbb{E}f}{|x|^{1-a}}\right\|^{\delta}_{L^p(\mathbb{G})} \\ &= \left|\frac{p}{Q-p(1-a)}\right|^{\delta}\left\|\frac{\mathcal{R}f}{|x|^{-a}}\right\|^{\delta}_{L^p(\mathbb{G})}, \quad 1<p<\infty.\end{aligned} \tag{3.139}$$

Putting this in (3.125), one has

$$\||x|^c f\|_{L^r(\mathbb{G})} \leq \left|\frac{p}{Q-p(1-a)}\right|^{\delta}\left\|\frac{\mathcal{R}f}{|x|^{-a}}\right\|^{\delta}_{L^p(\mathbb{G})}\left\|\frac{f}{|x|^{-b}}\right\|^{1-\delta}_{L^q(\mathbb{G})}.$$

We note that in the case of $p=q$ and $a-b=1$, Hölder's equality condition of the inequalities (3.137) and (3.139) holds true for functions $\frac{1}{|x|^c}$, $c \in \mathbb{R}\backslash\{0\}$. Moreover, in the case of $p \neq q$ and $p(1-a)+bq \neq 0$, Hölder's equality condition of the inequalities (3.137) and (3.139) holds true for the function $h(x)$ in (3.138). Therefore, the constant in (3.130) is sharp when $p=q$ and $a-b=1$, or when $p \neq q$ and $p(1-a)+bq \neq 0$.

Now let us consider the case $Q=p(1-a)$. Using Theorem 3.2.3, one has

$$\left\|\frac{f}{|x|^{1-a}}\right\|^{\delta}_{L^p(\mathbb{G})} \leq p^{\delta}\left\|\frac{\log|x|}{|x|^{1-a}}\mathbb{E}f\right\|^{\delta}_{L^p(\mathbb{G})}, \quad 1<p<\infty.$$

Then, putting this in (3.137), we obtain

$$\begin{aligned}\||x|^c f\|_{L^r(\mathbb{G})} &\leq p^{\delta}\left\|\frac{\log|x|}{|x|^{1-a}}\mathbb{E}f\right\|^{\delta}_{L^p(\mathbb{G})}\left\|\frac{f}{|x|^{-b}}\right\|^{1-\delta}_{L^q(\mathbb{G})} \\ &= p^{\delta}\left\|\frac{\log|x|}{|x|^{-a}}\mathcal{R}f\right\|^{\delta}_{L^p(\mathbb{G})}\left\|\frac{f}{|x|^{-b}}\right\|^{1-\delta}_{L^q(\mathbb{G})},\end{aligned}$$

completing the proof. □

4

A Study of Fractional Hardy Inequalities

In this chapter we present results concerning fractional forms of Hardy inequalities. Such a topic is well investigated in the Abelian Euclidean setting and we will be providing relevant references in the sequel. For a general survey of fractional Laplacians in the Euclidean setting see, e.g., [Gar17]. However, as usual, the general approach based on homogeneous groups allows one to get insights also in the Abelian case, for example, from the point of view of the possibility of choosing an arbitrary quasi-norm. Moreover, another application of the setting of homogeneous groups is that the results can be equally applied to both elliptic and subelliptic problems.

We start by discussing fractional Sobolev and Hardy inequalities on the homogeneous groups. As a consequence of these inequalities, we derive a Lyapunov type inequality for the fractional p-sub-Laplacian, which also implies an estimate of the first eigenvalue in a quasi-ball for the Dirichlet fractional p-sub-Laplacian. We also extend this analysis to systems of fractional p-sub-Laplacians and to Riesz potential operators.

4.1 Gagliardo seminorms and fractional p-sub-Laplacians

Throughout this chapter $\mathbb{G}$ will be a homogeneous group of homogeneous dimension Q. Let $|\cdot|$ be a homogeneous quasi-norm on $\mathbb{G}$. We start with the definition of the fractional p-sub-Laplacian.

Definition 4.1.1 (Fractional p-sub-Laplacian). Let $p > 1$ and let $s \in (0,1)$. For a measurable and compactly supported function u the *fractional p-sub-Laplacian* $(-\Delta_{p,|\cdot|})^s = (-\Delta_p)^s$ on $\mathbb{G}$ is defined by the formula

$$(-\Delta_p)^s u(x) := 2 \lim_{\delta \searrow 0} \int_{\mathbb{G}\setminus B(x,\delta)} \frac{|u(x)-u(y)|^{p-2}(u(x)-u(y))}{|y^{-1}x|^{Q+sp}} dy, \quad x \in \mathbb{G}, \quad (4.1)$$

where $B(x,\delta) = B_{|\cdot|}(x,\delta)$ is a quasi-ball with respect to the quasi-norm $|\cdot|$, with radius δ centred at $x \in \mathbb{G}$.

Definition 4.1.2 (Gagliardo seminorm and fractional Sobolev spaces)**.** For a measurable function $u : \mathbb{G} \to \mathbb{R}$, its *Gagliardo seminorm* is defined as

$$[u]_{s,p,|\cdot|} = [u]_{s,p} := \left(\int_{\mathbb{G}} \int_{\mathbb{G}} \frac{|u(x) - u(y)|^p}{|y^{-1}x|^{Q+sp}} dxdy \right)^{1/p}. \tag{4.2}$$

For $p \geq 1$ and $s \in (0,1)$, the functional space

$$W^{s,p}(\mathbb{G}) = \{u \in L^p(\mathbb{G}) : u \text{ is measurable and } [u]_{s,p} < +\infty\}, \tag{4.3}$$

endowed with the norm

$$\|u\|_{W^{s,p}(\mathbb{G})} := (\|u\|^p_{L^p(\mathbb{G})} + [u]^p_{s,p})^{1/p}, \quad u \in W^{s,p}(\mathbb{G}), \tag{4.4}$$

is called the *fractional Sobolev space on* $\mathbb{G}$. Sometimes, to emphasize the dependence on a particular quasi-norm, we may write $[u]_{s,p,|\cdot|}$ and $W^{s,p,|\cdot|}$ but we note that the space $W^{s,p,|\cdot|}$ is independent of a particular choice of a quasi-norm due to their equivalence, see Proposition 1.2.3.

Similarly, if $\Omega \subset \mathbb{G}$ is a Haar measurable set, we define the *fractional Sobolev space* $W^{s,p}(\Omega)$ on Ω by

$$W^{s,p}(\Omega) = \left\{ u \in L^p(\Omega) : u \text{ is measurable} \right.$$
$$\left. \text{and } \left(\int_{\Omega} \int_{\Omega} \frac{|u(x) - u(y)|^p}{|y^{-1}x|^{Q+sp}} dxdy \right)^{1/p} < +\infty \right\},$$

endowed with the norm

$$\|u\|_{W^{s,p}(\Omega)} := \left(\|u\|^p_{L^p(\Omega)} + \int_{\Omega} \int_{\Omega} \frac{|u(x) - u(y)|^p}{|y^{-1}x|^{Q+sp}} dxdy \right)^{1/p}. \tag{4.5}$$

Moreover, the Sobolev space $W^{s,p}_0(\Omega)$ is defined as the completion of $C^\infty_0(\Omega)$ with respect to the norm $\|u\|_{W^{s,p}(\Omega)}$.

4.2 Fractional Hardy inequalities on homogeneous groups

In the present section we establish fractional Hardy inequalities on homogeneous groups.

Theorem 4.2.1 (Fractional Hardy inequality). *Let $\mathbb{G}$ be a homogeneous group of homogeneous dimension Q with a homogeneous quasi-norm $|\cdot|$. Let $p > 1$, $s \in (0, 1)$ and $Q > sp$. Then for all $u \in C_0^\infty(\mathbb{G})$ we have*

$$2\mu(\gamma) \int_{\mathbb{G}} \frac{|u(x)|^p}{|x|^{ps}} dx \leq [u]^p_{s,p,|\cdot|}, \tag{4.6}$$

where $\mu(\gamma)$ is defined in (4.10).

Remark 4.2.2. In [AB17] the authors studied the weighted fractional p-Laplacian and established the following weighted fractional L^p-Hardy inequality:

$$C \int_{\mathbb{R}^N} \frac{|u(x)|^p}{|x|_E^{ps+2\beta}} dx \leq \int_{\mathbb{R}^N} \int_{\mathbb{R}^N} \frac{|u(x) - u(y)|^p}{|x-y|_E^{N+ps} |x|_E^\beta |y|_E^\beta} dxdy, \tag{4.7}$$

where $\beta < \frac{N-ps}{2}$, $u \in C_0^\infty(\mathbb{R}^N)$, $C > 0$ is a positive constant, and $|\cdot|_E$ is the Euclidean distance in $\mathbb{R}^N$.

Before we prove Theorem 4.2.1, let us establish the following two lemmas that will be instrumental in the proof.

Lemma 4.2.3. *We fix a homogeneous quasi-norm $|\cdot|$ on $\mathbb{G}$. Let $\omega \in W_0^{s,p}(\Omega)$ and assume that $w > 0$ in $\Omega \subset \mathbb{G}$. Assume that $(-\Delta_p)^s \omega = \nu > 0$ with $\nu \in L^1_{\text{loc}}(\Omega)$. Then for all $u \in C_0^\infty(\Omega)$, we have*

$$\frac{1}{2} \int_\Omega \int_\Omega \frac{|u(x) - u(y)|^p}{|y^{-1}x|^{Q+ps}} dxdy \geq \left\langle (-\Delta_p)^s \omega, \frac{|u|^p}{\omega^{p-1}} \right\rangle.$$

Proof of Lemma 4.2.3. Using the notations

$$v := \frac{|u|^p}{|\omega|^{p-1}} \text{ and } k(x,y) := \frac{1}{|y^{-1}x|^{Q+ps}},$$

we get

$$\begin{aligned}\langle (-\Delta_p)^s \omega(x), v(x) \rangle &= \int_\Omega v(x) dx \int_\Omega |\omega(x) - \omega(y)|^{p-2} (\omega(x) - \omega(y)) k(x,y) dy \\ &= \int_\Omega \frac{|u(x)|^p}{|\omega(x)|^{p-1}} dx \int_\Omega |\omega(x) - \omega(y)|^{p-2} (\omega(x) - \omega(y)) k(x,y) dy.\end{aligned}$$

Let us show that $k(x,y)$ is symmetric, that is, $k(x,y)=k(y,x)$ for all $x,y\in\mathbb{G}$. This readily follows, since by the definition of the quasi-norm we have $|x^{-1}|=|x|$ for all $x\in\mathbb{G}$. So, since $k(x,y)$ is symmetric, we obtain

$$\begin{aligned}&\langle(-\Delta_p)^s\omega(x),v(x)\rangle\\&=\frac{1}{2}\int_\Omega\int_\Omega\left(\frac{|u(x)|^p}{|\omega(x)|^{p-1}}-\frac{|u(y)|^p}{|\omega(y)|^{p-1}}\right)|\omega(x)-\omega(y)|^{p-2}(\omega(x)-\omega(y))k(x,y)dydx.\end{aligned}$$

Let $g:=\frac{u}{\omega}$ and

$$R(x,y):=|u(x)-u(y)|^p-(|g(x)|^p\omega(x)-|g(y)|^p\omega(y))|\omega(x)-\omega(y)|^{p-2}(\omega(x)-\omega(y)).$$

Then we have

$$\langle(-\Delta_p)^s\omega,v\rangle+\frac{1}{2}\int_\Omega\int_\Omega R(x,y)k(x,y)dydx=\frac{1}{2}\int_\Omega\int_\Omega|u(x)-u(y)|^pk(x,y)dydx.$$

By the symmetry argument, we can assume that $\omega(x)\geq\omega(y)$. By using the inequality (see, e.g., [FS08, Lemma 2.6])

$$|a-t|^p\geq(1-t)^{p-1}(|a|^p-t),\ p>1,\ t\in[0,1],\ a\in\mathbb{C},\tag{4.8}$$

with $t=\frac{\omega(y)}{\omega(x)}$ and $a=\frac{g(x)}{g(y)}$, we see that $R(x,y)\geq 0$. Therefore, we have proved the inequality

$$\langle(-\Delta_p)^s\omega,v\rangle\leq\frac{1}{2}\int_\Omega\int_\Omega\frac{|u(x)-u(y)|^p}{|y^{-1}x|^{Q+sp}}dydx,$$

completing the proof of Lemma 4.2.3. □

Lemma 4.2.4. *Let $p>1$ and $\gamma\in\left(0,\frac{Q-ps}{p-1}\right)$. Then there exists a positive constant $\mu(\gamma)>0$ such that*

$$(-\Delta_p)^s(|x|^{-\gamma})=\mu(\gamma)\frac{1}{|x|^{ps+\gamma(p-1)}}\quad a.e.\ in\ \mathbb{G}\setminus\{0\}.\tag{4.9}$$

Proof of Lemma 4.2.4. Let us denote $\omega(x):=|x|^{-\gamma}$. We set $r=|x|$ and $\rho=|y|$ with $x=rx'$ and $y=\rho y'$ where $|x'|=|y'|=1$. Then we have

$$\begin{aligned}(-\Delta_p)^s\omega&=\int_0^{+\infty}||x|^{-\gamma}-|y|^{-\gamma}|^{p-2}(|x|^{-\gamma}-|y|^{-\gamma})|y|^{Q-1}\left(\int_{|y'|=1}\frac{d\sigma(y)}{|y^{-1}x|^{Q+ps}}\right)d|y|\\&=\frac{1}{|x|^{ps+\gamma(p-1)}}\int_0^{+\infty}\left|1-\frac{|y|^{-\gamma}}{|x|^{-\gamma}}\right|^{p-2}\\&\quad\times\left(1-\frac{|y|^{-\gamma}}{|x|^{-\gamma}}\right)\frac{|y|^{Q-1}}{|x|^{Q-1}}\left(\int_{|y'|=1}\frac{d\sigma(y)}{\left|\left(\frac{|y|}{|x|}y'\right)^{-1}\circ x'\right|^{Q+ps}}\right)d|y|.\end{aligned}$$

Let $\rho = \frac{|y|}{|x|}$ and $L(\rho) = \int_{|y'|=1} \frac{d\sigma(y)}{|(\rho y')^{-1} \circ x'|^{Q+ps}}$. Then we have

$$(-\Delta_p)^s \omega = \frac{1}{|x|^{ps+\gamma(p-1)}} \int_0^{+\infty} |1-\rho^{-\gamma}|^{p-2}(1-\rho^{-\gamma}) L(\rho)\rho^{Q-1} d\rho.$$

We then have (4.9) with

$$\mu(\gamma) := \int_0^{+\infty} \phi(\rho) d\rho \tag{4.10}$$

for

$$\phi(\rho) = |1-\rho^{-\gamma}|^{p-2}(1-\rho^{-\gamma}) L(\rho)\rho^{Q-1}.$$

Now it remains to show that $\mu(\gamma)$ is positive and bounded. Firstly, let us show that $\mu(\gamma)$ is bounded. We have

$$\mu(\gamma) = \int_0^1 \phi(\rho) d\rho + \int_1^{+\infty} \phi(\rho) d\rho = I_1 + I_2. \tag{4.11}$$

Using the new variable $\zeta = \frac{1}{\rho}$ we have $L(\rho) = L\left(\frac{1}{\zeta}\right) = \zeta^{Q+ps} L(\zeta)$ for any $\zeta > 0$. Thus, we get that

$$\mu(\gamma) = \int_1^{+\infty} (\rho^{-\gamma} - 1)^{p-1} (\rho^{Q-1-\gamma(p-1)} - \rho^{ps-1}) L(\rho) d\rho. \tag{4.12}$$

For $\rho \to 1$ we have

$$(\rho^{-\gamma} - 1)^{p-1} (\rho^{Q-1-\gamma(p-1)} - \rho^{ps-1}) L(\rho) \simeq (\rho - 1)^{-1-ps+p} \in L^1(1,2). \tag{4.13}$$

Similarly, for $\rho \to \infty$ we get

$$(\rho^{-\gamma} - 1)^{p-1} (\rho^{Q-1-\gamma(p-1)} - \rho^{ps-1}) L(\rho) \simeq \rho^{-1-ps} \in L^1(2,\infty). \tag{4.14}$$

These properties show that $\mu(\gamma)$ is bounded. On the other hand, by (4.12) with $\gamma \in \left(0, \frac{Q-ps}{p-1}\right)$, we see that $\mu(\gamma)$ is positive.

Lemma 4.2.4 is proved. □

Proof of Theorem 4.2.1. Let $u \in C_0^\infty(\mathbb{G})$ and $\gamma < \frac{Q-ps}{p-1}$. By Lemma 4.2.4 and Lemma 4.2.3 we readily obtain that

$$\begin{aligned}\frac{1}{2}[u]^p_{s,p} &= \frac{1}{2}\int_{\mathbb{G}}\int_{\mathbb{G}} \frac{|u(x)-u(y)|^p}{|y^{-1}x|^{Q+ps}} dxdy \\ &\geq \left\langle (-\Delta_p)^s(|x|^{-\gamma}), \frac{|u(x)|^p}{|x|^{-\gamma(p-1)}} \right\rangle = \mu(\gamma) \int_{\mathbb{G}} \frac{|u(x)|^p}{|x|^{ps}} dx.\end{aligned}$$

This completes the proof of Theorem 4.2.1. □

4.3 Fractional Sobolev inequalities on homogeneous groups

In this section we establish fractional Sobolev inequalities on homogeneous groups.

Theorem 4.3.1 (Fractional Sobolev inequality). *Let $\mathbb{G}$ be a homogeneous group of homogeneous dimension Q. Let us fix a homogeneous quasi-norm $|\cdot|$ on $\mathbb{G}$. Let $p > 1$, $s \in (0,1)$ and $Q > sp$, and let us set $p^* := \frac{Qp}{Q-sp}$. Then there is a positive constant $C = C(Q, p, s, |\cdot|)$ such that we have*

$$\|u\|_{L^{p^*}(\mathbb{G})} \leq C[u]_{s,p,|\cdot|}, \tag{4.15}$$

for all measurable compactly supported functions $u : \mathbb{G} \to \mathbb{R}$.

Remark 4.3.2. In [DNPV12] the authors obtained the fractional Sobolev inequality in the case $N > sp$, $1 < p < \infty$, and $s \in (0,1)$. Namely, for all measurable and compactly supported functions u one has

$$\|u\|_{L^{p^*}(\mathbb{R}^N)} \leq C[u]_{s,p}, \tag{4.16}$$

where $p^* = \frac{Np}{N-sp}$, $C = C(N,p,s) > 0$ is a suitable constant independent of u, and

$$[u]^p_{s,p} = \int_{\mathbb{R}^N} \int_{\mathbb{R}^N} \frac{|u(x) - u(y)|^p}{|x-y|_E^{N+sp}} dxdy,$$

with $|\cdot|_E$ being the Euclidean distance in $\mathbb{R}^N$.

To prove the above analogue of the fractional Sobolev inequality, first we present the following two lemmas.

Lemma 4.3.3. *Let $p > 1$, $s \in (0,1)$, and let $K \subset \mathbb{G}$ be a Haar measurable set. Fix $x \in \mathbb{G}$ and a quasi-norm $|\cdot|$ on $\mathbb{G}$. Then we have*

$$\int_{K^c} \frac{dy}{|y^{-1}x|^{Q+sp}} \geq C|K|^{-sp/Q}, \tag{4.17}$$

where $C = C(Q, s, p, |\cdot|)$ is a positive constant, $K^c := \mathbb{G} \setminus K$, and $|K|$ is the Haar measure of K.

Proof of Lemma 4.3.3. Let $\delta := \left(\frac{|K|}{\omega_Q}\right)^{1/Q}$, where ω_Q is a surface measure of the unit quasi-ball on $\mathbb{G}$. For the corresponding quasi-ball $B(x,\delta) = B_{|\cdot|}(x,\delta)$ centred at x with radius δ, we have

$$\begin{aligned} |K^c \cap B(x,\delta)| &= |B(x,\delta)| - |K \cap B(x,\delta)| \\ &= |K| - |K \cap B(x,\delta)| = |K \cap B^c(x,\delta)|, \end{aligned} \tag{4.18}$$

where $|\cdot|$ (by abuse of notation, only in this proof) is the Haar measure on $\mathbb{G}$. Then,

$$\begin{aligned}\int_{K^c}\frac{dy}{|y^{-1}x|^{Q+sp}} &= \int_{K^c\cap B(x,\delta)}\frac{dy}{|y^{-1}x|^{Q+sp}}+\int_{K^c\cap B^c(x,\delta)}\frac{dy}{|y^{-1}x|^{Q+sp}}\\ &\geq \int_{K^c\cap B(x,\delta)}\frac{dy}{\delta^{Q+sp}}+\int_{K^c\cap B^c(x,\delta)}\frac{dy}{|y^{-1}x|^{Q+sp}}\\ &= \frac{|K^c\cap B(x,\delta)|}{\delta^{Q+sp}}+\int_{K^c\cap B^c(x,\delta)}\frac{dy}{|y^{-1}x|^{Q+sp}}.\end{aligned}$$

By using (4.18) we obtain

$$\begin{aligned}\int_{K^c}\frac{dy}{|y^{-1}x|^{Q+sp}} &\geq \frac{|K^c\cap B(x,\delta)|}{\delta^{Q+sp}}+\int_{K^c\cap B^c(x,\delta)}\frac{dy}{|y^{-1}x|^{Q+sp}}\\ &= \frac{|K\cap B^c(x,\delta)|}{\delta^{Q+sp}}+\int_{K^c\cap B^c(x,\delta)}\frac{dy}{|y^{-1}x|^{Q+sp}}\\ &\geq \int_{K\cap B^c(x,\delta)}\frac{dy}{|y^{-1}x|^{Q+sp}}+\int_{K^c\cap B^c(x,\delta)}\frac{dy}{|y^{-1}x|^{Q+sp}}\\ &= \int_{B^c(x,\delta)}\frac{dy}{|y^{-1}x|^{Q+sp}}.\end{aligned}$$

Now using the polar decomposition formula in Proposition 1.2.10 we obtain that

$$\int_{K^c}\frac{dy}{|y^{-1}x|^{Q+sp}}\geq C|K|^{-sp/Q}, \tag{4.19}$$

completing the proof. □

We now establish a useful technical estimate for the Gagliardo seminorm $[u]_{s,p}$ defined in (4.2).

Lemma 4.3.4. *Let $p>1$, $s\in(0,1)$ and $Q>sp$. Let $u\in L^\infty(\mathbb{G})$ be compactly supported and denote $a_k:=|\{|u|>2^k\}|$ for any $k\in\mathbb{Z}$. Then we have*

$$C\sum_{k\in\mathbb{Z},\,a_k\neq 0}a_{k+1}a_k^{-sp/Q}2^{kp}\leq [u]^p_{s,p}, \tag{4.20}$$

where $C=C(Q,p,s,|\cdot|)$ is a positive constant.

Proof of Lemma 4.3.4. We define

$$A_k:=\{|u|>2^k\},\ k\in\mathbb{Z}, \tag{4.21}$$

and

$$D_k:=A_k\setminus A_{k+1}=\{2^k<|u|\leq 2^{k+1}\}\text{ and }d_k:=|D_k|, \tag{4.22}$$

the Haar measure of D_k. Since $A_{k+1} \subseteq A_k$, it follows that

$$a_{k+1} \leq a_k. \tag{4.23}$$

By the assumption $u \in L^\infty(\mathbb{G})$ is compactly supported, a_k and d_k are bounded and vanish when k is large enough. Also, we notice that the D_k's are disjoint, therefore,

$$\bigcup_{l \in \mathbb{Z},\ l \leq k} D_l = A_{k+1}^c \tag{4.24}$$

and

$$\bigcup_{l \in \mathbb{Z},\ l \geq k} D_l = A_k. \tag{4.25}$$

From (4.25) it follows that

$$\sum_{l \in \mathbb{Z},\ l \geq k} d_l = a_k \tag{4.26}$$

and

$$d_k = a_k - \sum_{l \in \mathbb{Z},\ l \geq k+1} d_l. \tag{4.27}$$

Since a_k and d_k are bounded and vanish when k is large enough, (4.26) and (4.27) are convergent. We define the convergent series

$$S := \sum_{l \in \mathbb{Z},\ a_{l-1} \neq 0} 2^{lp} a_{l-1}^{-sp/Q} d_l. \tag{4.28}$$

We have that $D_k \subseteq A_k \subseteq A_{k-1}$, therefore, $a_{i-1}^{-sp/Q} d_l \leq a_{i-1}^{-sp/Q} a_{l-1}$. Thus,

$$\{(i,l) \in \mathbb{Z} \text{ s.t. } a_{i-1} \neq 0 \text{ and } a_{i-1}^{-sp/Q} d_l \neq 0\} \subseteq \{(i,l) \in \mathbb{Z} \text{ s.t. } a_{l-1} \neq 0\}. \tag{4.29}$$

By using (4.29) and (4.23), we can estimate

$$\begin{aligned}
\sum_{\substack{i \in \mathbb{Z}, \\ a_{i-1} \neq 0}} \sum_{\substack{l \in \mathbb{Z}, \\ l \geq i+1}} 2^{ip} a_{i-1}^{-sp/Q} d_l &= \sum_{\substack{i \in \mathbb{Z}, \\ a_{i-1} \neq 0}} \sum_{\substack{l \in \mathbb{Z}, \\ l \geq i+1, \\ a^{sp/Q} d_l \neq 0}} 2^{ip} a_{i-1}^{-sp/Q} d_l \\
\leq \sum_{i \in \mathbb{Z}} \sum_{\substack{l \in \mathbb{Z}, \\ l \geq i+1, \\ a_{l-1} \neq 0}} 2^{ip} a_{i-1}^{-sp/Q} d_l &= \sum_{\substack{l \in \mathbb{Z}, \\ a_{l-1} \neq 0}} \sum_{\substack{i \in \mathbb{Z}, \\ i \leq l-1}} 2^{ip} a_{i-1}^{-sp/Q} d_l \\
\leq \sum_{\substack{l \in \mathbb{Z}, \\ a_{l-1} \neq 0}} \sum_{\substack{i \in \mathbb{Z}, \\ i \leq l-1}} 2^{ip} a_{l-1}^{-sp/Q} d_l &= \sum_{\substack{l \in \mathbb{Z}, \\ a_{l-1} \neq 0}} \sum_{k=0}^{+\infty} 2^{p(l-1-k)} a_{l-1}^{-sp/Q} d_l \leq S.
\end{aligned} \tag{4.30}$$

Notice that

$$||u(x)| - |u(y)|| \leq |u(x) - u(y)|,$$

for any $x, y \in \mathbb{G}$. If we fix $i \in \mathbb{Z}$ and $x \in D_i$, then for any $j \in \mathbb{Z}$ with $j \leq i-2$, for any $y \in D_j$ using the above inequality, we obtain that

$$|u(x) - u(y)| \geq 2^i - 2^{j+1} \geq 2^i - 2^{i-1} \geq 2^{i-1}.$$

Then, using (4.24), we have

$$\begin{aligned}\sum_{j\in\mathbb{Z},\, j\leq i-2} \int_{D_j} \frac{|u(x)-u(y)|^p}{|y^{-1}x|^{Q+sp}} dy &\geq 2^{(i-1)p} \sum_{j\in\mathbb{Z},\, j\leq i-2} \int_{D_j} \frac{dy}{|y^{-1}x|^{Q+sp}} \\ &= 2^{(i-1)p} \int_{A_{i-1}^c} \frac{dy}{|y^{-1}x|^{Q+sp}}.\end{aligned} \tag{4.31}$$

Now using (4.31) and Lemma 4.3.3, we obtain that

$$\sum_{j\in\mathbb{Z},\, j\leq i-2} \int_{D_j} \frac{|u(x)-u(y)|^p}{|y^{-1}x|^{Q+sp}} dy \geq C 2^{ip} a_{i-1}^{-sp/Q},$$

with some positive constant C. That is, for any $i \in \mathbb{Z}$, we have

$$\sum_{j\in\mathbb{Z},\, j\leq i-2} \int_{D_i} \int_{D_j} \frac{|u(x)-u(y)|^p}{|y^{-1}x|^{Q+sp}} dx dy \geq C 2^{ip} a_{i-1}^{-sp/Q} d_i. \tag{4.32}$$

From (4.32) and (4.27) we get

$$\begin{aligned}&\sum_{j\in\mathbb{Z},\, j\leq i-2} \int_{D_i} \int_{D_j} \frac{|u(x)-u(y)|^p}{|y^{-1}x|^{Q+sp}} dx dy \\ &\geq C \left(2^{ip} a_{i-1}^{-sp/Q} a_i - \sum_{l\in\mathbb{Z},\, l\geq i+1} 2^{ip} a_{i-1}^{-sp/Q} d_l \right).\end{aligned} \tag{4.33}$$

By (4.32) and (4.28) it follows that

$$\begin{aligned}&\sum_{i\in\mathbb{Z}, a_{i-1}\neq 0} \sum_{j\in\mathbb{Z}, j\leq i-2} \int_{D_i} \int_{D_j} \frac{|u(x)-u(y)|^p}{|y^{-1}x|^{Q+sp}} dx dy \\ &\geq C \sum_{i\in\mathbb{Z}, a_{i-1}\neq 0} 2^{ip} a_{i-1}^{-sp/Q} d_i \geq C\, S.\end{aligned} \tag{4.34}$$

Then, by using (4.30), (4.33) and (4.34), we obtain that

$$\begin{aligned}&\sum_{i\in\mathbb{Z}, a_{i-1}\neq 0} \sum_{j\in\mathbb{Z}, j\leq i-2} \int_{D_i} \int_{D_j} \frac{|u(x)-u(y)|^p}{|y^{-1}x|^{Q+sp}} dx dy \\ &\geq C \sum_{i\in\mathbb{Z},\, a_{i-1}\neq 0} 2^{ip} a_{i-1}^{-sp/Q} a_i - C \sum_{i\in\mathbb{Z}, a_{i-1}\neq 0} \sum_{l\in\mathbb{Z}, l\geq i+1} 2^{ip} a_{i-1}^{-sp/Q} d_l\end{aligned}$$

$$\geq C \sum_{i\in\mathbb{Z},\, a_{i-1}\neq 0} 2^{ip} a_{i-1}^{-sp/Q} a_i - C\, S$$

$$\geq C \sum_{i\in\mathbb{Z}, a_{i-1}\neq 0} 2^{ip} a_{i-1}^{-sp/Q} a_i - \sum_{i\in\mathbb{Z}, a_{i-1}\neq 0} \sum_{j\in\mathbb{Z}, j\leq i-2} \int_{D_i}\int_{D_j} \frac{|u(x)-u(y)|^p}{|y^{-1}x|^{Q+sp}} dxdy.$$

This means that

$$\begin{aligned} \sum_{i\in\mathbb{Z}, a_{i-1}\neq 0} \sum_{j\in\mathbb{Z}, j\leq i-2} \int_{D_i}\int_{D_j} \frac{|u(x)-u(y)|^p}{|y^{-1}x|^{Q+sp}} dxdy \\ \geq \frac{C}{2} \sum_{i\in\mathbb{Z}, a_{i-1}\neq 0} 2^{ip} a_{i-1}^{-sp/Q} a_i, \end{aligned} \tag{4.35}$$

for some constant $C > 0$. By symmetry and using (4.35), we arrive at

$$\begin{aligned} [u]^p_{s,p,|\cdot|} &= \int_{\mathbb{G}}\int_{\mathbb{G}} \frac{|u(x)-u(y)|^p}{|y^{-1}x|^{Q+sp}} dxdy = \sum_{i,j\in\mathbb{Z}} \int_{D_i}\int_{D_j} \frac{|u(x)-u(y)|^p}{|y^{-1}x|^{Q+sp}} dxdy \\ &\geq 2 \sum_{i,j\in\mathbb{Z},\, j<i} \int_{D_i}\int_{D_j} \frac{|u(x)-u(y)|^p}{|y^{-1}x|^{Q+sp}} dxdy \\ &\geq 2 \sum_{i\in\mathbb{Z}, a_{i-1}\neq 0} \sum_{j\in\mathbb{Z}, j\leq i-2} \int_{Di}\int_{D_j} \frac{|u(x)-u(y)|^p}{|y^{-1}x|^{Q+sp}} dxdy \\ &\geq C \sum_{i\in\mathbb{Z},\, a_{i-1}\neq 0} 2^{ip} a_{i-1}^{-sp/Q} a_i, \end{aligned}$$

completing the proof of Lemma 4.3.4. □

Proof of Theorem 4.3.1. Assume that Gagliardo's seminorm $[u]_{s,p}$ is bounded, i.e., that

$$[u]^p_{s,p} = \int_{\mathbb{G}}\int_{\mathbb{G}} \frac{|u(x)-u(y)|^p}{|y^{-1}x|^{Q+sp}} dxdy < +\infty. \tag{4.36}$$

Suppose also that $u \in L^\infty(\mathbb{G})$.

If (4.36) is satisfied for bounded functions, it holds also for the function u_n, obtained from u by cutting at levels $-n$ and n, that is, for

$$u_n := \max\{\min\{u(x), n\}, -n\},$$

for any $n \in \mathbb{R}$ and $x \in \mathbb{G}$. Thus, using the fact that

$$\lim_{n\to+\infty} \|u_n\|_{L^p(\mathbb{G})} = \|u\|_{L^p(\mathbb{G})},$$

$1 < p < \infty$, and by using (4.36) with the dominated convergence theorem, we obtain that

$$\begin{aligned}\lim_{n\to+\infty}[u_n]^p_{s,p} &= \lim_{n\to+\infty}\int_{\mathbb{G}}\int_{\mathbb{G}}\frac{|u_n(x)-u_n(y)|^p}{|y^{-1}x|^{Q+sp}}dxdy \\ &= \int_{\mathbb{G}}\int_{\mathbb{G}}\frac{|u(x)-u(y)|^p}{|y^{-1}x|^{Q+sp}}dxdy = [u]^p_{s,p}.\end{aligned} \tag{4.37}$$

Defining a_k and A_k as in Lemma 4.3.4, we have

$$\begin{aligned}\|u\|_{L^{p^*}(\mathbb{G})} &= \left(\sum_{k\in\mathbb{Z}}\int_{A_k\setminus A_{k+1}}|u(x)|^{p^*}dx\right)^{1/p^*} \\ &\leq \left(\sum_{k\in\mathbb{Z}}\int_{A_k\setminus A_{k+1}}2^{(k+1)p^*}dx\right)^{1/p^*} \\ &\leq \left(\sum_{k\in\mathbb{Z}}2^{(k+1)p^*}a_k\right)^{1/p^*}.\end{aligned}$$

Recall the following fact from [DNPV12, Lemma 6.2]: let $T, p > 1$ and $s \in (0,1)$ be such that $Q > sp$, $m \in \mathbb{Z}$, and assume that a_k is a bounded, decreasing, non-negative sequence with $a_k = 0$ for any $k \geq m$; then we have

$$\sum_{k\in\mathbb{Z}}a_k^{(Q-sp)/Q}T^k \leq C\sum_{k\in\mathbb{Z},\, a_k\neq 0}a_{k+1}a_k^{-sp/Q}T^k, \tag{4.38}$$

for some positive constant $C = C(Q, s, p, T)$.

Then, with $p/p^* = 1 - sp/Q < 1$ and $T = 2^p$, this fact yields

$$\begin{aligned}\|u\|^p_{L^{p^*}(\mathbb{G})} &\leq 2^p\left(\sum_{k\in\mathbb{Z}}2^{kp^*}a_k\right)^{p/p^*} \leq 2^p\sum_{k\in\mathbb{Z}}2^{kp}a_k^{(Q-sp)/Q} \\ &\leq C\sum_{k\in\mathbb{Z},\, a_k\neq 0}2^{kp}a_k^{-sp/Q}a_{k+1}\end{aligned} \tag{4.39}$$

for a positive constant $C = C(Q, p, s, |\cdot|)$. Finally, using Lemma 4.3.4 we arrive at

$$\begin{aligned}\|u\|^p_{L^{p^*}(\mathbb{G})} &\leq C\sum_{k\in\mathbb{Z},\, a_k\neq 0}2^{kp}a_k^{-sp/Q}a_{k+1} \\ &\leq C\int_{\mathbb{G}}\int_{\mathbb{G}}\frac{|u(x)-u(y)|^p}{|y^{-1}x|^{Q+sp}}dxdy = C[u]^p_{s,p,|\cdot|}.\end{aligned}$$

Theorem 4.3.1 is proved. ∎

4.4 Fractional Gagliardo–Nirenberg inequalities

In this section we discuss an analogue of the fractional Gagliardo–Nirenberg inequality on homogeneous groups. As it can be partly expected, in its proof we will use an already established version of a fractional Sobolev inequality on the homogeneous groups.

Theorem 4.4.1 (Fractional Gagliargo–Nirenberg inequality). *Let $\mathbb{G}$ be a homogeneous group of homogeneous dimension Q with a quasi-norm $|\cdot|$. Assume that $Q \geq 2$, $s \in (0,1)$, $p > 1$, $\alpha \geq 1$, $\tau > 0$, $a \in (0,1]$, $Q > sp$ and*

$$\frac{1}{\tau} = a\left(\frac{1}{p} - \frac{s}{Q}\right) + \frac{1-a}{\alpha}.$$

Then there exists $C = C(s,p,Q,a,\alpha) > 0$ such that

$$\|u\|_{L^\tau(\mathbb{G})} \leq C[u]^a_{s,p,|\cdot|}\|u\|^{1-a}_{L^\alpha(\mathbb{G})} \tag{4.40}$$

holds for all $u \in C^1_c(\mathbb{G})$.

Remark 4.4.2. Theorem 4.4.1 was proved in [KRS18a]. In the Abelian case $(\mathbb{R}^N,+)$ with the standard Euclidean distance instead of the quasi-norm, Theorem 4.4.1 covers the fractional Gagliardo–Nirenberg inequality which was proved in [NS18a].

Proof of Theorem 4.4.1. By using Hölder's inequality, for every $\frac{1}{\tau} = a\left(\frac{1}{p} - \frac{s}{Q}\right) + \frac{1-a}{\alpha}$ we get

$$\|u\|^\tau_{L^\tau(\mathbb{G})} = \int_{\mathbb{G}} |u|^\tau dx = \int_{\mathbb{G}} |u|^{a\tau}|u|^{(1-a)\tau} dx \leq \|u\|^{a\tau}_{L^{p^*}(\mathbb{G})}\|u\|^{(1-a)\tau}_{L^\alpha(\mathbb{G})}, \tag{4.41}$$

where $p^* = \frac{Qp}{Q-sp}$. From (4.41), by using the fractional Sobolev inequality (Theorem 4.3.1), we obtain

$$\|u\|^\tau_{L^\tau(\mathbb{G})} \leq \|u\|^{a\tau}_{L^{p^*}(\mathbb{G})}\|u\|^{(1-a)\tau}_{L^\alpha(\mathbb{G})} \leq C[u]^{a\tau}_{s,p,|\cdot|}\|u\|^{(1-a)\tau}_{L^\alpha(\mathbb{G})},$$

that is,

$$\|u\|_{L^\tau(\mathbb{G})} \leq C[u]^a_{s,p,|\cdot|}\|u\|^{1-a}_{L^\alpha(\mathbb{G})}, \tag{4.42}$$

where C is a positive constant independent of u. Theorem 4.4.1 is proved. □

4.5 Fractional Caffarelli–Kohn–Nirenberg inequalities

In this section we discuss the weighted fractional Caffarelli–Kohn–Nirenberg inequalities on homogeneous groups. First, let us define a weighted version of fractional Sobolev spaces from Definition 4.1.2.

Definition 4.5.1 (Weighted fractional Sobolev spaces). We define the *weighted fractional Sobolev space* on a homogeneous group $\mathbb{G}$ with homogeneous dimension Q and homogeneous quasi-norm $|\cdot|$ by

$$W^{s,p,\beta}(\mathbb{G}) = \Big\{ u \in L^p(\mathbb{G}) : u \text{ is measurable}, \tag{4.43}$$

$$[u]_{s,p,\beta,|\cdot|} := \left(\int_{\mathbb{G}} \int_{\mathbb{G}} \frac{|x|^{\beta_1 p}|y|^{\beta_2 p}|u(x)-u(y)|^p}{|y^{-1}x|^{Q+sp}} dxdy \right)^{\frac{1}{p}} < +\infty \Big\},$$

where $\beta_1,\ \beta_2 \in \mathbb{R}$ with $\beta = \beta_1 + \beta_2$. We note that the space $W^{s,p,\beta}(\mathbb{G})$ depends on β_1 and β_2. At the same time, it is independent of a particular choice of a quasi-norm due to their equivalence, see Proposition 1.2.3.

For a Haar measurable set $\Omega \subset \mathbb{G}$, $p \geq 1$, $s \in (0,1)$ and $\beta_1,\ \beta_2 \in \mathbb{R}$ with $\beta = \beta_1 + \beta_2$, we define the weighted fractional Sobolev space on Ω by

$$W^{s,p,\beta}(\Omega) = \Big\{ u \in L^p(\Omega) : u \text{ is measurable}, \tag{4.44}$$

$$[u]_{s,p,\beta,|\cdot|,\Omega} := \left(\int_{\Omega} \int_{\Omega} \frac{|x|^{\beta_1 p}|y|^{\beta_2 p}|u(x)-u(y)|^p}{|y^{-1}x|^{Q+sp}} dxdy \right)^{\frac{1}{p}} < +\infty \Big\}.$$

Theorem 4.5.2 (Fractional Caffarelli–Kohn–Nirenberg inequality). *Let $\mathbb{G}$ be a homogeneous group of homogeneous dimension Q. Let $Q \geq 2$, $s \in (0,1)$, $p > 1$, $\alpha \geq 1$, $\tau > 0$, $a \in (0,1]$, β_1, β_2, β, μ, $\gamma \in \mathbb{R}$, $\beta_1 + \beta_2 = \beta$ and*

$$\frac{1}{\tau} + \frac{\gamma}{Q} = a\left(\frac{1}{p} + \frac{\beta - s}{Q}\right) + (1-a)\left(\frac{1}{\alpha} + \frac{\mu}{Q}\right). \tag{4.45}$$

In addition, assume that, $0 \leq \beta - \sigma$ with $\gamma = a\sigma + (1-a)\mu$. We also assume

$$\beta - \sigma \leq s \text{ only if } \frac{1}{\tau} + \frac{\gamma}{Q} = \frac{1}{p} + \frac{\beta - s}{Q}. \tag{4.46}$$

Then when $\frac{1}{\tau} + \frac{\gamma}{Q} > 0$ we have

$$\||x|^{\gamma}u\|_{L^{\tau}(\mathbb{G})} \leq C[u]^{a}_{s,p,\beta,|\cdot|}\||x|^{\mu}u\|^{1-a}_{L^{\alpha}(\mathbb{G})}, \tag{4.47}$$

for all $u \in C_c^1(\mathbb{G})$, and when $\frac{1}{\tau} + \frac{\gamma}{Q} < 0$ we have

$$\||x|^{\gamma}u\|_{L^{\tau}(\mathbb{G})} \leq C[u]^{a}_{s,p,\beta,|\cdot|}\||x|^{\mu}u\|^{1-a}_{L^{\alpha}(\mathbb{G})}, \tag{4.48}$$

for all $u \in C_c^1(\mathbb{G} \setminus \{0\})$.

Remark 4.5.3.

1. The critical case $\frac{1}{\tau}+\frac{\gamma}{Q}=0$ will be considered in Theorem 4.5.5.
2. In the Abelian (Euclidean) case $(\mathbb{R}^N,+)$ with the usual Euclidean distance instead of the quasi-norm in Theorem 4.5.2, we get the (Abelian) fractional Caffarelli–Kohn–Nirenberg inequality (see, e.g., [NS18a], Theorem 1.1). In (4.48) by setting $a=1$, $\tau=p$, $\beta_1=\beta_2=0$, and $\gamma=-s$, one has an analogue of the fractional Hardy inequality on homogeneous groups.
3. In the Abelian (Euclidean) case $(\mathbb{R}^N,+)$ again with the usual Euclidian distance instead of the quasi-norm and by taking in (4.48) the values $a=1$, $\tau=p$, $\beta_1=\beta_2=0$, and $\gamma=-s$, we get the fractional Hardy inequality (see [FS74, Theorem 1.1]).
4. The results of this section were obtained in [KRS18a] and we follow the presentation there in our proof.

To prove the fractional weighted Caffarelli–Kohn–Nirenberg inequality on $\mathbb{G}$ we will use Theorem 4.4.1 in the proof of the following lemma.

Lemma 4.5.4. *Let $Q\geq 2$, $s\in(0,1)$, $p>1$, $\alpha\geq 1$, $\tau>0$, $a\in(0,1]$ and*

$$\frac{1}{\tau}\geq a\left(\frac{1}{p}-\frac{s}{Q}\right)+\frac{1-a}{\alpha}.$$

Let $\lambda>0$ and $0<r<R$ and set

$$\Omega=\{x\in\mathbb{G}:\ \lambda r<|x|<\lambda R\}.$$

Then, for all $u\in C^1(\overline{\Omega})$, we have

$$\left(\fint_{\Omega}|u-u_{\Omega}|^{\tau}dx\right)^{\frac{1}{\tau}}\leq C_{r,R}\lambda^{\frac{a(sp-Q)}{p}}[u]^{a}_{s,p,|\cdot|,\Omega}\left(\fint_{\Omega}|u|^{\alpha}dx\right)^{\frac{1-a}{\alpha}},\tag{4.49}$$

where $C_{r,R}$ is a positive constant independent of u and λ.

Proof of Lemma 4.5.4. Without loss of generality, we assume that $0<s'\leq s$ and $\tau'\geq\tau$ are such that

$$\frac{1}{\tau'}=a\left(\frac{1}{p}-\frac{s'}{Q}\right)+\frac{1-a}{\alpha},$$

and $\lambda=1$ with

$$\Omega_1:=\{x\in\mathbb{G}:\ r<|x|<R\}.$$

By using Theorem 4.4.1, Jensen's inequality and $[u]_{s',p,|\cdot|,\Omega}\leq C[u]_{s,p,|\cdot|,\Omega}$, we compute

$$\left(\fint_{\Omega_1}|u-u_{\Omega_1}|^{\tau}dx\right)^{1/\tau}=\frac{1}{|\Omega_1|^{\frac{1}{\tau}}}\|u-u_{\Omega_1}\|_{\tau}\leq C_{r,R}\|u-u_{\Omega_1}\|_{L^{\tau'}(\Omega_1)}$$

$$
\begin{aligned}
&\leq C_{r,R}[u-u_{\Omega_1}]^a_{s',p,|\cdot|,\Omega_1}\|u\|^{1-a}_{L^\alpha(\Omega_1)}\\
&\leq C_{r,R}\left(\int_{\Omega_1}\int_{\Omega_1}\frac{|u(x)-u_{\Omega_1}-u(y)+u_{\Omega_1}|^p}{|y^{-1}x|^{Q+s'p}}dxdy\right)^{a/p}\|u\|^{1-a}_{L^\alpha(\Omega_1)}\\
&\leq C_{r,R}[u]^a_{s,p,|\cdot|,\Omega_1}\|u\|^{1-a}_{L^\alpha(\Omega_1)}\leq C_{r,R}[u]^a_{s,p,|\cdot|,\Omega_1}\left(\fint_{\Omega_1}|u|^\alpha dx\right)^{(1-a)/\alpha},
\end{aligned}
$$

where $C_{r,R}>0$. Let us apply the above inequality to $u(\lambda x)$ instead of $u(x)$. This yields

$$
\begin{aligned}
&\left(\fint_{\Omega_1}\left|u(\lambda x)-\fint_{\Omega_1}u(\lambda x)dx\right|^\tau dx\right)^{1/\tau}\\
&\quad\leq C_{r,R}\left(\int_{\Omega_1}\int_{\Omega_1}\frac{|u(\lambda x)-u(\lambda y)|^p}{|y^{-1}x|^{Q+sp}}dxdy\right)^{a/p}\left(\frac{1}{|\Omega_1|}\int_{\Omega_1}|u(\lambda x)|^\alpha dx\right)^{(1-a)/\alpha}.
\end{aligned}
$$

Therefore, we have

$$
\begin{aligned}
&\left(\fint_{\Omega}\left|u(x)-\fint_{\Omega}u(x)dx\right|^\tau dx\right)^{\frac{1}{\tau}}=\left(\frac{1}{|\Omega|}\int_{\Omega}\left|u(x)-\frac{1}{|\Omega|}\int_{\Omega}u(x)dx\right|^\tau dx\right)^{\frac{1}{\tau}}\\
&=\left(\frac{1}{|\Omega|}\int_{\Omega}\left|u(\lambda y)-\frac{1}{|\Omega|}\int_{\Omega}u(\lambda y)d(\lambda y)\right|^\tau d(\lambda y)\right)^{\frac{1}{\tau}}\\
&=\left(\frac{1}{|\Omega_1|}\int_{\Omega_1}\frac{\lambda^Q}{\lambda^Q}\left|u(\lambda y)-\frac{\lambda^Q}{\lambda^Q|\Omega_1|}\int_{\Omega_1}u(\lambda y)dy\right|^\tau dy\right)^{\frac{1}{\tau}}\\
&=\left(\frac{1}{|\Omega_1|}\int_{\Omega_1}\left|u(\lambda y)-\frac{1}{|\Omega_1|}\int_{\Omega_1}u(\lambda y)dy\right|^\tau dy\right)^{\frac{1}{\tau}}\\
&\leq C_{r,R}\left(\int_{\Omega_1}\int_{\Omega_1}\frac{|u(\lambda x)-u(\lambda y)|^p}{|y^{-1}x|^{Q+sp}}dxdy\right)^{\frac{a}{p}}\left(\frac{1}{|\Omega_1|}\int_{\Omega_1}|u(\lambda x)|^\alpha dx\right)^{\frac{1-a}{\alpha}}\\
&=C_{r,R}\left(\int_{\Omega_1}\int_{\Omega_1}\frac{\lambda^{2Q}\lambda^{Q+sp}|u(\lambda x)-u(\lambda y)|^p}{\lambda^{2Q}\lambda^{Q+sp}|y^{-1}x|^{Q+sp}}dxdy\right)^{\frac{a}{p}}\left(\frac{1}{|\Omega_1|}\int_{\Omega_1}\frac{\lambda^Q}{\lambda^Q}|u(\lambda x)|^\alpha dx\right)^{\frac{1-a}{\alpha}}\\
&=C_{r,R}\left(\int_{\Omega}\int_{\Omega}\frac{\lambda^{sp-Q}|u(\lambda x)-u(\lambda y)|^p}{|(\lambda y)^{-1}\lambda x|^{Q+sp}}d(\lambda x)d(\lambda y)\right)^{\frac{a}{p}}\left(\frac{1}{|\Omega|}\int_{\Omega}|u(\lambda x)|^\alpha d(\lambda x)\right)^{\frac{1-a}{\alpha}}\\
&=C_{r,R}\left(\int_{\Omega}\int_{\Omega}\frac{\lambda^{sp-Q}|u(x)-u(y)|^p}{|y^{-1}x|^{Q+sp}}dxdy\right)^{\frac{a}{p}}\left(\frac{1}{|\Omega|}\int_{\Omega}|u(x)|^\alpha dx\right)^{\frac{1-a}{\alpha}}\\
&=C_{r,R}\lambda^{\frac{a(sp-Q)}{p}}[u]^a_{s,p,|\cdot|,\Omega}\left(\frac{1}{|\Omega|}\int_{\Omega}|u(x)|^\alpha dx\right)^{\frac{1-a}{\alpha}}.
\end{aligned}
$$

This completes the proof. □

Proof of Theorem 4.5.2. First let us consider the case (4.46), that is, $\beta - \sigma \leq s$ and $\frac{1}{\tau} + \frac{\gamma}{Q} = \frac{1}{p} + \frac{\beta - s}{Q}$. By using Lemma 4.5.4 with $\lambda = 2^k$, $r = 1$, $R = 2$ and $\Omega = A_k$, we get

$$\left(\fint_{A_k} |u - u_{A_k}|^\tau dx \right)^{1/\tau} \leq C 2^{\frac{ak(sp-Q)}{p}} [u]^a_{s,p,|\cdot|,A_k} \left(\fint_{A_k} |u|^\alpha dx \right)^{(1-a)/\alpha}. \quad (4.50)$$

Here and below A_k is the quasi-annulus defined by

$$A_k := \{x \in \mathbb{G} : \ 2^k \leq |x| < 2^{k+1}\},$$

for $k \in \mathbb{Z}$. Now by using (4.50) we obtain

$$\begin{aligned} \int_{A_k} |u|^\tau dx &= \int_{A_k} |u - u_{A_k} + u_{A_k}|^\tau dx \\ &\leq C \left(\int_{A_k} |u_{A_k}|^\tau dx + \int_{A_k} |u - u_{A_k}|^\tau dx \right) \\ &= C \left(\int_{A_k} |u_{A_k}|^\tau dx + \frac{|A_k|}{|A_k|} \int_{A_k} |u - u_{A_k}|^\tau dx \right) \\ &= C \left(|A_k| |u_{A_k}|^\tau + |A_k| \fint_{A_k} |u - u_{A_k}|^\tau dx \right) \qquad (4.51) \\ &\leq C \left(|A_k| |u_{A_k}|^\tau + 2^{\frac{ak(sp-Q)\tau}{p}} |A_k| [u]^{a\tau}_{s,p,|\cdot|,A_k} \left(\frac{1}{|A_k|} \int_{A_k} |u|^\alpha dx \right)^{\frac{(1-a)\tau}{\alpha}} \right) \\ &\leq C \left(2^{Qk} |u_{A_k}|^\tau + 2^{\frac{ak(sp-Q)\tau}{p}} 2^{kQ} 2^{-\frac{Q(1-a)\tau k}{\alpha}} [u]^{a\tau}_{s,p,|\cdot|,A_k} \|u\|^{(1-a)\tau}_{L^\alpha(A_k)} \right). \end{aligned}$$

Then, from (4.51) we get

$$\begin{aligned} \int_{A_k} &|x|^{\gamma\tau} |u|^\tau dx \leq 2^{(k+1)\gamma\tau} \int_{A_k} |u|^\tau dx \\ &\leq C 2^{(Q+\gamma\tau)k} |u_{A_k}|^\tau + C 2^{\gamma\tau k} 2^{kQ} 2^{\frac{ak(sp-Q)\tau}{p}} 2^{-\frac{Q(1-a)\tau k}{\alpha}} [u]^{a\tau}_{s,p,|\cdot|,A_k} \|u\|^{(1-a)\tau}_{L^\alpha(A_k)} \\ &= C 2^{(Q+\gamma\tau)k} |u_{A_k}|^\tau \\ &\quad + C 2^{\left(\gamma\tau + Q + \frac{a(sp-Q)\tau}{p} - \frac{Q(1-a)\tau}{\alpha}\right)k} \left(\int_{A_k} \int_{A_k} \frac{2^{kp\beta_1} 2^{kp\beta_2} |u(x) - u(y)|^p}{2^{kp\beta} |y^{-1}x|^{Q+sp}} dx dy \right)^{\frac{a\tau}{p}} \\ &\quad \times \left(\int_{A_k} \frac{2^{k\alpha\mu}}{2^{k\alpha\mu}} |u(x)|^\alpha dx \right)^{\frac{(1-a)\tau}{\alpha}} \\ &\leq C 2^{(Q+\gamma\tau)k} |u_{A_k}|^\tau \\ &\quad + C 2^{\left(\gamma\tau + Q + \frac{a(sp-Q)\tau}{p} - \frac{Q(1-a)\tau}{\alpha} - a\beta\tau - \mu\tau(1-a)\right)k} \\ &\quad \times \left(\int_{A_k} \int_{A_k} \frac{|x|^{p\beta_1} |y|^{p\beta_2} |u(x) - u(y)|^p}{|y^{-1}x|^{Q+sp}} dx dy \right)^{\frac{a\tau}{p}} \end{aligned}$$

$$\begin{aligned}
&\times \left(\int_{A_k} |x|^{\alpha\mu}|u(x)|^\alpha dx\right)^{\frac{(1-a)\tau}{\alpha}} \\
&\leq C2^{(Q+\gamma\tau)k}|u_{A_k}|^\tau \qquad (4.52)\\
&+ C2^{\left(\gamma\tau+Q+\frac{a(sp-Q)\tau}{p}-\frac{Q(1-a)\tau}{\alpha}-a\beta\tau-\mu\tau(1-a)\right)k}[u]^{a\tau}_{s,p,\beta,|\cdot|,A_k}\||x|^\mu u\|^{(1-a)\tau}_{L^\alpha(A_k)}.
\end{aligned}$$

Here by (4.45), we have

$$\begin{aligned}
&\gamma\tau + Q + \frac{a(sp-Q)\tau}{p} - \frac{Q(1-a)\tau}{\alpha} - a\beta\tau - \mu\tau(1-a) \\
&= Q\tau\left(\frac{\gamma}{Q} + \frac{1}{\tau} + \frac{a(sp-Q)}{Qp} - \frac{(1-a)}{\alpha} - \frac{a\beta}{Q} - \frac{\mu(1-a)}{Q}\right) \\
&= Q\tau\left(a\left(\frac{1}{p} + \frac{\beta-s}{Q}\right) + (1-a)\left(\frac{1}{\alpha} + \frac{\mu}{Q}\right)\right) \\
&+ Q\tau\left(\frac{a(sp-Q)}{Qp} - \frac{(1-a)}{\alpha} - \frac{a\beta}{Q} - \frac{\mu(1-a)}{Q}\right) = 0. \qquad (4.53)
\end{aligned}$$

Thus, we obtain

$$\int_{A_k} |x|^{\gamma\tau}|u|^\tau dx \leq C2^{(\gamma\tau+Q)k}|u_{A_k}|^\tau + C[u]^{a\tau}_{s,p,\beta,|\cdot|,A_k}\||x|^\mu u\|^{(1-a)\tau}_{L^\alpha(A_k)}, \qquad (4.54)$$

and by summing over k from m to n, we get

$$\begin{aligned}
\int_{\cup_{k=m}^n A_k} |x|^{\gamma\tau}|u|^\tau dx &= \int_{\{2^m<|x|<2^{n+1}\}} |x|^{\gamma\tau}|u|^\tau dx \qquad (4.55)\\
&\leq C\sum_{k=m}^n 2^{(\gamma\tau+Q)k}|u_{A_k}|^\tau + C\sum_{k=m}^n [u]^{a\tau}_{s,p,\beta,|\cdot|,A_k}\||x|^\mu u\|^{(1-a)\tau}_{L^\alpha(A_k)},
\end{aligned}$$

where $k, m, n \in \mathbb{Z}$ and $m \leq n-2$.

To prove (4.47) let us choose n such that

$$\operatorname{supp} u \subset B_{2^n}, \qquad (4.56)$$

where B_{2^n} is a quasi-ball of $\mathbb{G}$ with the radius 2^n.

Let us consider the following integral

$$\begin{aligned}
&⨍_{A_{k+1}\cup A_k}\left|u - ⨍_{A_{k+1}\cup A_k} u\right|^\tau dx = \frac{1}{|A_{k+1}|+|A_k|}\int_{A_{k+1}\cup A_k}\left|u - ⨍_{A_{k+1}\cup A_k} u\right|^\tau dx \\
&= \frac{1}{|A_{k+1}|+|A_k|}\left(\int_{A_{k+1}}\left|u - ⨍_{A_{k+1}\cup A_k} u\right|^\tau dx + \int_{A_k}\left|u - ⨍_{A_{k+1}\cup A_k} u\right|^\tau dx\right).
\end{aligned}$$

On the other hand, a direct calculation gives

$$\begin{aligned}
&\fint_{A_{k+1}\cup A_k}\left|u-\fint_{A_{k+1}\cup A_k}u\right|^\tau dx\\
&=\frac{1}{|A_{k+1}|+|A_k|}\left(\int_{A_{k+1}}\left|u-\fint_{A_{k+1}\cup A_k}u\right|^\tau dx+\int_{A_k}\left|u-\fint_{A_{k+1}\cup A_k}u\right|^\tau dx\right)\\
&\geq\frac{1}{|A_{k+1}|+|A_k|}\int_{A_k}\left|u-\fint_{A_{k+1}\cup A_k}u\right|^\tau dx\\
&\geq\frac{1}{|A_{k+1}|+|A_k|}\left|\int_{A_k}\left(u-\fint_{A_{k+1}\cup A_k}u\right)dx\right|^\tau\\
&=\frac{1}{|A_{k+1}|+|A_k|}\left|\int_{A_k}udx-\frac{|A_k|}{|A_{k+1}|+|A_k|}\int_{A_k}udx-\frac{|A_k|}{|A_{k+1}|+|A_k|}\int_{A_{k+1}}udx\right|^\tau\\
&=\frac{1}{|A_{k+1}|+|A_k|}\left|\frac{|A_{k+1}|}{|A_{k+1}|+|A_k|}\int_{A_k}udx-\frac{|A_k|}{|A_{k+1}|+|A_k|}\int_{A_{k+1}}udx\right|^\tau\\
&=\frac{1}{(|A_{k+1}|+|A_k|)^{\tau+1}}\left||A_{k+1}|\int_{A_k}udx-|A_k|\int_{A_{k+1}}udx\right|^\tau\\
&=\frac{|A_{k+1}|^\tau|A_k|^\tau}{(|A_{k+1}|+|A_k|)^{\tau+1}}\left|\frac{1}{|A_k|}\int_{A_k}udx-\frac{1}{|A_{k+1}|}\int_{A_{k+1}}udx\right|^\tau\\
&=\frac{|A_{k+1}|^\tau|A_k|^\tau}{(|A_{k+1}|+|A_k|)^{\tau+1}}|u_{A_{k+1}}-u_{A_k}|^\tau\\
&\geq C|u_{A_{k+1}}-u_{A_k}|^\tau. \qquad (4.57)
\end{aligned}$$

From (4.57) and Lemma 4.5.4, we obtain

$$\begin{aligned}
|u_{A_{k+1}}-u_{A_k}|^\tau &\leq C\fint_{A_{k+1}\cup A_k}\left|u-\fint_{A_{k+1}\cup A_k}u\right|^\tau dx\\
&\leq C2^{\frac{ak(sp-Q)}{p}}[u]^{\tau a}_{s,p,|\cdot|,A_{k+1}\cup A_k}\left(\fint_{A_{k+1}\cup A_k}|u|^\alpha dx\right)^{\frac{(1-a)\tau}{\alpha}}. \qquad (4.58)
\end{aligned}$$

By using this fact, taking $\tau=1$ we have

$$\begin{aligned}
|u_{A_k}| &\leq |u_{A_{k+1}}-u_{A_k}|+|u_{A_{k+1}}|\\
&\leq |u_{A_{k+1}}|+C2^{\frac{ak(sp-Q)}{p}}[u]^{a}_{s,p,|\cdot|,A_{k+1}\cup A_k}\left(\fint_{A_{k+1}\cup A_k}|u|^\alpha dx\right)^{\frac{(1-a)}{\alpha}}. \qquad (4.59)
\end{aligned}$$

On the other hand (see, e.g., [NS18b, Lemma 2.2]), there exists a positive constant C depending $\xi > 1$ and $\eta > 1$ such that $1 < \zeta < \xi$,

$$(|a| + |b|)^\eta \leq \zeta |a|^\eta + \frac{C}{(\zeta - 1)^{\eta - 1}} |b|^\eta, \quad \forall\, a, b \in \mathbb{R}. \tag{4.60}$$

Thus, by using with $\eta = \tau$, $\zeta = 2^{\gamma\tau + Q} c$, where $c = \frac{2}{1 + 2^{\gamma\tau + Q}} < 1$, since $\gamma\tau + Q > 0$, from the previous inequality we have

$$2^{(\gamma\tau+Q)k} |u_{A_k}|^\tau \leq c 2^{(k+1)(\gamma\tau+Q)} |u_{A_{k+1}}|^\tau + C[u]^{\tau a}_{s,p,\beta,|\cdot|,A_{k+1}\cup A_k} \| |x|^\mu u \|^{(1-a)\tau}_{L^\alpha(A_{k+1}\cup A_k)}.$$

By summing over k from m to n and by using (4.56) we have

$$\begin{aligned} \sum_{k=m}^{n} 2^{(\gamma\tau+Q)k} |u_{A_k}|^\tau &\leq \sum_{k=m}^{n} c 2^{(k+1)(\gamma\tau+Q)} |u_{A_{k+1}}|^\tau \\ &\quad + C \sum_{k=m}^{n} [u]^{\tau a}_{s,p,\beta,|\cdot|,A_{k+1}\cup A_k} \| |x|^\mu u \|^{(1-a)\tau}_{L^\alpha(A_{k+1}\cup A_k)}. \end{aligned} \tag{4.61}$$

By using (4.61), we compute

$$\begin{aligned} (1-c) \sum_{k=m}^{n} 2^{(\gamma\tau+Q)k} |u_{A_k}|^\tau &\leq 2^{(\gamma\tau+Q)m} |u_{A_m}|^\tau + (1-c) \sum_{k=m+1}^{n} 2^{(\gamma\tau+Q)k} |u_{A_k}|^\tau \\ &\leq C \sum_{k=m}^{n} [u]^{\tau a}_{s,p,\beta,|\cdot|,A_{k+1}\cup A_k} \| |x|^\mu u \|^{(1-a)\tau}_{L^\alpha(A_{k+1}\cup A_k)}. \end{aligned}$$

This yields

$$\sum_{k=m}^{n} 2^{(\gamma\tau+Q)k} |u_{A_k}|^\tau \leq C \sum_{k=m}^{n} [u]^{\tau a}_{s,p,\beta,|\cdot|,A_{k+1}\cup A_k} \| |x|^\mu u \|^{(1-a)\tau}_{L^\alpha(A_{k+1}\cup A_k)}. \tag{4.62}$$

From (4.55) and (4.62), we have

$$\int_{\{2^m < |x| < 2^{n+1}\}} |x|^{\gamma\tau} |u|^\tau dx \leq C \sum_{k=m}^{n} [u]^{\tau a}_{s,p,\beta,|\cdot|,A_{k+1}\cup A_k} \| |x|^\mu u \|^{(1-a)\tau}_{L^\alpha(A_{k+1}\cup A_k)}. \tag{4.63}$$

Let $s, t \geq 0$ be such that $s + t \geq 1$. Then for any $x_k, y_k \geq 0$, we have

$$\sum_{k=m}^{n} x_k^s y_k^t \leq \left(\sum_{k=m}^{n} x_k \right)^s \left(\sum_{k=m}^{n} y_k \right)^t. \tag{4.64}$$

By using this inequality in (4.63) with $s = \frac{\tau a}{p}$, $t = \frac{(1-a)\tau}{\alpha}$, $\frac{a}{p} + \frac{1-a}{\alpha} \geq \frac{1}{\tau}$ and $s \geq \beta - \sigma$, we obtain

$$\int_{\{|x| > 2^m\}} |x|^{\gamma\tau} |u|^\tau dx \leq C [u]^{a\tau}_{s,p,\beta,|\cdot|,\cup_{k=m}^{\infty} A_k} \| |x|^\mu u \|^{(1-a)\tau}_{L^\alpha(\cup_{k=m}^{\infty} A_k)}.$$

Inequality (4.47) is proved.

Let us prove (4.48). The strategy of the proof is similar to the previous case. Choose m such that

$$\operatorname{supp} u \cap B_{2^m} = \emptyset. \tag{4.65}$$

From Lemma 4.5.4 we have

$$|u_{A_{k+1}} - u_{A_k}|^\tau \leq C 2^{\frac{a\tau k(sp-Q)}{p}} [u]^{\tau a}_{s,p,|\cdot|,A_{k+1}\cup A_k} \left(\fint_{A_{k+1}\cup A_k} |u|^\alpha dx \right)^{\frac{(1-a)\tau}{\alpha}}.$$

By (4.60) and choosing $c = \frac{1+2^{\gamma\tau+Q}}{2} < 1$, since $\gamma\tau + Q < 0$, we have

$$2^{(\gamma\tau+Q)(k+1)} |u_{A_{k+1}}|^\tau \leq c 2^{k(\gamma\tau+Q)} |u_{A_k}|^\tau + C[u]^{\tau a}_{s,p,\beta,|\cdot|,A_{k+1}\cup A_k} \| |x|^\mu u \|^{(1-a)\tau}_{L^\alpha(A_{k+1}\cup A_k)},$$

and by summing over k from m to n and by using (4.65) we obtain

$$\sum_{k=m}^{n} 2^{(\gamma\tau+Q)k} |u_{A_k}|^\tau \leq C \sum_{k=m-1}^{n-1} [u]^{\tau a}_{s,p,\beta,|\cdot|,A_{k+1}\cup A_k} \| |x|^\mu u \|^{(1-a)\tau}_{L^\alpha(A_{k+1}\cup A_k)}. \tag{4.66}$$

From (4.55) and (4.66), we establish that

$$\int_{\{2^m<|x|<2^{n+1}\}} |x|^{\gamma\tau} |u|^\tau dx \leq C \sum_{k=m-1}^{n-1} [u]^{\tau a}_{s,p,\beta,|\cdot|,A_{k+1}\cup A_k} \| |x|^\mu u \|^{(1-a)\tau}_{L^\alpha(A_{k+1}\cup A_k)}.$$

Now by using (4.64) we get

$$\int_{\{|x|<2^{n+1}\}} |x|^{\gamma\tau} |u|^\tau dx \leq C[u]^{\tau a}_{s,p,\beta,|\cdot|,\cup_{k=-\infty}^{n} A_k} \| |x|^\mu u \|^{(1-a)\tau}_{L^\alpha(\cup_{k=-\infty}^{n} A_k)}.$$

The proof of the case $s \geq \beta - \sigma$ is complete.

Let us prove the case of $\beta - \sigma > s$. Without loss of generality, we assume that

$$[u]_{s,p,\beta,|\cdot|} = \|u\|_{L^\alpha(\mathbb{G})} = 1,$$

where

$$\frac{1}{p} + \frac{\beta - s}{Q} \neq \frac{1}{\alpha} + \frac{\mu}{Q}.$$

We also assume that $a_1 > 0$, $1 > a_2$ and $\tau_1, \tau_2 > 0$ with

$$\frac{1}{\tau_2} = \frac{a_2}{p} + \frac{1-a_2}{\alpha},$$

and

$$\begin{aligned} &\text{if } \frac{a}{p} + \frac{1-a}{\alpha} - \frac{as}{Q} > 0, \quad \text{then } \frac{1}{\tau_1} = \frac{a_1}{p} + \frac{1-a_1}{\alpha} - \frac{a_1 s}{Q}, \\ &\text{if } \frac{a}{p} + \frac{1-a}{\alpha} - \frac{as}{Q} \leq 0, \quad \text{then } \frac{1}{\tau} > \frac{1}{\tau_1} \geq \frac{a_1}{p} + \frac{1-a_1}{\alpha} - \frac{a_1 s}{Q}. \end{aligned} \tag{4.67}$$

Taking $\gamma_1 = a_1\beta + (1-a_1)\mu$ and $\gamma_2 = a_2(\beta - s) + (1-a_2)\mu$, we obtain

$$\frac{1}{\tau_1} + \frac{\gamma_1}{Q} \geq a_1\left(\frac{1}{p} + \frac{\beta - s}{Q}\right) + (1-a_1)\left(\frac{1}{\alpha} + \frac{\mu}{Q}\right) \tag{4.68}$$

and

$$\frac{1}{\tau_2} + \frac{\gamma_2}{Q} = a_2\left(\frac{1}{p} + \frac{\beta - s}{Q}\right) + (1-a_2)\left(\frac{1}{\alpha} + \frac{\mu}{Q}\right). \tag{4.69}$$

Let a_1 and a_2 be such that

$$|a - a_1| \text{ and } |a - a_2| \text{ are small enough,} \tag{4.70}$$

$$\begin{aligned} &a_2 < a < a_1, \text{ if } \frac{1}{p} + \frac{\beta - s}{Q} > \frac{1}{\alpha} + \frac{\mu}{Q}, \\ &a_1 < a < a_2, \text{ if } \frac{1}{p} + \frac{\beta - s}{Q} < \frac{1}{\alpha} + \frac{\mu}{Q}. \end{aligned} \tag{4.71}$$

By using (4.70)–(4.71) in (4.68), (4.69) and (4.45), we establish

$$\frac{1}{\tau_1} + \frac{\gamma_1}{Q} > \frac{1}{\tau} + \frac{\gamma}{Q} > \frac{1}{\tau_2} + \frac{\gamma_2}{Q} > 0.$$

From (4.67) in the case $\frac{a}{p} + \frac{1-a}{\alpha} - \frac{as}{Q} > 0$ with $a > 0$, $\beta - \sigma > s$ and (4.70), we get

$$\frac{1}{\tau} - \frac{1}{\tau_1} = (a - a_1)\left(\frac{1}{p} - \frac{s}{Q} - \frac{1}{\alpha}\right) + \frac{a}{Q}(\beta - \sigma) > 0, \tag{4.72}$$

and

$$\frac{1}{\tau} - \frac{1}{\tau_2} = (a - a_2)\left(\frac{1}{p} - \frac{1}{\alpha}\right) + \frac{a}{Q}(\beta - \sigma - s) > 0. \tag{4.73}$$

From (4.67), (4.72) and (4.73), we have

$$\tau_1 > \tau, \quad \tau_2 > \tau.$$

Thus, using this, (4.70) and Hölder's inequality, we obtain

$$\||x|^{\gamma}u\|_{L^{\tau}(\mathbb{G}\setminus B_1)} \leq C\||x|^{\gamma_1}u\|_{L^{\tau_1}(\mathbb{G})},$$

and

$$\||x|^{\gamma}u\|_{L^{\tau}(B_1)} \leq C\||x|^{\gamma_2}u\|_{L^{\tau_2}(\mathbb{G})},$$

where B_1 is the unit quasi-ball. By using the previous case, we establish

$$\||x|^{\gamma_1}u\|_{L^{\tau_1}(\mathbb{G})} \leq C[u]^{a_1}_{s,p,\beta,|\cdot|}\||x|^{\mu}u\|^{1-a_1}_{L^{\alpha}(\mathbb{G})} \leq C,$$

and

$$\||x|^{\gamma_2}u\|_{L^{\tau_2}(\mathbb{G})} \leq C[u]^{a_2}_{s,p,\beta,|\cdot|}\||x|^{\mu}u\|^{1-a_2}_{L^{\alpha}(\mathbb{G})} \leq C.$$

The proof of Theorem 4.5.2 is complete. □

Now we consider the critical case $\frac{1}{\tau} + \frac{\gamma}{Q} = 0$.

Theorem 4.5.5 (Fractional critical Caffarelli–Kohn–Nirenberg inequality). *Let $\mathbb{G}$ be a homogeneous group of homogeneous dimension Q. Let $Q \geq 2$, $s \in (0,1)$, $p > 1$, $\alpha \geq 1$, $\tau > 1$, $a \in (0,1]$, β_1, β_2, β, μ, $\gamma \in \mathbb{R}$, $\beta_1 + \beta_2 = \beta$,*

$$\frac{1}{\tau} + \frac{\gamma}{Q} = a\left(\frac{1}{p} + \frac{\beta - s}{Q}\right) + (1-a)\left(\frac{1}{\alpha} + \frac{\mu}{Q}\right). \tag{4.74}$$

In addition, assume that $0 \leq \beta - \sigma \leq s$ with $\gamma = a\sigma + (1-a)\mu$.

If $\frac{1}{\tau} + \frac{\gamma}{Q} = 0$ and $\operatorname{supp} u \subset B_R = \{x \in \mathbb{G} : |x| < R\}$, then we have

$$\left\| \frac{|x|^{\gamma}}{\ln \frac{2R}{|x|}} u \right\|_{L^{\tau}(\mathbb{G})} \leq C[u]^{a}_{s,p,\beta,|\cdot|} \||x|^{\mu} u\|^{1-a}_{L^{\alpha}(\mathbb{G})}, \tag{4.75}$$

for all $u \in C^1_c(\mathbb{G})$.

Proof of Theorem 4.5.5. The proof is similar to the proof of Theorem 4.5.2. In (4.54), summing over k from m to n and fixing $\varepsilon > 0$, we have

$$\begin{aligned}\int_{\{|x|>2^m\}} \frac{|x|^{\gamma\tau}}{\ln^{1+\varepsilon}\left(\frac{2R}{|x|}\right)} |u|^{\tau} dx &\leq C \sum_{k=m}^{n} \frac{1}{(n+1-k)^{1+\varepsilon}} |u_{A_k}|^{\tau} \\ &\quad + C \sum_{k=m}^{n} [u]^{a\tau}_{s,p,\beta,|\cdot|,A_k} \||x|^{\mu} u\|^{(1-a)\tau}_{L^{\alpha}(A_k)}.\end{aligned} \tag{4.76}$$

From Lemma 4.5.4, we have

$$|u_{A_{k+1}} - u_{A_k}| \leq C 2^{\frac{ak(sp-Q)}{p}} [u]^{a}_{s,p,|\cdot|,A_{k+1}\cup A_k} \left(\int_{A_{k+1}\cup A_k} |u|^{\alpha} dx \right)^{\frac{1-a}{\alpha}}.$$

By using (4.60) with $\zeta = \frac{(n+1-k)^{\varepsilon}}{(n+\frac{1}{2}-k)^{\varepsilon}}$ we get

$$\begin{aligned}\frac{|u_{A_k}|^{\tau}}{(n+1-k)^{\varepsilon}} &\leq \frac{|u_{A_{k+1}}|^{\tau}}{(n+\frac{1}{2}-k)^{\varepsilon}} \\ &\quad + C(n+1-k)^{\tau-1-\varepsilon} [u]^{a\tau}_{s,p,\beta,|\cdot|,A_{k+1}\cup A_k} \||x|^{\mu} u\|^{(1-a)\tau}_{L^{\alpha}(A_{k+1}\cup A_k)}.\end{aligned} \tag{4.77}$$

For $\varepsilon > 0$ and $n \geq k$, we have

$$\frac{1}{(n-k+1)^{\varepsilon}} - \frac{1}{(n-k+\frac{3}{2})^{\varepsilon}} \sim \frac{1}{(n-k+1)^{1+\varepsilon}}. \tag{4.78}$$

By using this fact, (4.77), (4.78) and $\varepsilon = \tau - 1$, we obtain

$$\sum_{k=m}^{n} \frac{|u_{A_k}|^{\tau}}{(n+1-k)^{\tau}} \leq C \sum_{k=m}^{n} [u]^{a\tau}_{s,p,\beta,|\cdot|,A_{k+1}\cup A_k} \|q^{\mu}(x) u\|^{(1-a)\tau}_{L^{\alpha}(A_{k+1}\cup A_k)}. \tag{4.79}$$

From (4.76) and (4.79), we establish

$$\int_{\{|x|>2^m\}} \frac{|x|^{\gamma\tau}}{\ln^\tau \frac{2R}{|x|}} |u|^\tau dx \leq C \sum_{k=m}^{n} [u]^{a\tau}_{s,p,\beta,|\cdot|,A_{k+1}\cup A_k} \||x|^\mu u\|^{(1-a)\tau}_{L^\alpha(A_{k+1}\cup A_k)}.$$

By using (4.64) with (4.74) and $0 \leq \beta - \sigma \leq s$, where $s = \frac{\tau a}{p}$; $t = \frac{(1-a)\tau}{\alpha}$, we have $s + t \geq 1$ and we arrive at

$$\int_{\{|x|>2^m\}} \frac{|x|^{\gamma\tau}}{\ln^\tau \frac{2R}{|x|}} |u|^\tau dx \leq C \sum_{k=m}^{n} [u]^{a\tau}_{s,p,\beta,|\cdot|,\cup_{k=m}^{\infty} A_k} \||x|^\mu u\|^{(1-a)\tau}_{L^\alpha(\cup_{k=m}^{\infty} A_k)}.$$

Theorem 4.5.5 is proved. □

4.6 Lyapunov inequalities on homogeneous groups

In this section we give an application of the preceding results to derive a Lyapunov type inequality for the fractional p-sub-Laplacian with homogeneous Dirichlet boundary condition on homogeneous groups. First, we summarize the basic results concerning the classical Lyapunov inequality.

Remark 4.6.1 (Euclidean Lyapunov inequalities).

1. In [Lya07], Lyapunov obtained the following result for the one-dimensional homogeneous Dirichlet boundary value problem. Consider the second-order ordinary differential equation

$$\begin{cases} u''(x) + \omega(x)u(x) = 0, \ x \in (a,b), \\ u(a) = u(b) = 0. \end{cases} \tag{4.80}$$

Then, if (4.80) has a non-trivial solution u, and $\omega = \omega(x)$ is a real-valued and continuous function on $[a,b]$, then we must have

$$\int_a^b |\omega(x)|dx > \frac{4}{b-a}. \tag{4.81}$$

Inequality (4.81) is called a (classical) *Lyapunov inequality*.

2. Nowadays, there are many extensions of the Lyapunov inequality (4.81). For example, in [Elb81] the Lyapunov inequality for the one-dimensional Dirichlet p-Laplacian was obtained: if

$$\begin{cases} (|u'(x)|^{p-2}u'(x))' + \omega(x)u(x) = 0, \ x \in (a,b), \ 1 < p < \infty, \\ u(a) = u(b) = 0, \end{cases} \tag{4.82}$$

has a non-trivial solution u for $\omega \in L^1(a,b)$, then

$$\int_a^b |\omega(x)|dx > \frac{2^p}{(b-a)^{p-1}}, \quad 1 < p < \infty. \tag{4.83}$$

Obviously, taking $p=2$ in (4.83), we recover (4.81).

3. In [JKS17], the following Lyapunov inequality was obtained for the multi-dimensional fractional p-Laplacian $(-\Delta_p)^s$, $1<p<\infty$, $s \in (0,1)$, with a homogeneous Dirichlet boundary condition, that is, for the equation

$$\begin{cases} (-\Delta_p)^s u = \omega(x)|u|^{p-2}u, \ x \in \Omega, \\ u(x) = 0, \ x \in \mathbb{R}^N \setminus \Omega, \end{cases} \tag{4.84}$$

where $\Omega \subset \mathbb{R}^N$ is a measurable set, $1<p<\infty$, and $s \in (0,1)$. More precisely, let $\omega \in L^\theta(\Omega)$ with $N > sp$, $\frac{N}{sp} < \theta < \infty$, be a non-negative weight. Suppose that the problem (4.84) has a non-trivial weak solution $u \in W_0^{s,p}(\Omega)$. Then we have

$$\left(\int_\Omega \omega^\theta(x)\, dx\right)^{1/\theta} > \frac{C}{r_\Omega^{sp-\frac{N}{\theta}}}, \tag{4.85}$$

where $C>0$ is a universal constant and r_Ω is the inner radius of Ω.

The appearance of the inner radius in (4.85) motivates one to define its analogue also in the setting of homogeneous groups.

Definition 4.6.2 (Inner quasi-radius)**.** Let $p>1$ and $s \in (0,1)$ be such that $Q > sp$. Let $\Omega \subset \mathbb{G}$ be a Haar measurable set. We denote by $r_{\Omega,q}$ the *inner quasi-radius* of Ω, that is,

$$r_\Omega = r_{\Omega,|\cdot|} := \sup\{|x| : \ x \in \Omega\}. \tag{4.86}$$

Clearly, the exact values depend on the choice of a homogeneous quasi-norm $|\cdot|$. As before, if the quasi-norm is fixed, we can omit it from the notation.

4.6.1 Lyapunov type inequality for fractional p-sub-Laplacians

We now turn our attention to the Lyapunov inequalities for the fractional p-sub-Laplacian $(-\Delta_p)^s$ from Definition 4.1.1. Let us consider the boundary value problem

$$\begin{cases} (-\Delta_p)^s u(x) = \omega(x)|u(x)|^{p-2}u(x), \ x \in \Omega, \\ u(x) = 0, \quad x \in \mathbb{G} \setminus \Omega, \end{cases} \tag{4.87}$$

where $\omega \in L^\infty(\Omega)$. A function $u \in W_0^{s,p}(\Omega)$ is called a *weak solution of the problem* (4.87) if we have

$$\int_\Omega \int_\Omega \frac{|u(x)-u(y)|^{p-2}(u(x)-u(y))(v(x)-v(y))}{|y^{-1}x|^{Q+sp}} dxdy$$
$$= \int_\Omega \omega(x)|u(x)|^{p-2}u(x)v(x)dx \quad \text{for all } v \in W_0^{s,p}(\Omega).$$

Theorem 4.6.3 (Lyapunov inequality for fractional p-sub-Laplacian). *Let $\mathbb{G}$ be a homogeneous group of homogeneous dimension Q. Let $\Omega \subset \mathbb{G}$ be a Haar measurable set. Let $\omega \in L^\theta(\Omega)$ be a non-negative function with $\frac{Q}{sp} < \theta < \infty$, and with $Q > ps$. Suppose that the boundary value problem (4.87) has a non-trivial weak solution $u \in W_0^{s,p}(\Omega)$. Then we have*

$$\|\omega\|_{L^\theta(\Omega)} \geq C \Big/ r_{\Omega,|\cdot|}^{sp-Q/\theta}, \tag{4.88}$$

for some $C = C(Q, p, s, |\cdot|) > 0$.

Proof of Theorem 4.6.3. We fix a homogeneous quasi-norm $|\cdot|$ thus also eliminating it from the notation. Let $\alpha := \frac{\theta-\theta/sp}{\theta-1} \in (0,1)$ and let p^* be the Sobolev conjugate exponent as in Theorem 4.3.1. Let us define

$$\beta := \alpha p + (1-\alpha)p^*.$$

Let $\beta = p\theta'$ with $1/\theta + 1/\theta' = 1$. Then we have

$$\int_\Omega \frac{|u(x)|^\beta}{r_\Omega^{\alpha sp}} dx \leq \int_\Omega \frac{|u(x)|^\beta}{|x|^{\alpha sp}} dx. \tag{4.89}$$

On the other hand, the Hölder inequality with exponents $\nu = \alpha^{-1}$ and its conjugate $1/\nu + 1/\nu' = 1$ implies

$$\begin{aligned}\int_\Omega \frac{|u(x)|^\beta}{|x|^{\alpha sp}} dx &\leq \int_\Omega \frac{|u(x)|^{\alpha p}|u(x)|^{(1-\alpha)p^*}}{|x|^{\alpha sp}} dx \\ &\leq \left(\int_\Omega \frac{|u(x)|^p}{|x|^{sp} dx}\right)^\alpha \left(\int_\Omega |u(x)|^{p^*} dx\right)^{1-\alpha}.\end{aligned}$$

Further, by using Theorem 4.3.1 and Theorem 4.2.1, we get

$$\begin{aligned}\int_\Omega \frac{|u(x)|^\beta}{|x|^{\alpha sp}} dx &\leq C_1^\alpha \left(\int_\Omega \int_\Omega \frac{|u(x)-u(y)|^p}{|y^{-1}x|^{Q+sp}} dx dy\right)^{\alpha/p} C_2^{(1-\alpha)p^*/p} [u]_{s,p}^{(1-\alpha)p^*/p} \\ &\leq C_1^\alpha [u]_{s,p}^\alpha C_2^{(1-\alpha)p^*/p} [u]_{s,p}^{(1-\alpha)p^*/p} \\ &= C\left([u]_{s,p}^p\right)^{(\alpha p + (1-\alpha)p^*)/p} = C\left(\int_\Omega \omega(x)|u(x)|^p dx\right)^{\theta'} \\ &\leq C\left(\int_\Omega \omega^\theta(x) dx\right)^{\theta'/\theta} \int_\Omega |u(x)|^{p\theta'} dx = C\|\omega\|_{L^\theta(\Omega)}^{\theta'} \int_\Omega |u(x)|^\beta dx.\end{aligned}$$

That is, we obtain

$$\int_\Omega \frac{|u(x)|^\beta}{|x|^{\alpha sp}} dx \leq C\|\omega\|_{L^\theta(\Omega)}^{\theta'} \int_\Omega |u(x)|^\beta dx.$$

Thus, from (4.89) we have

$$\frac{1}{r_\Omega^{\alpha sp}} \int_\Omega |u(x)|^\beta dx \leq \int_\Omega \frac{|u(x)|^\beta}{|x|^{\alpha sp}} dx \leq C\|\omega\|_{L^\theta(\Omega)}^{\theta'} \int_\Omega |u(x)|^\beta dx.$$

Finally, we arrive at

$$\frac{C}{r_{\Omega}^{sp-Q/\theta}} \leq \|\omega\|_{L^{\theta}(\Omega)}.$$

Theorem 4.6.3 is proved. □

As an application of the Lyapunov inequality, let us consider the spectral problem for the (nonlinear) fractional p-sub-Laplacian $(-\Delta_p)^s$, $1 < p < \infty$, $s \in (0,1)$, with the Dirichlet boundary condition:

$$\begin{cases} (-\Delta_p)^s u = \lambda |u|^{p-2} u, \ x \in \Omega, \\ u(x) = 0, \ x \in \mathbb{G} \setminus \Omega. \end{cases} \tag{4.90}$$

We define the corresponding *Rayleigh quotient* by

$$\lambda_1 := \inf_{u \in W_0^{s,p}(\Omega),\ u \neq 0} \frac{[u]_{s,p}^p}{\|u\|_{L^p(\mathbb{G})}^p}. \tag{4.91}$$

Clearly, its precise value may depend on the choice of the homogeneous quasi-norm $|\cdot|$. As a consequence of Theorem 4.6.3 we have

Theorem 4.6.4 (First eigenvalue for the fractional p-sub-Laplacian). *Let $\mathbb{G}$ be a homogeneous group of homogeneous dimension Q. Let λ_1 be the first eigenvalue of problem* (4.90) *given by* (4.91). *Let $Q > sp$, $s \in (0,1)$ and $1 < p < \infty$. Then we have*

$$\lambda_1 \geq \sup_{\frac{Q}{sp} < \theta < \infty} \frac{C}{|\Omega|^{\frac{1}{\theta}} r_{\Omega,|\cdot|}^{sp-Q/\theta}}, \tag{4.92}$$

where C is a positive constant given in Theorem 4.6.3 *and $|\Omega|$ is the Haar measure of Ω.*

Proof of Theorem 4.6.4. In Theorem 4.6.3, taking $\omega = \lambda \in L^{\theta}(\Omega)$ and using Lyapunov type inequality (4.88), we get that

$$\|\omega\|_{L^{\theta}(\Omega)} = \|\lambda\|_{L^{\theta}(\Omega)} = \left(\int_{\Omega} \lambda^{\theta} dx\right)^{1/\theta} \geq \frac{C}{r_{\Omega,|\cdot|}^{sp-Q/\theta}}.$$

For every $\theta > \frac{Q}{sp}$, we have

$$\lambda_1 \geq \frac{C}{|\Omega|^{\frac{1}{\theta}} r_{\Omega,|\cdot|}^{sp-Q/\theta}}.$$

Thus, we obtain

$$\lambda_1 \geq \sup_{\frac{Q}{sp} < \theta < \infty} \frac{C}{|\Omega|^{\frac{1}{\theta}} r_{\Omega,|\cdot|}^{sp-Q/\theta}},$$

for all $\frac{Q}{sp} < \theta < \infty$. Theorem 4.6.4 is proved. □

4.6.2 Lyapunov type inequality for systems

In the previous section we have presented the Lyapunov type inequality for the fractional p-sub-Laplacian with the homogeneous Dirichlet condition. Now we discuss the Lyapunov type inequality for the fractional p-sub-Laplacian system for the homogeneous Dirichlet problem.

Let $\Omega \subset \mathbb{G}$ be a Haar measurable set, let $\omega_i \in L^1(\Omega)$, $\omega_i \geq 0$, $s_i \in (0,1)$, $p_i \in (1,\infty)$. As before in Definition 4.1.1, for each p_i we denote by $(-\Delta_{p_i})^{s_i}$ the fractional p-sub-Laplacian on $\mathbb{G}$ defined by

$$(-\Delta_{p_i})^{s_i} u_i(x) = 2 \lim_{\delta \searrow 0} \int_{\mathbb{G}\setminus B(x,\delta)} \frac{|u_i(x)-u_i(y)|^{p_i-2}(u_i(x)-u_i(y))}{|y^{-1}x|^{Q+s_ip_i}} dy, \quad x \in \mathbb{G},$$
$$i = 1,\ldots,n.$$

Here $B(x,\delta)$ is a quasi-ball with respect to a fixed quasi-norm $|\cdot|$, with radius δ, centred at $x \in \mathbb{G}$. Let α_i be positive parameters such that

$$\sum_{i=1}^{n} \frac{\alpha_i}{p_i} = 1. \tag{4.93}$$

Then, we consider the following system of the fractional p-sub-Laplacians:

$$\begin{cases} (-\Delta_{p_1})^{s_1} u_1(x) = \omega_1(x)|u_1(x)|^{\alpha_1-2}u_1(x)|u_2(x)|^{\alpha_2}\cdots|u_n(x)|^{\alpha_n}, \ x \in \Omega, \\ (-\Delta_{p_2})^{s_2} u_2(x) = \omega_2(x)|u_1(x)|^{\alpha_1}|u_2(x)|^{\alpha_2-2}u_2(x)\cdots|u_n(x)|^{\alpha_n}, \ x \in \Omega, \\ \cdots \\ (-\Delta_{p_n})^{s_n} u_n(x) = \omega_n(x)|u_1(x)|^{\alpha_1}|u_2(x)|^{\alpha_2}\cdots|u_n(x)|^{\alpha_n-2}u_n(x), \ x \in \Omega, \end{cases} \tag{4.94}$$

with homogeneous Dirichlet conditions

$$u_i(x) = 0, \quad x \in \mathbb{G}\setminus\Omega, \quad i = 1,\ldots,n. \tag{4.95}$$

We denote by r_Ω the inner quasi-radius of Ω, that is,

$$r_\Omega = r_{\Omega,|\cdot|} := \max\{|x| : x \in \Omega\}.$$

Definition 4.6.5 (Weak solutions of the p-sub-Laplacian system). We will say that $(u_1,\ldots,u_n) \in \prod_{i=1}^n W_0^{s_i,p_i}(\Omega)$ is a *weak solution* of (4.94)–(4.95) if for all $(v_1,\ldots,v_n) \in \prod_{i=1}^n W_0^{s_i,p_i}(\Omega)$, we have

$$\int_{\mathbb{G}}\int_{\mathbb{G}} \frac{|u_i(x)-u_i(y)|^{p_i-2}(u_i(x)-u_i(y))(v_i(x)-v_i(y))}{|y^{-1}x|^{Q+s_ip_i}} dxdy \tag{4.96}$$
$$= \int_\Omega \omega_i(x) \left(\prod_{j=1}^{i-1} |u_j(x)|^{\alpha_j}\right)\left(\prod_{j=i+1}^{n} |u_j(x)|^{\alpha_j}\right) |u_i(x)|^{\alpha_i-2}u_i(x)v_i(x)dx,$$

for every $i = 1,\ldots,n$.

Now we present the following analogue of the Lyapunov type inequality for the fractional p-sub-Laplacian system on $\mathbb{G}$.

Theorem 4.6.6 (Lyapunov inequality for fractional p-sub-Laplacian system). *Let $\mathbb{G}$ be a homogeneous group of homogeneous dimension Q. Let $s_i \in (0,1)$ and $p_i \in (1,\infty)$ be such that $Q > s_i p_i$ with $i = 1,\ldots,n$. Let $\omega_i \in L^\theta(\Omega)$ be a non-negative weight and assume that*

$$1 < \max_{i=1,\ldots,n} \left\{ \frac{Q}{s_i p_i} \right\} < \theta < \infty.$$

If (4.94)–(4.95) *admits a non-trivial weak solution, then we have*

$$\prod_{i=1}^{n} \|\omega_i\|_{L^\theta(\Omega)}^{\frac{\theta\alpha_i}{p_i}} \geq C r_{\Omega,q}^{Q-\theta\sum_{j=1}^{n} s_j\alpha_j}, \tag{4.97}$$

where C is a positive constant.

Proof of Theorem 4.6.6. Set

$$\xi_i := \gamma_i p_i + (1-\gamma_i)p_i^*, \quad i = 1,\ldots,n,$$

and

$$\gamma_i := \frac{\theta - \frac{Q}{s_i p_i}}{\theta - 1}, \tag{4.98}$$

where $p_i^* = \frac{Q}{Q - s_i p_i}$ is the Sobolev conjugate exponent as in Theorem 4.3.1. Notice that for all $i = 1,\ldots,n$ we have $\gamma_i \in (0,1)$ and $\xi_i = p_i\theta'$, where $\theta' = \frac{\theta}{\theta-1}$. Then for every i we have

$$\int_\Omega \frac{|u_i(x)|^{\xi_i}}{r_{\Omega,q}^{\gamma_i s_i p_i}} dx \leq \int_\Omega \frac{|u_i(x)|^{\xi_i}}{|x|^{\gamma_i s_i p_i}} dx,$$

and by using the Hölder inequality with $\nu_i = \frac{1}{\gamma_i}$ and $\frac{1}{\nu_i} + \frac{1}{\nu_i'} = 1$, we get

$$\begin{aligned} \int_\Omega \frac{|u_i(x)|^{\xi_i}}{|x|^{\gamma_i s_i p_i}} dx &= \int_\Omega \frac{|u_i(x)|^{\gamma_i p_i} |u_i(x)|^{(1-\gamma_i)p_i^*}}{|x|^{\gamma_i s_i p_i}} dx \\ &\leq \left(\int_\Omega \frac{|u_i(x)|^{p_i}}{|x|^{s_i p_i}} dx \right)^{\gamma_i} \left(\int_\Omega |u_i(x)|^{p_i^*} dx \right)^{1-\gamma_i}. \end{aligned} \tag{4.99}$$

On the other hand, from Theorem 4.3.1, we obtain

$$\left(\int_\Omega |u_i(x)|^{p_i^*} dx \right)^{1-\gamma_i} \leq C[u_i]_{s_i,p_i,|\cdot|}^{p_i^*(1-\gamma_i)},$$

and from Theorem 4.2.1, we have

$$\left(\int_\Omega \frac{|u_i(x)|^{p_i}}{|x|^{s_i p_i}} dx \right)^{\gamma_i} \leq C[u_i]_{s_i,p_i,|\cdot|}^{p_i\gamma_i}.$$

Thus, from (4.99) and by setting $u_i(x) = v_i(x)$ in (4.96), we get

$$\int_\Omega \frac{|u_i(x)|^{\xi_i}}{|x|^{\gamma_i s_i p_i}} dx \le C([u_i]^{p_i}_{s_i,p_i,|\cdot|,\Omega})^{\frac{\xi_i}{p_i}} \le C([u_i]^{p_i}_{s_i,p_i,|\cdot|})^{\frac{\xi_i}{p_i}}$$

$$= C\left(\int_\Omega \omega_i(x) \prod_{j=1}^n |u_j|^{\alpha_j} dx\right)^{\frac{\xi_i}{p_i}} = C\left(\int_\Omega \omega_i(x) \prod_{j=1}^n |u_j|^{\alpha_j} dx\right)^{\theta'},$$

for every $i = 1, \ldots, n$. Therefore, by using the Hölder inequality with exponents θ and θ', we obtain

$$\int_\Omega \frac{|u_i(x)|^{\xi_i}}{|x|^{\gamma_i s_i p_i}} dx \le C\|\omega_i\|^{\frac{\theta}{\theta-1}}_{L^\theta(\Omega)} \int_\Omega \prod_{j=1}^n |u_j(x)|^{\alpha_j \theta'} dx.$$

Again by using the Hölder inequality and (4.93), we get

$$\int_\Omega \prod_{j=1}^n |u_j(x)|^{\alpha_j \theta'} dx \le \prod_{j=1}^n \left(\int_\Omega |u_j|^{\theta' p_j} dx\right)^{\frac{\alpha_j}{p_j}}.$$

It yields that

$$\int_\Omega \frac{|u_i(x)|^{\xi_i}}{|x|^{\gamma_i s_i p_i}} dx \le C\|\omega_i\|^{\frac{\theta}{\theta-1}}_{L^\theta(\Omega)} \prod_{j=1}^n \left(\int_\Omega |u_j|^{\theta' p_j} dx\right)^{\frac{\alpha_j}{p_j}}.$$

That is, we have

$$\int_\Omega \frac{|u_i(x)|^{\xi_i}}{r_\Omega^{\gamma_i s_i p_i}} dx \le \int_\Omega \frac{|u_i(x)|^{\xi_i}}{|x|^{\gamma_i s_i p_i}} dx \le C\|\omega_i\|^{\frac{\theta}{\theta-1}}_{L^\theta(\Omega)} \prod_{j=1}^n \left(\int_\Omega |u_j|^{\theta' p_j} dx\right)^{\frac{\alpha_j}{p_j}}.$$

Thus, for every $e_i > 0$ we have

$$\left(\int_\Omega \frac{|u_i(x)|^{\xi_i}}{r_\Omega^{\gamma_i s_i p_i}} dx\right)^{e_i} = \frac{1}{r_\Omega^{e_i \gamma_i s_i p_i}} \left(\int_\Omega |u_i(x)|^{\xi_i} dx\right)^{e_i}$$

$$\le C\|\omega_i\|^{\frac{e_i\theta}{\theta-1}}_{L^\theta(\Omega)} \prod_{j=1}^n \left(\int_\Omega |u_j|^{\theta' p_j} dx\right)^{\frac{e_i \alpha_j}{p_j}},$$

so that

$$\frac{1}{r_\Omega^{\sum_{j=1}^n \gamma_j s_j p_j e_j}} \prod_{i=1}^n \left(\int_\Omega |u_i(x)|^{\theta' p_i} dx\right)^{e_i}$$

$$\le C\left(\prod_{i=1}^n \|\omega_i\|^{\frac{e_i\theta}{\theta-1}}_{L^\theta(\Omega)}\right)\left(\prod_{i=1}^n \left(\int_\Omega |u_i(x)|^{\theta' p_i} dx\right)^{\frac{\alpha_i \sum_{j=1}^n e_j}{p_i}}\right).$$

This implies

$$\frac{1}{r_{\Omega}^{\sum_{j=1}^{n}\gamma_j s_j p_j e_j}} \leq C\left(\prod_{i=1}^{n}\|\omega_i\|_{L^{\theta}(\Omega)}^{\frac{e_i\theta}{\theta-1}}\right)\left(\prod_{i=1}^{n}\left(|u_i(x)|^{\theta' p_i}dx\right)^{\frac{\alpha_i\sum_{j=1}^{n}e_j}{p_i}-e_i}\right), \quad (4.100)$$

where C is a positive constant. Then, we choose e_i, $i = 1,\ldots,n$, such that $\frac{\alpha_i\sum_{j=1}^{n}e_j}{p_i} - e_i = 0$. Consequently, from (4.93) we have the solution of this system

$$e_i = \frac{\alpha_i}{p_i},\ i = 1,\ldots,n. \quad (4.101)$$

Combining (4.100), (4.98) and (4.101) we arrive at

$$\prod_{i=1}^{n}\|\omega_i\|_{L^{\theta}(\Omega)}^{\frac{\theta\alpha_i}{p_i}} \geq C r_{\Omega}^{Q-\theta\sum_{j=1}^{n}s_j\alpha_j}. \quad (4.102)$$

Theorem 4.6.6 is proved. □

In order to discuss an application of the Lyapunov type inequality for the fractional p-sub-Laplacian system on $\mathbb{G}$, we consider the spectral problem for the system of fractional p-sub-Laplacians:

$$\begin{cases} (-\Delta_{p_1,q})^{s_1}u_1(x) = \lambda_1\alpha_1\varphi(x)|u_1(x)|^{\alpha_1-2}u_1(x)|u_2(x)|^{\alpha_2}\cdots|u_n(x)|^{\alpha_n}, & x\in\Omega,\\ (-\Delta_{p_2,q})^{s_2}u_2(x) = \lambda_2\alpha_2\varphi(x)|u_1(x)|^{\alpha_1}|u_2(x)|^{\alpha_2-2}u_2(x)\cdots|u_n(x)|^{\alpha_n}, & x\in\Omega,\\ \quad\cdots \\ (-\Delta_{p_n,q})^{s_n}u_n(x) = \lambda_n\alpha_n\varphi(x)|u_1(x)|^{\alpha_1}|u_2(x)|^{\alpha_2}\cdots|u_n(x)|^{\alpha_n-2}u_n(x), & x\in\Omega, \end{cases} \quad (4.103)$$

with

$$u_i(x) = 0,\quad x\in\mathbb{G}\setminus\Omega,\ i = 1,\ldots,n, \quad (4.104)$$

where $\Omega\subset\mathbb{G}$ is a Haar measurable set, $\varphi\in L^1(\Omega)$, $\varphi\geq 0$ and $s_i\in(0,1)$, $p_i\in(1,\infty)$, $i = 1,\ldots,n$.

Definition 4.6.7 (Eigenvalues of p-sub-Laplacian system (4.103)–(4.104)). We say that $\lambda = (\lambda_1,\ldots,\lambda_n)$ is an eigenvalue if the problem (4.103)–(4.104) admits at least one non-trivial weak solution $(u_1,\ldots,u_n)\in\prod_{i=1}^{n}W_0^{s_i,p_i}(\Omega)$.

Theorem 4.6.8 (Eigenvalue estimates for p-sub-Laplacian system). *Let $s_i\in(0,1)$ and $p_i\in(1,\infty)$ be such that $Q > s_i p_i$, for all $i = 1,\ldots,n$, and*

$$1 < \max_{i=1,\ldots,n}\left\{\frac{Q}{s_i p_i}\right\} < \theta < \infty.$$

Let $\varphi\in L^{\theta}(\Omega)$ with $\|\varphi\|_{L^{\theta}(\Omega)}\neq 0$. Then we have

$$\lambda_k \geq \frac{C}{\alpha_k}\left(\frac{1}{\prod_{i=1,i\neq k}^{n}\lambda_i^{\frac{\alpha_i}{p_i}}}\right)^{\frac{p_k}{\alpha_k}}\left(\frac{1}{r_{\Omega}^{\theta\sum_{i=1}^{n}\alpha_i s_i - Q}\prod_{i=1,i\neq k}^{n}\alpha_i^{\frac{\theta\alpha_i}{p_i}}\int_{\Omega}\varphi^{\theta}(x)dx}\right)^{\frac{p_k}{\theta\alpha_k}},$$

where C is a positive constant and $k = 1,\ldots,n$.

Proof of Theorem 4.6.8. In Theorem 4.6.6 by setting $\omega_k = \lambda_k \alpha_k \varphi(x)$, $k = 1, \dots, n$, we get

$$\alpha_k^{\frac{\theta \alpha_k}{p_k}} \lambda_k^{\frac{\theta \alpha_k}{p_k}} \prod_{i=1, i\neq k}^{n} (\alpha_i \lambda_i)^{\frac{\theta \alpha_i}{p_i}} \prod_{i=1}^{n} \|\varphi\|_{L^\theta(\Omega)}^{\frac{\theta \alpha_i}{p_i}} \geq C r_{\Omega,q}^{Q-\theta \sum_{j=1}^{n} s_j \alpha_j}.$$

So from (4.93) we have

$$\alpha_k^{\frac{\theta \alpha_k}{p_k}} \lambda_k^{\frac{\theta \alpha_k}{p_k}} \prod_{i=1, i\neq k}^{n} (\alpha_i \lambda_i)^{\frac{\theta \alpha_i}{p_i}} \int_\Omega \varphi^\theta(x) dx \geq C r_{\Omega,q}^{Q-\theta \sum_{j=1}^{n} s_j \alpha_j}.$$

This yields

$$\lambda_k^{\frac{\theta \alpha_k}{p_k}} \geq \frac{C}{\alpha_k^{\frac{\theta \alpha_k}{p_k}} r_{\Omega,q}^{\theta \sum_{j=1}^{n} s_j \alpha_j - Q} \prod_{i=1, i\neq k}^{n} (\alpha_i \lambda_i)^{\frac{\theta \alpha_i}{p_i}} \int_\Omega \varphi^\theta(x) dx}, \quad k = 1, \dots, n.$$

Therefore, we have

$$\lambda_k \geq \frac{C}{\alpha_k} \left(\frac{1}{\prod_{i=1, i\neq k}^{n} \lambda_i^{\frac{\alpha_i}{p_i}}} \right)^{\frac{p_k}{\alpha_k}} \left(\frac{1}{r_\Omega^{\theta \sum_{i=1}^{n} \alpha_i s_i - Q} \prod_{i=1, i\neq k}^{n} \alpha_i^{\frac{\theta \alpha_i}{p_i}} \int_\Omega \varphi^\theta(x) dx} \right)^{\frac{p_k}{\theta \alpha_k}},$$
$$k = 1, \dots, n,$$

completing the proof. □

4.6.3 Lyapunov type inequality for Riesz potentials

In this section we discuss the Lyapunov type inequality for the Riesz potential operators on homogeneous groups. As an application, we discuss a two-sided estimate for the first eigenvalue of the Riesz potential.

Definition 4.6.9 (Riesz potentials). Let $\mathbb{G}$ be a homogeneous group of homogeneous dimension Q with a quasi-norm $|\cdot|$. The *Riesz potential* on a Haar measurable set $\Omega \subset \mathbb{G}$ is the operator given by the formula

$$\mathfrak{R}u(x) = \int_\Omega \frac{u(y)}{|y^{-1}x|^{Q-2s}} dy, \quad 0 < 2s < Q. \tag{4.105}$$

The (*weighted*) *Riesz potential* is defined by

$$\mathfrak{R}(\omega u)(x) = \int_\Omega \frac{\omega(y) u(y)}{|y^{-1}x|^{Q-2s}} dy, \quad 0 < 2s < Q. \tag{4.106}$$

A Lyapunov type inequality for the weighted Riesz potentials can be formulated as follows.

Theorem 4.6.10 (Lyapunov type inequality for weighted Riesz potentials). *Let $\mathbb{G}$ be a homogeneous group of homogeneous dimension Q with a quasi-norm $|\cdot|$. Let $\Omega \subset \mathbb{G}$ be a Haar measurable set, let $Q \geq 2 > 2s > 0$ and $1 < p < 2$. Assume that $\omega \in L^{\frac{p}{2-p}}(\Omega)$, $\frac{1}{|y^{-1}x|^{Q-2s}} \in L^{\frac{p}{p-1}}(\Omega \times \Omega)$ and*

$$C_0 = \left\| \frac{1}{|y^{-1}x|^{Q-2s}} \right\|_{L^{\frac{p}{p-1}}(\Omega\times\Omega)} < \infty.$$

Let $u \in L^{\frac{p}{p-1}}(\Omega)$, $u \neq 0$, satisfy

$$\mathfrak{R}(\omega u)(x) = \int_\Omega \frac{\omega(y)u(y)}{|y^{-1}x|^{Q-2s}} dy = u(x), \text{ for a.e. } x \in \Omega. \tag{4.107}$$

Then

$$\|\omega\|_{L^{\frac{p}{2-p}}(\Omega)} \geq \frac{1}{C_0}. \tag{4.108}$$

Proof of Theorem 4.6.10. In (4.107), by using Hölder's inequality for $p, \theta > 1$ with $\frac{1}{p} + \frac{1}{p'} = 1$ and $\frac{1}{\theta} + \frac{1}{\theta'} = 1$, we get

$$\begin{aligned}
|u(x)| &= \left| \int_\Omega \frac{\omega(y)u(y)}{|y^{-1}x|^{Q-2s}} dy \right| \\
&\leq \left(\int_\Omega |\omega(y)u(y)|^p dy \right)^{\frac{1}{p}} \left(\int_\Omega \left| \frac{1}{|y^{-1}x|^{Q-2s}} \right|^{p'} dy \right)^{\frac{1}{p'}} \\
&\leq \left(\int_\Omega |\omega(y)|^{p\theta} dy \right)^{\frac{1}{p\theta}} \left(\int_\Omega |u(y)|^{\theta' p} dy \right)^{\frac{1}{\theta' p}} \left(\int_\Omega \left| \frac{1}{|y^{-1}x|^{Q-2s}} \right|^{p'} dy \right)^{\frac{1}{p'}} \\
&= \|\omega\|_{L^{p\theta}(\Omega)} \|u\|_{L^{p\theta'}(\Omega)} \left(\int_\Omega \left| \frac{1}{|y^{-1}x|^{Q-2s}} \right|^{p'} dy \right)^{\frac{1}{p'}}.
\end{aligned}$$

Let p' be such that $p' = p\theta'$, so that $\theta = \frac{1}{2-p}$. Then we have

$$|u(x)| \leq \|\omega\|_{L^{\frac{p}{2-p}}(\Omega)} \|u\|_{L^{\frac{p}{p-1}}(\Omega)} \left(\int_\Omega \left| \frac{1}{|y^{-1}x|^{Q-2s}} \right|^{\frac{p}{p-1}} dy \right)^{\frac{p-1}{p}}. \tag{4.109}$$

From (4.109) we calculate

$$\begin{aligned}
\|u\|_{L^{\frac{p}{p-1}}(\Omega)} &\leq \|\omega\|_{L^{\frac{p}{2-p}}(\Omega)} \|u\|_{L^{\frac{p}{p-1}}(\Omega)} \left(\int_\Omega \int_\Omega \left| \frac{1}{|y^{-1}x|^{Q-2s}} \right|^{\frac{p}{p-1}} dx dy \right)^{\frac{p-1}{p}} \\
&= C_0 \|\omega\|_{L^{\frac{p}{2-p}}(\Omega)} \|u\|_{L^{\frac{p}{p-1}}(\Omega)}.
\end{aligned}$$

Finally, since $u \neq 0$, this implies

$$\|\omega\|_{L^{\frac{p}{2-p}}(\Omega)} \geq \frac{1}{C_0},$$

completing the proof. □

Let us now consider the following spectral problem for the Riesz potential:

$$\mathfrak{R}u(x) = \int_\Omega \frac{u(y)}{|y^{-1}x|^{Q-2s}} dy = \lambda u(x), \quad x \in \Omega, \quad 0 < 2s < Q. \tag{4.110}$$

We recall the Rayleigh quotient for the Riesz potential:

$$\lambda_1(\Omega) = \sup_{u \neq 0} \frac{\int_\Omega \int_\Omega \frac{u(x)u(y)}{|y^{-1}x|^{Q-2s}} dxdy}{\|u\|^2_{L^2(\Omega)}}, \tag{4.111}$$

where $\lambda_1(\Omega)$ is the first eigenvalue of the Riesz potential. A direct consequence of Theorem 4.6.10 is the following estimate for this first eigenvalue.

Theorem 4.6.11 (First eigenvalue of the Riesz potential). *Let $\mathbb{G}$ be a homogeneous group of homogeneous dimension Q. Let $\Omega \subset \mathbb{G}$ be a Haar measurable set, let $Q \geq 2 > 2s > 0$ and $1 < p < 2$ with $\frac{1}{|y^{-1}x|^{Q-2s}} \in L^{\frac{p}{p-1}}(\Omega \times \Omega)$ and set*

$$C_0 := \left\| \frac{1}{|y^{-1}x|^{Q-2s}} \right\|_{L^{\frac{p}{p-1}}(\Omega\times\Omega)}.$$

Then for the spectral problem (4.110), *we have*

$$\lambda_1(\Omega) \leq C_0 |\Omega|^{\frac{2-p}{p}}. \tag{4.112}$$

Proof of Theorem 4.6.11. By using (4.111), Theorem 4.6.10 and $\omega = \frac{1}{\lambda_1(\Omega)}$, we obtain

$$\lambda_1(\Omega) \leq C_0 |\Omega|^{\frac{2-p}{p}}. \tag{4.113}$$

Theorem 4.6.11 is proved. □

Euclidean case. Let us now record several applications of the above constructions in the case of the Abelian group $(\mathbb{R}^N, +)$. With the Euclidean distance $|\cdot|_E$, the Riesz potential is given by

$$\mathfrak{R}u(x) = \int_\Omega \frac{u(y)}{|x-y|_E^{N-2s}} dy, \quad 0 < 2s < N, \quad \Omega \subset \mathbb{R}^N, \tag{4.114}$$

and the weighted Riesz potential is

$$\mathfrak{R}(\omega u)(x) = \int_\Omega \frac{\omega(y)u(y)}{|x-y|_E^{N-2s}} dy, \quad 0 < 2s < N, \quad \Omega \subset \mathbb{R}^N. \tag{4.115}$$

Then, in Theorem 4.6.10, setting $\mathbb{G} = (\mathbb{R}^N, +)$ and taking the standard Euclidean distance instead of the quasi-norm, we obtain

Theorem 4.6.12 (Euclidean Lyapunov inequality for the Riesz potential). *Let $\Omega \subset \mathbb{R}^N$ be a measurable set with $|\Omega| < \infty$, $1 < p < 2$ and let $N \geq 2 > 2s > 0$. Let $\omega \in L^{\frac{p}{2-p}}(\Omega)$, $\frac{1}{|x-y|_E^{N-2s}} \in L^{\frac{p}{p-1}}(\Omega \times \Omega)$ and set*

$$S := \left\| \frac{1}{|x-y|_E^{N-2s}} \right\|_{L^{\frac{p}{p-1}}(\Omega\times\Omega)}.$$

In addition, assume that $u \in L^{\frac{p}{p-1}}(\Omega)$, $u \neq 0$, satisfies

$$\mathfrak{R}(\omega u)(x) = u(x), \quad x \in \Omega.$$

Then we have

$$\|\omega\|_{L^{\frac{p}{2-p}}(\Omega)} \geq \frac{1}{S}.$$

Now let us consider the spectral problem for the Euclidean Riesz potential from (4.114):

$$\mathfrak{R}u(x) = \int_\Omega \frac{u(y)}{|x-y|_E^{N-2s}} dy = \lambda u(x), \quad 0 < 2s < N. \tag{4.116}$$

Theorem 4.6.13 (Isoperimetric inequality for the Euclidean Riesz potential). *Let $\Omega \subset \mathbb{R}^N$ be a set with $|\Omega| < \infty$, $1 < p < 2$ and $N \geq 2 > 2s > 0$ and $1 < p < 2$. Assume that $\omega \in L^{\frac{p}{2-p}}(\Omega)$, $\frac{1}{|x-y|_E^{N-2s}} \in L^{\frac{p}{p-1}}(\Omega \times \Omega)$ and set*

$$S := \left\| \frac{1}{|x-y|_E^{N-2s}} \right\|_{L^{\frac{p}{p-1}}(\Omega\times\Omega)}.$$

Then, for the spectral problem (4.116), *we have,*

$$\lambda_1(\Omega) \leq \lambda_1(B) \leq S|B|^{\frac{2-p}{p}},$$

where $B \subset \mathbb{R}^N$ is an open ball, $\lambda_1(\Omega)$ is the first eigenvalue of the spectral problem (4.116), *for all sets Ω with $|\Omega| = |B|$.*

Proof of Theorem 4.6.13. The proof of $\lambda_1(B) \leq S|B|^{\frac{2-p}{p}}$ is the same as the proof of Theorem 4.6.11. From [RRS16] we have

$$\lambda_1(B) \geq \lambda_1(\Omega),$$

which completes the proof of Theorem 4.6.13. □

4.7 Hardy inequalities for fractional sub-Laplacians on stratified groups

In this section we discuss the Hardy inequalities involving fractional powers of the sub-Laplacian from a different point of view. First, we observe another way of writing the well-known one-dimensional L^p-Hardy inequality

$$\int_0^\infty \left|\frac{f(x)}{x}\right|^p dx \leq \left(\frac{p}{p-1}\right)^p \int_0^\infty \left|\frac{df(x)}{dx}\right|^p dx, \quad p > 1.$$

Consequently, one can replace $f(x)$ by $xf(x)$ and restate this inequality in terms of the boundedness of the operator

$$Tf(x) = \frac{d(xf(x))}{dx},$$

or of its dual operator

$$T^*f(x) = x\frac{df(x)}{dx}.$$

In this section, we discuss this point of view and its extension to the subelliptic setting. For fractional powers of the sub-Laplacian on stratified groups this has been analysed in [CCR15], with subsequent extensions to more general hypoelliptic operators and more general graded groups in [RY18a]. We discuss this matter in the spirit of the former.

4.7.1 Riesz kernels on stratified Lie groups

In this section we briefly recapture some properties of the Riesz kernels of the sub-Laplacians on stratified Lie groups. Historically, these have been consistently developed in [Fol75]. For a detailed unifying description containing also higher-order hypoelliptic operators on general graded groups, their fractional powers, and the corresponding Riesz and Bessel kernels we refer to the exposition in [FR16, Section 4.3] where one can find proofs of most of the properties described in this section.

The analysis of second-order hypoelliptic operators, including the sub-Laplacian, is significantly simpler than that of higher-order operators. This difference is based on the Hunt theorem [Hun56] which asserts that the semigroup $\{e^{-t\mathcal{L}}\}_{t>0}$ generated by the sub-Laplacian $\mathcal{L}$ (or more precisely, by its unique self-adjoint extension) consists of convolutions with probability measures. Moreover, the hypoellipticity of $\mathcal{L}$ implies that these measures are absolutely continuous with respect to the Haar measure and have densities in the Schwartz space [FS82], which are strictly positive by the Bony maximum principle [Bon69]. Therefore, for all $t \in \mathbb{R}^+$, there exists the heat kernel of $\mathcal{L}$, that is, a function $p_t \in \mathcal{S}(\mathbb{G})$ such that

$$e^{-t\mathcal{L}}f(x) = f * p_t(x) = \int_{\mathbb{G}} f(xy^{-1})p_t(y)dy$$

for all $x \in \mathbb{G}$ and all $f \in L^2(\mathbb{G})$. Since $\mathcal{L}$ is homogeneous of order two with respect to the dilations D_r of a stratified group $\mathbb{G}$, we also have

$$p_t(x) = t^{-Q/2} P(D_{t^{-1/2}}x)$$

for all $x \in \mathbb{G}$, where $P = p_1$, which this is a strictly positive function in the Schwartz class $\mathcal{S}(\mathbb{G})$ and

$$\int_{\mathbb{G}} P(x)dx = 1.$$

Let us define the following fractional integral kernel F_α on $\mathbb{G}\backslash\{0\}$ by

$$F_\alpha(x) := \int_0^\infty t^{\alpha/2} p_t(x) \frac{dt}{t} = \int_0^\infty t^{(\alpha-Q)/2} P(D_{t^{-1/2}}x)\frac{dt}{t}, \quad \operatorname{Re}\alpha < Q, \tag{4.117}$$

which converges absolutely and uniformly on compact subsets of $\mathbb{G}\backslash\{0\}$ to a smooth function, homogeneous of degree $\alpha - Q$. Since P is positive F_α is positive when α is real. Thus, the function $|\cdot|_\alpha$, given by the formula

$$|x|_\alpha := \begin{cases} (F_0(x))^{-1/(Q-\alpha)} & \text{if } x \in \mathbb{G}\backslash\{0\}, \\ 0 & \text{if } x = 0, \end{cases} \tag{4.118}$$

is non-negative and homogeneous of degree 1, and vanishes only at the origin. So $|\cdot|_\alpha$ is a homogeneous norm, and for further discussions it will be convenient to use the norm $|\cdot|_0$, which is a limit of the norms $|\cdot|_\alpha$.

The fractional integral $F_\alpha(x)$ in (4.117) is holomorphic with respect to α in $\mathcal{H}_Q := \{\alpha \in \mathbb{C} : \operatorname{Re}\alpha < Q\}$, and its derivatives

$$F_\alpha^{(k)}(x) = \frac{1}{2^k}\int_0^\infty t^{\alpha/2}(\log t)^k p_t(x)\frac{dt}{t}$$

are the absolutely convergent integrals. The functions F_α are smooth and homogeneous of degree $\alpha - Q$ and locally integrable on $\mathbb{G}$ when $0 < \operatorname{Re}\alpha < Q$. Thus, the associated distributions are defined by

$$\begin{aligned} \langle F_\alpha, \phi\rangle &:= \int_{\mathbb{G}} F_\alpha(x)\phi(x)dx = \int_{\mathbb{G}}\int_0^\infty \phi(x)t^{(\alpha-Q)/2-1}P(D_{t^{-1/2}}x)dtdx \\ &= 2\int_0^\infty \int_{\mathbb{G}} s^{\alpha-1}\phi(D_s x)P(x)dxds \end{aligned} \tag{4.119}$$

for all $\phi \in C_0^\infty(\mathbb{G})$. Now it is clear that

$$I_\alpha := \Gamma(\alpha/2)^{-1}F_\alpha$$

is the *Riesz kernel of order* α, i.e., the convolution kernel of $\mathcal{L}^{-\alpha/2}$, that is,

$$\mathcal{L}^{-\alpha/2}f = f * I_\alpha, \tag{4.120}$$

with

$$I_\alpha|_{\alpha=0} = \delta_0. \tag{4.121}$$

The family of Riesz kernels can be analytically continued as distributions to the half-plane $\mathcal{H}_Q$ by the identity

$$I_\alpha = \mathcal{L}(I_{\alpha+2}).$$

The analytic continuation to the strip $\{\alpha \in \mathbb{C} : -1 < \operatorname{Re}\alpha < Q\}$ will be enough for our purpose. An explicit expression is obtained by rewriting (4.119) in the form

$$\begin{aligned}\langle I_\alpha, \phi\rangle = &\frac{2}{\Gamma\left(\frac{\alpha}{2}\right)} \int_{\mathbb{G}} \int_0^1 s^{\alpha-1}(\phi(D_s x) - \phi(0)) ds P(x) dx + \frac{2\phi(0)}{\alpha\Gamma(\frac{\alpha}{2})} \\ &+ \frac{2}{\Gamma(\frac{\alpha}{2})} \int_{\mathbb{G}} \int_1^\infty s^{\alpha-1}\phi(D_s x) ds P(x) dx.\end{aligned} \tag{4.122}$$

Since $\Gamma'(1) = -\gamma$ and hence

$$\frac{1}{\Gamma(t)} = \frac{t}{\Gamma(t+1)} = t - \gamma t^2 + O(t^3) \text{ as } t \to 0, \tag{4.123}$$

and

$$s^t = 1 + \log s \int_0^t s^u du,$$

the equality (4.122) implies that for α near 0, for $\phi \in C_0^\infty(\mathbb{G})$ we have

$$c\langle I_\alpha, \phi\rangle = \phi(0) + \langle \Lambda, \phi\rangle\alpha + O(\alpha^2), \tag{4.124}$$

where Λ is the distribution defined by

$$\begin{aligned}\langle \Lambda, \phi\rangle := &\int_{\mathbb{G}} \int_0^1 \frac{1}{s}(\phi(D_s x) - \phi(0)) ds P(x) dx + \frac{\gamma\phi(0)}{2} \\ &+ \int_{\mathbb{G}} \int_1^\infty \frac{1}{s}\phi(D_s x) ds P(x) dx.\end{aligned} \tag{4.125}$$

Thus, (4.124) is the Taylor expansion of I_α around 0. By the analytic continuation of (4.122) one obtains the following representation of the distribution Λ in terms of the homogeneous norm $|\cdot|_0$ defined in (4.118): There exists $a > 0$ such that

$$\langle \Lambda, \phi\rangle = \frac{1}{2}\left(\int_{\{x\in\mathbb{G}:|x|_0<a\}} (\phi(x) - \phi(0))|x|_0^{-Q} dx + \int_{\{x\in\mathbb{G}:|x|_0>a\}} \phi(x)|x|_0^{-Q} dx\right), \tag{4.126}$$

for all $\phi \in C_0^\infty$.

Let $0 < \operatorname{Re}\alpha$, $0 < \operatorname{Re}\beta$ and $\operatorname{Re}(\alpha+\beta) < Q$. Then the convolution of the Riesz potentials of orders α and β is defined pointwise by absolutely convergent integrals as well as

$$I_\alpha * I_\beta = I_{\alpha+\beta} \tag{4.127}$$

is satisfied pointwise and in the sense of distributions. This identity is equivalent to the functional equality $\mathcal{L}^{-\alpha/2}\mathcal{L}^{-\beta/2} = \mathcal{L}^{-(\alpha+\beta)/2}$ and can be obtained from (4.117) by using the properties of the heat kernel. This fact can be also extended by using the analytical continuation. That is,

$$(\phi * I_\alpha) * I_\beta = \phi * I_{\alpha+\beta}$$

holds if $\operatorname{Re}\alpha > -1$, $\operatorname{Re}\beta > -1$ and $\operatorname{Re}(\alpha+\beta) < Q$. These distributions will be useful to present a family of Hardy type inequalities in the following sections. We refer to [FR16, Section 4.3] for the detailed discussion of these and other properties of the Riesz kernels and fractional powers of left invariant hypoelliptic operators.

4.7.2 Hardy inequalities for fractional powers of sub-Laplacians

As in this whole section, let $\mathcal{L}$ be the sub-Laplacian on the stratified Lie group $\mathbb{G}$ of homogeneous dimension Q. Let us define the operator T_α on $C_0^\infty(\mathbb{G})$ by the formula

$$T_\alpha f := |\cdot|^{-\alpha}\mathcal{L}^{-\alpha/2} f = |\cdot|^{-\alpha}(f * I_\alpha), \quad -1 < \operatorname{Re}\alpha < Q.$$

Consequently, the operator

$$T_\alpha^* g = (|\cdot|^{-\overline{\alpha}} g) * I_{\overline{\alpha}}$$

satisfies

$$\langle f, T_\alpha^* g\rangle = \langle T_\alpha f, g\rangle \tag{4.128}$$

for all $f, g \in C_0^\infty(\mathbb{G})$. It turns out that the operator T_α is bounded on $L^p(\mathbb{G})$ when $1 < p < \infty$ and $0 < \alpha < Q/p$ which, in turn, can be formulated as a version of a Hardy inequality:

Theorem 4.7.1 (Hardy inequality for fractional powers of sub-Laplacian). *Let $\mathbb{G}$ be a stratified Lie group of homogeneous dimension Q. Let $1 < p < \infty$ and $0 < \alpha < Q/p$. Then the operator T_α extends uniquely to a bounded operator on $L^p(\mathbb{G})$.*

Applying this to $\mathcal{L}^{\alpha/2} f$ rather than f we get that for all $f \in C_0^\infty(\mathbb{G})$, we have

$$\left\|\frac{f}{|x|^\alpha}\right\|_{L^p(\mathbb{G})} \le \|T_\alpha\|_{L^p(\mathbb{G})\to L^p(\mathbb{G})}\|\mathcal{L}^{\alpha/2} f\|_{L^p(\mathbb{G})}, \quad 1 < p < \infty. \tag{4.129}$$

Remark 4.7.2 (Hardy–Sobolev inequalities).

1. Theorem 4.7.1 was established in [CCR15]. It is easy to see that combined with the Sobolev inequality, it implies the following *Hardy–Sobolev inequality*:

   Let $0 \le b < Q$ and $\frac{a}{Q} = \frac{1}{p} - \frac{1}{q} + \frac{b}{qQ}$. Then there exists a positive constant $C > 0$ such that we have

$$\left\|\frac{f}{|x|^{\frac{b}{q}}}\right\|_{L^q(\mathbb{G})} \le C\|(-\mathcal{L})^{\frac{a}{2}} f\|_{L^p(\mathbb{G})}. \tag{4.130}$$

Such an inequality can be interpreted as a weighted Sobolev embedding of the homogeneous Sobolev space $\dot{L}^p_a(\mathbb{G})$ over L^p of order a, based on the sub-Laplacian $\mathcal{L}$. The theory of such spaces has been extensively developed by Folland [Fol75]. We refer to [RY18a] for the inequality (4.130).

2. The asymptotic behaviour of $\|T_\alpha\|_{L^p(\mathbb{G})\to L^p(\mathbb{G})}$ with respect to α will be given in Theorem 4.7.3; in its proof we will follow the original proof in [CCR15], as well as for Theorem 4.7.4.
3. (Critical global Hardy inequality for $a = Q/p$) Let $1 < p < r < \infty$ and $p < q < (r-1)p'$, where $1/p + 1/p' = 1$. Then there exists a positive constant $C = C(p, q, r, Q) > 0$ such that we have

$$\left\| \frac{f}{\left(\log\left(e + \frac{1}{|x|}\right)\right)^{\frac{r}{q}} |x|^{\frac{Q}{q}}} \right\|_{L^q(\mathbb{G})} \leq C(\|f\|_{L^p(\mathbb{G})} + \|(-\mathcal{L})^{\frac{Q}{2p}} f\|_{L^p(\mathbb{G})}). \tag{4.131}$$

4. (Critical local Hardy inequality for $a = Q/p$) Let $1 < p < \infty$ and $\beta \in [0, Q)$. Let $r > 0$ be given and let x_0 be any point of $\mathbb{G}$. Then for any $p \leq q < \infty$ there exists a positive constant $C = C(p, Q, \beta, r, q)$ such that we have

$$\left\| \frac{f}{|x|^{\frac{\beta}{q}}} \right\|_{L^q(B(x_0,r))} \leq Cq^{1-1/p}(\|f\|_{L^p(B(x_0,r))} + \|(-\mathcal{L})^{\frac{Q}{2p}} f\|_{L^p(B(x_0,r))}), \tag{4.132}$$

and such that

$$\lim_{q\to\infty} \sup C(p, Q, \beta, r, q) < \infty.$$

Inequalities (4.131) and (4.132) have been obtained in [RY18a], to which we refer for their proofs as well as for the expressions for the asymptotically sharp constants in these inequalities.

5. (Trudinger–Moser inequality on stratified groups) Let $Q \geq 3$ and let $|\cdot|$ be a homogeneous quasi-norm on $\mathbb{G}$. Then there exists a constant $\alpha_Q > 0$ such that we have

$$\sup_{\|f\|_{L_1^Q(\mathbb{G})}\leq 1} \int_{\mathbb{G}} \frac{1}{|x|^\beta} \left(\exp(\alpha|f(x)|^{Q'}) - \sum_{k=0}^{Q-2} \frac{\alpha^k |f(x)|^{kQ'}}{k!} \right) dx < \infty \tag{4.133}$$

for any $\beta \in [0, Q)$ and $\alpha \in (0, \alpha_Q(1 - \beta/Q))$, where $Q' = Q/(Q-1)$. When $\alpha > \alpha_Q(1 - \beta/Q)$, the integral in (4.133) is still finite for any $f \in L_1^Q(\mathbb{G})$, but the supremum is infinite. The space $L_1^Q(\mathbb{G})$ is the Sobolev space with the norm $\|f\|_{L_1^Q(\mathbb{G})} = \|f\|_{L^Q(\mathbb{G})} + \|\nabla_H f\|_{L^Q(\mathbb{G})}$.

The inequality (4.133) was obtained in [RY18a], also with a rather explicit expression for the constant α_Q. We refer to the above paper also for an extensive history of this subject.

6. (Hardy–Sobolev inequalities on graded groups) In fact, Hardy–Sobolev inequalities (4.130), (4.131) and (4.132), Trudinger–Moser inequalities (4.133) and their local versions, have been obtained in [RY18a] for general homogeneous left invariant hypoelliptic differential operators (the so-called Rockland operators) on **general graded Lie groups**. The methods of the proof, however, are rather different, and somewhat more involved, than the proofs for the sub-Laplacians presented in this section. Therefore, here we do not present them in full generality: these results, their proofs, and expressions for (asymptotically) best constants can be found in [RY18a].

Proof of Theorem 4.7.1. As usual here we denote the conjugate number to p by p'. Since the integral kernel of the operator T_α is positive, its L^p-boundedness follows from the following *Schur test* (see [FR75]):

Suppose that there exist a positive function g and constants $A_{\alpha,p}$ and $B_{\alpha,p}$ such that we have

$$T_\alpha(g^{p'})(x) \leq A_{\alpha,p} g^{p'}(x) \text{ and } T_\alpha^*(g^p)(x) \leq B_{\alpha,p} u^p(x) \tag{4.134}$$

for almost all $x \in \mathbb{G}$. Then the estimate

$$\|T_\alpha f\|_{L^p(\mathbb{G})} \leq A_{\alpha,p}^{1/p'} B_{\alpha,p}^{1/p} \|f\|_{L^p(\mathbb{G})} \tag{4.135}$$

holds for all $f \in L^p(\mathbb{G})$.

In order to produce g as in (4.134), let us consider the family of functions $g_\gamma := |\cdot|^{\gamma-Q}$, $\gamma > 0$, and consider the convolutions $g_\gamma^{p'} * I_\alpha$ and $(|\cdot|^{-\alpha} g_\gamma^p) * I_\alpha$ related to the computations of $T_\alpha(g_\gamma^{p'})$ and $T_\alpha^*(g_\gamma^p)$. Let

$$\beta' = Q + (\gamma - Q)p' \text{ and } \beta = Q - \alpha + (\gamma - Q)p, \tag{4.136}$$

so that $g_\gamma^{p'}$ and $|\cdot|^{-\alpha} g_\gamma^p$ have the same homogeneity as $I_{\beta'}$ and I_β, respectively. As in the case of (4.127) these convolution integrals converge absolutely in $\mathbb{G}\backslash\{0\}$ if and only if $0 < \beta < Q - \alpha$ and $0 < \beta' < Q - \alpha$, i.e., for γ such that

$$\max\left(\frac{Q}{p}, \frac{\alpha}{p} + \frac{Q}{p'}\right) < \gamma < Q - \frac{\alpha}{p'}. \tag{4.137}$$

Thus, if this condition is satisfied, then both $T_\alpha(g_\gamma^{p'})$ and $T_\alpha^*(g_\gamma^p)$ are positive functions, continuous away from the origin, and of the same homogeneity as $I_{\beta'}$ and I_β, respectively. The pointwise estimates (4.134) follow directly from the homogeneity. Finally, for γ the condition (4.137) is nontrivial if and only if $0 < \alpha < Q/p$, completing the proof of Theorem 4.7.1. □

By Theorem 4.7.1, the operator T_α is L^p-bounded and the constant in the inequality (4.129) depends on the choice of a homogeneous quasi-norm. While the inequality itself holds true for any homogeneous quasi-norm (due to their equivalence), in the case of particular homogeneous norm $|\cdot|_0$ from the family (4.118) one can give the following estimate for it.

Theorem 4.7.3 (Estimate for L^p-operator norm). *Let $\mathbb{G}$ be a stratified Lie group of homogeneous dimension Q. Let $1 < p < \infty$ and $0 < \alpha < Q/p$. For the particular homogeneous norm $|\cdot|_0$, the operator norm $\|T_\alpha\|_{L^p(\mathbb{G})\to L^p(\mathbb{G})}$ satisfies*

$$\|T_\alpha\|_{L^p(\mathbb{G})\to L^p(\mathbb{G})} \leq 1 + C\alpha + O(\alpha^2).$$

For the rest of this subsection, we assume that $|\cdot| = |\cdot|_0$ and that the operator T_α is defined using $|\cdot|_0$.

Proof of Theorem 4.7.3. The proof of Theorem 4.7.1, and especially (4.135), show that, if

$$0 < \alpha < \frac{Q}{p} \quad \text{and} \quad \frac{\alpha}{p} + \frac{Q}{p'} < \gamma < Q - \frac{\alpha}{p'}, \tag{4.138}$$

then

$$\|T_\alpha\|_{L^p(\mathbb{G})\to L^p(\mathbb{G})} \leq A^{1/p'}_{\alpha,\gamma,p} B^{1/p}_{\alpha,\gamma,p}, \tag{4.139}$$

where

$$\begin{aligned} A_{\alpha,\gamma,p} &:= \sup_{|y|=1} \left(|\cdot|_0^{(\gamma-Q)p'} * I_\alpha\right)(y) &&= \sup_{|y|=1} \left(|\cdot|_0^{\beta'-Q} * I_\alpha\right)(y), \\ B_{\alpha,\gamma,p} &:= \sup_{|y|=1} \left(|\cdot|_0^{(\gamma-Q)p-\alpha} * I_\alpha\right)(y) &&= \sup_{|y|=1} \left(|\cdot|_0^{\beta-Q} * I_\alpha\right)(y). \end{aligned} \tag{4.140}$$

First let us show that there is a constant C_p such that

$$\|T_\alpha f\|_{L^p(\mathbb{G})\to L^p(\mathbb{G})} \leq 1 + C_p\alpha + O(\alpha^2), \quad 1 < p < \infty, \tag{4.141}$$

holds for all sufficiently small positive α. Assume that $Q/p' < \gamma < Q$. Then (4.138) is valid for α in a neighborhood of 0^+. Moreover, with β and β' as in (4.136), the constants $A_{\alpha,\gamma,p}$ and $B_{\alpha,\gamma,p}$ in (4.140) are bounded by $1 + L_{\beta'}\alpha + O(\alpha^2)$ and $1 + L_\beta\alpha + O(\alpha^2)$, respectively. By (4.121) we obtain

$$\|T_\alpha\|_{L^p(\mathbb{G})\to L^p(\mathbb{G})} \leq 1 + \left(\frac{L_{\beta'}}{p'} + \frac{L_\beta}{p}\right)\alpha + O(\alpha^2),$$

which confirms (4.141). Combining this with Theorem 4.7.1 completes the proof of Theorem 4.7.3. □

Theorem 4.7.4 (A logarithmic version of uncertainty principle). *Let $\mathbb{G}$ be a stratified Lie group and let $1 < p < \infty$. Then we have*

$$\int_{\mathbb{G}} (\log|x|)|f(x)|^p dx + \int_{\mathbb{G}} \operatorname{Re}((\log \mathcal{L}^{1/2} f)(x)\overline{f(x)})|f(x)|^{p-2} dx \geq -C_p \|f\|^p_{L^p(\mathbb{G})}.$$

Proof of Theorem 4.7.4. Now let us recall the distribution Λ introduced in (4.125). We also define the operator

$$T_0' f := \left(\frac{d}{d\alpha} T_\alpha f\right)\bigg|_{\alpha=0} = -\log \mathcal{L}^{1/2} f - (\log|\cdot|_0) f = -f * \Lambda - (\log|\cdot|_0) f. \tag{4.142}$$

Setting $\phi = |y \cdot|_0^{\beta-Q}$ in (4.126), the integrals converge absolutely for any $|y|=1$, and so (4.124) also extends to

$$|\cdot|_0^{\beta-Q} * I_\alpha(y) = 1 + \alpha|\cdot|_0^{\beta-Q} * \Lambda(y) + O(\alpha^2);$$

the term $O(\alpha^2)$ is uniform in $|y| = 1$. Taking the supremum over y yields the logarithmic uncertainty inequality for T_0', that is, this shows that if α is small and $0 < \alpha < Q - \beta$, then

$$\sup_{|y|=1} \left(|\cdot|_0^{\beta-Q} * I_\alpha\right)(y) \leq 1 + L_\beta\alpha + O(\alpha^2), \quad 0 < \beta < Q,$$

where

$$L_\beta = \sup_{|y|=1} \left(|\cdot|_0^{\beta-Q} * \Lambda(y)\right). \tag{4.143}$$

Let $1 < p < \infty$. Taking $f \in C_0^\infty(\mathbb{G})$ with $\|f\|_{L^p(\mathbb{G})} = 1$ and restricting ourselves to positive values of α for which (4.141) holds and $1 + C_p\alpha > 0$, we consider the function

$$\Phi_\epsilon(\alpha) := (1 + (C_p + \epsilon)\alpha)^p - \|T_\alpha f\|^p_{L^p(\mathbb{G})},$$

where $\epsilon \geq 0$. The expression $\|T_\alpha f\|^p_{L^p(\mathbb{G})}$ can be differentiated in α at 0, and

$$\frac{d}{d\alpha}\left(\|T_\alpha f\|^p_{L^p(\mathbb{G})}\right)\Bigg|_{\alpha=0} = p\int_{\mathbb{G}} \operatorname{Re}\left(T_0' f(x)\overline{f(x)}\right)|f(x)|^{p-2}dx.$$

Hence Φ_ϵ is differentiable at 0. Now $\Phi_\epsilon(0) = 0$ and $\Phi_\epsilon(\alpha) \geq 0$ for all sufficiently small α, by (4.141), and so $\Phi_\epsilon'(0) \geq 0$. Now we let ϵ tend to 0, and deduce that $\Phi_\epsilon(0) \geq 0$. Thus, the statement of Theorem 4.7.4 follows from (4.142). □

4.7.3 Landau–Kolmogorov inequalities on stratified groups

In this section we discuss the stratified groups version of the Landau–Kolmogorov inequality, and some of its consequences. We start with a related simpler version of such an inequality.

Proposition 4.7.5 (Two weighted inequalities).

(1) *Let $\mathbb{G}$ be a homogeneous group. If $1 \leq p, q, r \leq \infty$, and*

$$\frac{1}{p} = \frac{\theta}{q} + \frac{1-\theta}{r}, \quad 0 \leq \theta \leq 1,$$

then

$$\||\cdot|^\alpha f\|_{L^p(\mathbb{G})} \leq \||\cdot|^{\alpha/\theta} f\|^\theta_{L^q(\mathbb{G})}\|f\|^{1-\theta}_{L^r(\mathbb{G})}, \quad \alpha \geq 0,$$

holds for all $f \in C_0^\infty(\mathbb{G})$.

(2) *Let $\mathbb{G}$ be a stratified group and let $\mathcal{L}$ be a sub-Laplacian on $\mathbb{G}$. Let $\beta > 0$, $\gamma > 0$, $p > 1$, $q \geq 1$, $r > 1$ be such that*

$$\gamma < \frac{Q}{r} \quad \text{and} \quad \frac{\beta+\gamma}{p} = \frac{\gamma}{q} + \frac{\beta}{r}.$$

Then

$$\|f\|_{L^p(\mathbb{G})} \leq C \| |\cdot|^\beta f\|_{L^q(\mathbb{G})}^{\gamma/(\beta+\gamma)} \|\mathcal{L}^{\gamma/2} f\|_{L^r(\mathbb{G})}^{\beta/(\beta+\gamma)} \tag{4.144}$$

holds for all $f \in C_0^\infty(\mathbb{G})$.

Proof. Part (1). Writing

$$(|\cdot|^\alpha |f|)^p = \left(|\cdot|^\alpha |f|^\theta\right)^p \left(|f|^{1-\theta}\right)^p,$$

then applying the Hölder inequality (with index s) we have

$$\left(\int_{\mathbb{G}} (|x|^\alpha |f(x)|^p) dx\right)^{\frac{1}{p}} \leq \left(\int_{\mathbb{G}} \left(|x|^\alpha |f(x)|^\theta\right)^{ps} dx\right)^{\frac{1}{ps}} \left(\int_{\mathbb{G}} (|f(x)|^{1-\theta})^{ps'} dx\right)^{\frac{1}{ps'}}.$$

If the index s is chosen so that $q = \theta ps$, then $r = (1-\theta)ps'$, finishing the proof.

Part (2). By using the Hölder inequality and (4.129) with (4.141) we have

$$\begin{aligned}\|f\|_{L^p(\mathbb{G})} &\leq \| |\cdot|^\beta f\|_{L^q(\mathbb{G})}^{\gamma/(\beta+\gamma)} \| |\cdot|^{-\gamma} f\|_{L^r(\mathbb{G})}^{\beta/(\beta+\gamma)} \\ &\leq C \| |\cdot|^\beta f\|_{L^q(\mathbb{G})}^{\gamma/(\beta+\gamma)} \|\mathcal{L}^{\gamma/2} f\|_{L^r(\mathbb{G})}^{\beta/(\beta+\gamma)},\end{aligned}$$

yielding (4.144). □

Now we present the following Landau–Kolmogorov type inequality.

Theorem 4.7.6 (Landau–Kolmogorov type inequality). *Let $\mathbb{G}$ be a stratified group and let $\mathcal{L}$ be a sub-Laplacian on $\mathbb{G}$. If $1 < p, q, r < \infty$, and*

$$\frac{1}{p} = \frac{\theta}{q} + \frac{1-\theta}{r}, \quad 0 \leq \theta \leq 1,$$

then we have

$$\|\mathcal{L}^{\alpha/2} f\|_{L^p(\mathbb{G})} \leq C \|\mathcal{L}^{\alpha/2\theta} f\|_{L^q(\mathbb{G})}^\theta \|f\|_{L^r(\mathbb{G})}^{1-\theta}, \quad \alpha \geq 0, \tag{4.145}$$

for all $f \in C_0^\infty(\mathbb{G})$.

Proof of Theorem 4.7.6. To prove this theorem we will use the well-known complex interpolation methods (see, e.g., [BL76] or [Ste56]). First, we have to see that the operator $\mathcal{L}^{iy}$, where $y \in \mathbb{R}$, is bounded on $L^s(\mathbb{G})$ with $1 < s < \infty$, and that

$$\|\mathcal{L}^{iy/2}\|_{L^s \to L^s} \leq C(s)\phi(y) \text{ with } \phi(y) = e^{\gamma|y|}.$$

In the case of the stratified groups this follows from the Mihlin–Hörmander multiplier theorem (see, e.g., [Chr91], [MM90]).

Let $g \in L^{p'}(\mathbb{G})$ be a compactly supported simple function of norm one. For z in the strip $S := \{z \in \mathbb{C} : 0 \leq \operatorname{Re} z \leq \frac{\alpha}{\theta}\}$, let us define

$$g_z(x) := \|g\|_{L^{p'}(\mathbb{G})}^{az+b} g(x)|g(x)|^{cx+d}$$

for all $x \in \mathbb{G}$, where

$$a = \frac{\theta p'}{\alpha}\left(\frac{1}{q} - \frac{1}{p}\right), \; b = -\frac{p'}{r'}, \; c = \frac{\theta p'}{\alpha}\left(\frac{1}{q} - \frac{1}{p}\right), \; d = \frac{p'}{r'} - 1.$$

For $\operatorname{Re} z = \eta$ and

$$\frac{1}{s} = \frac{1}{z} - \frac{\theta\eta}{\alpha}\left(\frac{1}{q} - \frac{1}{p}\right),$$

it follows that

$$\|g_z\|_{L^{s'}(\mathbb{G})}^{s'} = \int_{\mathbb{G}} |g_z(x)|^{s'} dx = \|g\|_{L^{p'}(\mathbb{G})}^{s'(a\eta+b)+p'} = 1.$$

Further, let us fix $f \in C_0^\infty(\mathbb{G})$ and set

$$h(z) := e^{z^2} \int_{\mathbb{G}} \mathcal{L}^{z/2} f(x) g_z(x) dx.$$

By the assumptions on f, for each $z \in S$, the function $\mathcal{L}^{z/2} f$ on $\mathbb{G}$ is smooth, while g_z is a simple function with compact support, and so $h(z)$ is well defined. In addition, if $z = \eta + iy$, then

$$\begin{aligned} |h(z)| &\leq e^{\eta^2 - y^2} \|\mathcal{L}^{z/2} f\|_{L^s(\mathbb{G})} \|g_z\|_{L^{s'}(\mathbb{G})} \\ &= e^{\eta^2 - y^2} \|\mathcal{L}^{iy/2} \mathcal{L}^{\eta/2} f\|_{L^s(\mathbb{G})} \leq C e^{-y^2} \phi(y) \|\mathcal{L}^{\eta/2} f\|_{L^s(\mathbb{G})}, \end{aligned}$$

hence

$$|h(\eta + iy)| \leq C \|\mathcal{L}^{\eta/2} f\|_{L^s(\mathbb{G})}.$$

Moreover, if $\operatorname{Re} z = 0$, then $|h(z)| \leq C\|f\|_{L^r(\mathbb{G})}$, while if $\operatorname{Re} z = \alpha/\theta$, then $|h(z)| \leq C\|\mathcal{L}^{\alpha/2\theta} f\|_{L^q(\mathbb{G})}$. On the other hand, the Phragmen–Lindelölf theorem implies that

$$|h(\alpha)| \leq C \|\mathcal{L}^{\alpha/2\theta} f\|_{L^q(\mathbb{G})}^{\theta} \|f\|_{L^r(\mathbb{G})}^{1-\theta}.$$

Since

$$h(\alpha) = \int_{\mathbb{G}} \mathcal{L}^{\alpha/2} f(x) g_\alpha(x) dx,$$

and g_α is an arbitrary simple function on $\mathbb{G}$ with compact support and $L^{p'}(\mathbb{G})$-norm equal to 1, this completes the proof. $\square$

Let us present the following consequence of the Landau–Kolmogorov inequality.

Corollary 4.7.7 (Consequence of the Landau–Kolmogorov inequality). *Let $\beta > 0$, $\delta > 0$, $p > 1$, $s \geq 1$, $r > 1$ be such that*

$$\frac{\beta+\delta}{p} = \frac{\delta}{s} + \frac{\beta}{r}.$$

Then we have

$$\|f\|_{L^p(\mathbb{G})} \leq C\||\cdot|^{\beta} f\|_{L^s(\mathbb{G})}^{\delta/(\beta+\delta)} \|\mathcal{L}^{\delta/2} f\|_{L^r(\mathbb{G})}^{\beta/(\beta+\delta)}$$

for all $f \in C_0^{\infty}(\mathbb{G})$.

Proof of Corollary 4.7.7. Obviously, if $\delta < Q/r$, there is nothing to prove. Otherwise, we use Proposition 4.7.5, Part (2), and the version of the Landau–Kolmogorov inequality (4.145). Let $0 < \theta < Q/(r\delta)$ and $\gamma = \theta\delta$. Then Proposition 4.7.5, Part (2), gives that

$$\|f\|_{L^p(\mathbb{G})} \leq C\||\cdot|^{\beta} f\|_{L^q(\mathbb{G})}^{\gamma/(\beta+\gamma)} \|\mathcal{L}^{\gamma/2} f\|_{L^r(\mathbb{G})}^{\beta/(\beta+\gamma)}$$

for all $f \in C_0^{\infty}(\mathbb{G})$. By Theorem 4.7.6, with r, s and p instead of p, q, and r. we have

$$\|\mathcal{L}^{\gamma/2} f\|_{L^r(\mathbb{G})} \leq C\|\mathcal{L}^{\delta/2} f\|_{L^s(\mathbb{G})}^{\theta} \|f\|_{L^p(\mathbb{G})}^{1-\theta},$$

that is,

$$\|f\|_{L^p(\mathbb{G})} \leq C\||\cdot|_0^{\beta} f\|_{L^q(\mathbb{G})}^{\gamma/(\beta+\gamma)} \|\mathcal{L}^{\gamma/2} f\|_{L^s(\mathbb{G})}^{\theta\beta/(\beta+\gamma)} \|f\|_{L^p(\mathbb{G})}^{(1-\theta)\beta/(\beta+\gamma)}.$$

This completes the proof. ☐

5

Integral Hardy Inequalities

In this chapter we discuss the integral form of Hardy inequalities where instead of estimating a function by its gradient, we estimate the integral by the function itself. This stems from the original version of Hardy's inequality [Har20]:

$$\int_b^\infty \left(\frac{\int_b^x f(t)dt}{x}\right)^p dx \leq \left(\frac{p}{p-1}\right)^p \int_b^\infty f(x)^p dx, \tag{5.1}$$

where $p > 1$, $b > 0$, and $f \geq 0$ is a non-negative function. We analyse the weighted versions of such inequalities in the setting of general homogeneous groups. Most of the results of this chapter have been obtained in [RY18a] and here we follow the presentation of this paper.

5.1 Two-weight integral Hardy inequalities

Here we discuss the weighted Hardy inequalities in the integral form extending that in (5.1). It turns out that one can actually derive the necessary and sufficient conditions on weights for these inequalities to hold.

Theorem 5.1.1 (Integral Hardy inequalities for $p \leq q$). *Let $\mathbb{G}$ be a homogeneous group of homogeneous dimension Q and let $1 < p \leq q < \infty$. Let $\phi_1 > 0$, $\phi_2 > 0$ be positive functions on $\mathbb{G}$. Then we have the following properties:*

(1) *The inequality*

$$\left(\int_{\mathbb{G}} \left(\int_{B(0,|x|)} f(z)dz\right)^q \phi_1(x)dx\right)^{1/q} \leq C_1 \left(\int_{\mathbb{G}} (f(x))^p \psi_1(x)dx\right)^{1/p} \tag{5.2}$$

holds for all $f \geq 0$ *a.e. on* $\mathbb{G}$ *if and only if*

$$A_1 := \sup_{R>0} \left(\int_{\{|x| \geq R\}} \phi_1(x) dx \right)^{1/q} \left(\int_{\{|x| \leq R\}} (\psi_1(x))^{-(p'-1)} dx \right)^{1/p'} < \infty. \tag{5.3}$$

(2) *The inequality*

$$\left(\int_{\mathbb{G}} \left(\int_{\mathbb{G} \setminus B(0,|x|)} f(z) dz \right)^q \phi_2(x) dx \right)^{1/q} \leq C_2 \left(\int_{\mathbb{G}} (f(x))^p \psi_2(x) dx \right)^{1/p} \tag{5.4}$$

holds for all $f \geq 0$ *a.e. on* $\mathbb{G}$ *if and only if*

$$A_2 := \sup_{R>0} \left(\int_{\{|x| \leq R\}} \phi_2(x) dx \right)^{1/q} \left(\int_{\{|x| \geq R\}} (\psi_2(x))^{-(p'-1)} dx \right)^{1/p'} < \infty. \tag{5.5}$$

(3) *If* $\{C_i\}_{i=1}^2$ *are the smallest constants for which* (5.2) *and* (5.4) *hold, then*

$$A_i \leq C_i \leq (p')^{\frac{1}{p'}} p^{\frac{1}{q}} A_i, \quad i = 1, 2. \tag{5.6}$$

Before we prove this theorem let us give a few comments.

Remark 5.1.2.

1. In the Abelian case $\mathbb{G} = (\mathbb{R}^n, +)$ and $Q = n$, if we take $p = q > 1$, and

$$\phi_1(x) = |B(0, |x|)|^{-p} \text{ and } \psi_1(x) = 1$$

in (5.2), then we have $A_1 = (p-1)^{-1/p}$ and

$$\left(\int_{\mathbb{R}^n} \left| \frac{1}{|B(0,|x|)|} \int_{B(0,|x|)} f(z) dz \right|^p dx \right)^{1/p} \leq \frac{p}{p-1} \left(\int_{\mathbb{R}^n} |f(x)|^p dx \right)^{1/p}, \tag{5.7}$$

where $|B(0, |x|)|$ is the volume of the ball $B(0, |x|)$. The inequality (5.7) was obtained in [CG95].

2. Theorem 5.1.1 was obtained in [RY18a] and here we follow the proof from that paper. However, due to the fact that the formulations do not make use of the differential structure, the statement can be actually extended to general metric measure spaces with polar decomposition. More specifically, consider a metric space $\mathbb{X}$ with a Borel measure dx allowing for the following *polar decomposition* at $a \in \mathbb{X}$: we assume that there is a locally integrable function $\lambda \in L^1_{\text{loc}}$ such that for all $f \in L^1(\mathbb{X})$ we have

$$\int_{\mathbb{X}} f(x) dx = \int_0^{\infty} \int_{\Sigma} f(r, \omega) \lambda(r, \omega) d\omega dr, \tag{5.8}$$

for some set $\Sigma \subset \mathbb{X}$ with a measure on it denoted by $d\omega$, and $(r, \omega) \to a$ as $r \to 0$. In the case of homogeneous groups such a polar decomposition is given in Proposition 1.2.10.

Let us denote by $B(a, r)$ the ball in $\mathbb{X}$ with centre a and radius r, i.e.,

$$B(a, r) := \{x \in \mathbb{X} : d(x, a) < r\},$$

where d is the metric on $\mathbb{X}$. Once and for all we will fix some point $a \in \mathbb{X}$, and we will write

$$|x|_a := d(a, x).$$

Then the following result was obtained in [RV18] which we record here without proof.

Theorem 5.1.3 (Integral Hardy inequality in metric measure spaces). *Let $1 < p \leq q < \infty$ and let $s > 0$. Let $\mathbb{X}$ be a metric measure space with a polar decomposition* (5.8) *at a. Let $u, v > 0$ be measurable functions positive a.e. in $\mathbb{X}$ such that $u \in L^1(\mathbb{X}\backslash\{a\})$ and $v^{1-p'} \in L^1_{\text{loc}}(\mathbb{X})$. Denote*

$$U(x) := \int_{\mathbb{X}\backslash B(a,|x|_a)} u(y)dy \qquad \textit{and} \qquad V(x) := \int_{B(a,|x|_a)} v^{1-p'}(y)dy$$

Then the inequality

$$\left(\int_{\mathbb{X}}\left(\int_{B(a,|x|_a)} |f(y)|dy\right)^q u(x)dx\right)^{1/q} \leq C\left\{\int_{\mathbb{X}} |f(x)|^p v(x)dx\right\}^{1/p} \tag{5.9}$$

holds for all measurable functions $f : \mathbb{X} \to \mathbb{C}$ if and only if any of the following equivalent conditions hold:

1. $\mathcal{D}_1 := \sup_{x \neq a}\left\{U^{\frac{1}{q}}(x)V^{\frac{1}{p'}}(x)\right\} < \infty.$

2. $\mathcal{D}_2 := \sup_{x \neq a}\left\{\int_{\mathbb{X}\backslash B(a,|x|_a)} u(y)V^{q(\frac{1}{p'}-s)}(y)dy\right\}^{1/q} V^s(x) < \infty.$

3. $\mathcal{D}_3 := \sup_{x \neq a}\left\{\int_{B(a,|x|_a)} u(y)V^{q(\frac{1}{p'}+s)}(y)dy\right\}^{1/q} V^{-s}(x) < \infty,$ *provided that $u, v^{1-p'} \in L^1(\mathbb{X})$.*

4. $\mathcal{D}_4 := \sup_{x \neq a}\left\{\int_{B(a,|x|_a)} v^{1-p'}(y)U^{p'(\frac{1}{q}-s)}(y)dy\right\}^{1/p'} U^s(x) < \infty.$

5. $\mathcal{D}_5 := \sup_{x \neq a}\left\{\int_{\mathbb{X}\backslash B(a,|x|_a)} v^{1-p'}(y)U^{p'(\frac{1}{q}+s)}(y)dy\right\}^{1/p'} U^{-s}(x) < \infty,$ *provided that $u, v^{1-p'} \in L^1(\mathbb{X})$.*

Moreover, the constant C for which (5.9) *holds and quantities $\mathcal{D}_1$–$\mathcal{D}_5$ are related by*

$$\mathcal{D}_1 \leq C \leq \mathcal{D}_1 (p')^{\frac{1}{p'}} p^{\frac{1}{q}}, \tag{5.10}$$

and

$$\mathcal{D}_1 \leq (\max(1, p's))^{\frac{1}{q}} \mathcal{D}_2, \ \mathcal{D}_2 \leq \left(\max(1, \frac{1}{p's})\right)^{1/q} \mathcal{D}_1,$$

$$\left(\frac{sp'}{1+p's}\right)^{1/q} \mathcal{D}_3 \leq \mathcal{D}_1 \leq (1+sp')^{\frac{1}{q}} \mathcal{D}_3,$$

$$\mathcal{D}_1 \leq (\max(1, qs))^{\frac{1}{p'}} \mathcal{D}_4, \ \mathcal{D}_4 \leq \left(\max(1, \frac{1}{qs})\right)^{1/p'} \mathcal{D}_1,$$

$$\left(\frac{sq}{1+qs}\right)^{1/p'} \mathcal{D}_5 \leq \mathcal{D}_1 \leq (1+sq)^{\frac{1}{p'}} \mathcal{D}_5.$$

3. As such, Theorem 5.1.3 is an extension of (5.1) to the setting of metric measures spaces $\mathbb{X}$ with the polar decomposition (5.8): in particular, for $p = q$ and real-valued non-negative measurable $f \geq 0$, inequality (5.9) becomes

$$\int_{\mathbb{X}} \left(\int_{B(a,|x|_a)} f(y)dy \right)^p u(x)dx \leq C \int_{\mathbb{X}} f(x)^p v(x)dx,$$

as an extension of (5.1). Indeed, in this case we can take $u(x) = \frac{1}{x^p}$, $v(x) = 1$, $\mathbb{X} = [b, \infty)$, $a = b$, so that Theorem 5.1.3 implies (5.1).

4. Let us give an application of Theorem 5.1.3 in the setting of homogeneous groups, recovering a two-weighted result obtained in [RV18]:

Corollary 5.1.4 (Characterization for homogeneous weights). *Let $\mathbb{G}$ be a homogeneous group of homogeneous dimension Q, equipped with a homogeneous quasi-norm $|\cdot|$. Let $1 < p \leq q < \infty$ and let $\alpha, \beta \in \mathbb{R}$. Then the inequality*

$$\left(\int_{\mathbb{G}} \left(\int_{B(0,|x|)} |f(y)|dy \right)^q |x|^\alpha dx \right)^{\frac{1}{q}} \leq C \left(\int_{\mathbb{G}} |f(x)|^p |x|^\beta dx \right)^{\frac{1}{p}} \tag{5.11}$$

holds for all measurable functions $f : \mathbb{G} \to \mathbb{C}$ if and only if $\alpha + Q < 0$, $\beta(1-p') + Q > 0$ and $\frac{\alpha+Q}{q} + \frac{\beta(1-p')+Q}{p'} = 0$. Moreover, the best constant C for (5.11) *satisfies*

$$\frac{\sigma^{\frac{1}{q}+\frac{1}{p'}}}{|\alpha+Q|^{\frac{1}{q}} (\beta(1-p')+Q)^{\frac{1}{p'}}} \leq C \leq (p')^{\frac{1}{p'}} p^{\frac{1}{q}} \frac{\sigma^{\frac{1}{q}+\frac{1}{p'}}}{|\alpha+Q|^{\frac{1}{q}} (\beta(1-p')+Q)^{\frac{1}{p'}}},$$

where σ is the area of the unit sphere in $\mathbb{G}$ with respect to the quasi-norm $|\cdot|$.

Let us show how Theorem 5.1.3 implies Corollary 5.1.4. If we take $a = 0$, and the power weights

$$u(x) = |x|^{\alpha} \qquad \text{and} \qquad v(x) = |x|^{\beta},$$

then the inequality (5.9) holds for $1 < p \leq q < \infty$ if and only if

$$\mathcal{D}_1 = \sup_{r>0} \left(\sigma \int_r^{\infty} \rho^{\alpha} \rho^{Q-1} d\rho\right)^{1/q} \left(\sigma \int_0^r \rho^{\beta(1-p')} \rho^{Q-1} d\rho\right)^{1/p'} < \infty,$$

where σ is the area of the unit sphere in $\mathbb{G}$ with respect to the quasi-norm $|\cdot|$. For this supremum to be well defined we need to have $\alpha + Q < 0$ and $\beta(1-p') + Q > 0$. Consequently, we can calculate

$$\begin{aligned} \mathcal{D}_1 &= \sigma^{\left(\frac{1}{q}+\frac{1}{p'}\right)} \sup_{r>0} \left(\int_r^{\infty} \rho^{\alpha+Q-1} d\rho\right)^{1/q} \left(\int_0^r \rho^{\beta(1-p')+Q-1} d\rho\right)^{1/p'} \\ &= \sigma^{\left(\frac{1}{q}+\frac{1}{p'}\right)} \sup_{r>0} \frac{r^{\frac{\alpha+Q}{q}}}{|\alpha+Q|^{\frac{1}{q}}} \frac{r^{\frac{\beta(1-p')+Q}{p'}}}{(\beta(1-p')+Q)^{\frac{1}{p'}}}, \end{aligned}$$

which is finite if and only if the power of r is zero. Consequently, Corollary 5.1.4 follows from Theorem 5.1.3.

Proof of Theorem 5.1.1. We will prove Part (1) of the theorem since the proof of Part (2) is similar. The obtained estimates will also show the corresponding part of the statement in Part (3).

Thus, let us first show that (5.3) implies (5.2). Using the polar decomposition in Proposition 1.2.10 and denoting $r = |x|$, we write

$$\begin{aligned} &\int_{\mathbb{G}} \phi_1(x) \left[\int_{B(0,r)} f(z) dz\right]^q dx \\ &= \int_0^{\infty} \int_{\wp} r^{Q-1} \phi_1(ry) \left[\int_0^r \int_{\wp} s^{Q-1} f(sy) d\sigma(y) ds\right]^q d\sigma(y) dr. \end{aligned} \tag{5.12}$$

Denoting

$$g(r) := \left\{\int_{\wp} \int_0^r s^{Q-1} (\psi_1(sy))^{1-p'} ds d\sigma(y)\right\}^{1/(pp')}, \tag{5.13}$$

and using Hölder's inequality, we can estimate

$$\begin{aligned} &\int_0^r \int_{\wp} s^{Q-1} f(sy) d\sigma(y) ds \\ &= \int_{\wp} \int_0^r s^{(Q-1)/p} f(sy) (\psi_1(sy))^{1/p} g(s) s^{(Q-1)/p'} \\ &\quad \times \left((\psi_1(sy))^{1/p} g(s)\right)^{-1} ds d\sigma(y) \end{aligned}$$

$$\leq \left(\int_{\wp} \int_{0}^{r} s^{Q-1} \left[f(sy)(\psi_1(sy))^{1/p} g(s) \right]^p dsd\sigma(y) \right)^{1/p}$$
$$\times \left(\int_{\wp} \int_{0}^{r} s^{Q-1} \left[(\psi_1(sy))^{1/p} g(s) \right]^{-p'} dsd\sigma(y) \right)^{1/p'}. \quad (5.14)$$

Let us introduce the following notations:

$$U(s) := \int_{\wp} s^{Q-1} \left(f(sy)(\psi_1(sy))^{1/p} g(s) \right)^p d\sigma(y), \quad (5.15)$$

$$V(r) := \int_0^r \int_{\wp} s^{Q-1} \left((\psi_1(sy))^{1/p} g(s) \right)^{-p'} d\sigma(y) ds, \quad (5.16)$$

$$W_1(r) := \int_{\wp} r^{Q-1} \phi_1(ry) d\sigma(y), \quad (5.17)$$

for $s, r > 0$. Plugging (5.14) into (5.12) we obtain

$$\int_{\mathbb{G}} \phi_1(x) \left(\int_{B(0,r)} f(z)dz \right)^q dx$$
$$\leq \int_0^{\infty} W_1(r) \left(\int_0^r U(s)ds \right)^{q/p} (V(r))^{q/p'} dr. \quad (5.18)$$

We now recall the following continuous version of the *Minkowski inequality* (see, e.g., [DHK97, Formula 2.1]):

Let $\theta \geq 1$. Then for all $f_1(x), f_2(x) \geq 0$ on $(0, \infty)$, we have

$$\int_0^{\infty} f_1(x) \left(\int_0^x f_2(z)dz \right)^{\theta} dx \leq \left(\int_0^{\infty} f_2(z) \left(\int_z^{\infty} f_1(x)dx \right)^{1/\theta} dz \right)^{\theta}. \quad (5.19)$$

Using this inequality with $\theta = q/p \geq 1$ in the right-hand side of (5.18), we can estimate

$$\int_{\mathbb{G}} \phi_1(x) \left(\int_{B(0,r)} f(z)dz \right)^q dx \leq \left(\int_0^{\infty} U(s) \left(\int_s^{\infty} W_1(r)(V(r))^{q/p'} dr \right)^{p/q} ds \right)^{q/p}. \quad (5.20)$$

Let us introduce one more temporary notation

$$T(s) := \int_{\wp} s^{Q-1} (\psi_1(sy))^{1-p'} d\sigma(y).$$

Using (5.13), (5.16), the integration by parts, (5.3) and (5.17), we compute

$$
\begin{aligned}
V(r) &= \int_{\wp}\int_0^r s^{Q-1}(\psi_1(sy))^{1-p'}\left(\int_0^s\int_{\wp} t^{Q-1}(\psi_1(tw))^{1-p'}d\sigma(w)dt\right)^{-1/p} dsd\sigma(y) \\
&= \int_0^r T(s)\left(\int_0^s T(t)dt\right)^{-1/p} ds = p'\int_0^r \frac{d}{ds}\left(\int_0^s T(t)dt\right)^{1/p'} ds \\
&= p'\left(\int_0^r T(s)ds\right)^{1/p'} = p'\left(\int_0^r\int_{\wp} s^{Q-1}(\psi_1(sy))^{1-p'}d\sigma(y)ds\right)^{1/p'} \\
&\leq p'A_1\left(\int_r^\infty s^{Q-1}\int_{\wp}\phi_1(sw)d\sigma(w)ds\right)^{-1/q} = p'A_1\left(\int_r^\infty W_1(s)ds\right)^{-1/q}.
\end{aligned}
$$

Similarly, applying the integration by parts and (5.3), this implies

$$
\begin{aligned}
\int_s^\infty W_1(r)(V(r))^{q/p'}dr &= (p'A_1)^{q/p'}\int_s^\infty W_1(r)\left(\int_r^\infty W_1(s)ds\right)^{-1/p'} dr \\
&= (p'A_1)^{q/p'}p\left(\int_s^\infty W_1(r)dr\right)^{1/p} \\
&= (p'A_1)^{q/p'}p\left(\int_s^\infty\int_{\wp} r^{Q-1}\phi_1(ry)d\sigma(y)dr\right)^{1/p} \\
&\leq (p'A_1)^{q/p'}pA_1^{q/p}\left(\int_0^s r^{Q-1}\int_{\wp}(\psi_1(ry))^{1-p'}d\sigma(y)dr\right)^{-q/(p'p)} \\
&= A_1^q(p')^{q/p'}p(g(s))^{-q}, \qquad (5.21)
\end{aligned}
$$

where we have used (5.13) in the last line. Putting (5.21) in (5.20) and recalling (5.15), we obtain

$$
\begin{aligned}
\int_{\mathbb{G}}\phi_1(x)\left(\int_{B(0,r)} f(z)dz\right)^q dx &\leq \left(\int_0^\infty U(s)A_1^p(p')^{p-1}p^{p/q}(g(s))^{-p}ds\right)^{q/p} \\
&= A_1^q(p')^{q/p'}p\left(\int_0^\infty U(s)(g(s))^{-p}ds\right)^{q/p} \\
&= A_1^q(p')^{q/p'}p\left(\int_0^\infty\int_{\wp} s^{Q-1}(f(sy))^p\psi_1(sy)d\sigma(y)ds\right)^{q/p} \\
&= A_1^q(p')^{q/p'}p\left(\int_{\mathbb{G}}\psi_1(x)(f(x))^p dx\right)^{q/p}, \qquad (5.22)
\end{aligned}
$$

yielding (5.2) with $C_1 = A_1(p')^{1/p'}p^{1/q}$.

We now show the converse, namely, that (5.2) implies (5.3). For that, we set

$$f(x) := (\psi_1(x))^{1-p'}\chi_{(0,R)}(|x|),$$

with $R > 0$. For this f we observe the equality

$$\begin{aligned}
&\left(\int_{\mathbb{G}} \psi_1(x)(f(x))^p dx\right)^{1/p}\left(\int_{|x|\le R} (\psi_1(x))^{1-p'}dx\right)^{-1/p} \\
&= \left(\int_{|x|\le R} (\psi_1(x))^{1-p'}dx\right)^{1/p}\left(\int_{|x|\le R} (\psi_1(x))^{1-p'}dx\right)^{-1/p} = 1.
\end{aligned} \tag{5.23}$$

Consequently, by (5.2) we have

$$\begin{aligned}
C &= C\left(\int_{\mathbb{G}} \psi_1(x)(f(x))^p dx\right)^{1/p}\left(\int_{|x|\le R} (\psi_1(x))^{1-p'}dx\right)^{-1/p} \\
&\ge \left(\int_{\mathbb{G}} \phi_1(x)\left(\int_{|z|\le|x|} f(z)dz\right)^q dx\right)^{1/q}\left(\int_{|x|\le R} (\psi_1(x))^{1-p'}dx\right)^{-1/p} \\
&\ge \left(\int_{|x|\ge R} \phi_1(x)\left(\int_{|z|\le|x|} f(z)dz\right)^q dx\right)^{1/q}\left(\int_{|x|\le R} (\psi_1(x))^{1-p'}dx\right)^{-1/p} \\
&= \left(\int_{|x|\ge R} \phi_1(x)dx\right)^{1/q}\left(\int_{|z|\le R} (\psi_1(z))^{1-p'}dz\right)^{1/p'}.
\end{aligned} \tag{5.24}$$

Combining (5.23) and (5.24), we obtain (5.3) with $C \ge A_1$. □

The next case is the version of Theorem 5.1.1 for the indices $p > q$:

Theorem 5.1.5 (Integral Hardy inequalities for $p > q$). *Let $\mathbb{G}$ be a homogeneous group of homogeneous dimension Q and let $1 < q < p < \infty$ and $1/\delta = 1/q - 1/p$. Let ϕ_3 and ϕ_4 be positive functions on $\mathbb{G}$. Then we have the following properties:*

(1) *The inequality*

$$\left(\int_{\mathbb{G}}\left(\int_{B(0,|x|)} f(z)dz\right)^q \phi_3(x)dx\right)^{1/q} \le C_1\left(\int_{\mathbb{G}}(f(x))^p\psi_3(x)dx\right)^{1/p} \tag{5.25}$$

holds for all $f \ge 0$ if and only if

$$\int_{\mathbb{G}}\left(\int_{\mathbb{G}\setminus B(0,|x|)} \phi_3(z)dz\right)^{\delta/q}\left(\int_{B(0,|x|)} (\psi_3(z))^{1-p'}dz\right)^{\delta/q'}(\psi_3(x))^{1-p'}dx < \infty. \tag{5.26}$$

(2) *The inequality*

$$\left(\int_{\mathbb{G}}\left(\int_{\mathbb{G}\backslash B(0,|x|)} f(z)dz\right)^{q}\phi_4(x)dx\right)^{1/q} \leq C_2\left(\int_{\mathbb{G}}(f(x))^p\psi_4(x)dx\right)^{1/p} \tag{5.27}$$

holds for all $f \geq 0$ if and only if

$$\int_{\mathbb{G}}\left(\int_{B(0,|x|)}\phi_4(z)dz\right)^{\delta/q}\left(\int_{\mathbb{G}\backslash B(0,|x|)}(\psi_4(z))^{1-p'}dz\right)^{\delta/q'}(\psi_4(x))^{1-p'}dx < \infty. \tag{5.28}$$

Proof of Theorem 5.1.5. We will prove Part (1) of the theorem since Part (2) is similar. We will denote

$$A_3 := \int_{\mathbb{G}}\left(\int_{\mathbb{G}\backslash B(0,|x|)}\phi_3(z)dz\right)^{\delta/q}\left(\int_{B(0,|x|)}(\psi_3(z))^{1-p'}dz\right)^{\delta/q'}(\psi_3(x))^{1-p'}dx.$$

First we prove that if $A_3 < \infty$, then we have inequality (5.25). Denote

$$W_2(r) := \int_{\wp} r^{Q-1}\phi_3(ry)d\sigma(y) \tag{5.29}$$

and

$$G(s) := \int_{\wp} s^{Q-1}h(sy)(\psi_3(sy))^{1-p'}d\sigma(y) \tag{5.30}$$

for $h \geq 0$ on $\mathbb{G}$ to be chosen later. Using the polar decomposition in Proposition 1.2.10 we have the following equalities:

$$\begin{aligned}
&\int_{\mathbb{G}}\phi_3(x)\left(\int_{B(0,|x|)}h(z)(\psi_3(z))^{1-p'}dz\right)^{q}dx \\
&= \int_0^{\infty}\int_{\wp} r^{Q-1}\phi_3(rw)d\sigma(w)\left(\int_0^r\int_{\wp} s^{Q-1}h(sy)(\psi_3(sy))^{1-p'}d\sigma(y)ds\right)^{q}dr \\
&= \int_0^{\infty}W_2(r)\left(\int_0^r G(s)ds\right)^{q}dr \\
&= q\int_0^{\infty}G(s)\left(\int_0^s G(r)dr\right)^{q-1}\left(\int_s^{\infty}W_2(r)dr\right)ds \\
&= q\int_{\wp}\int_0^{\infty}s^{Q-1}h(sy)(\psi_3(sy))^{1-p'}\left(\int_0^s\int_{\wp}r^{Q-1}h(rw)(\psi_3(rw))^{1-p'}d\sigma(w)dr\right)^{q-1} \\
&\quad\times\left(\int_s^{\infty}W_2(r)dr\right)dsd\sigma(y)
\end{aligned}$$

$$
\begin{aligned}
&= q \int_{\wp} \int_0^\infty s^{Q-1} h(sy)(\psi_3(sy))^{(1-p')(\frac{1}{p}+\frac{q-1}{p}+\frac{p-q}{p})} \\
&\times \left(\frac{\int_{\wp} \int_0^s r^{Q-1} h(rw)(\psi_3(rw))^{1-p'} dr d\sigma(w)}{\int_{\wp} \int_0^s r^{Q-1} (\psi_3(rw))^{1-p'} dr d\sigma(w)} \right)^{q-1} \\
&\times \left(\left(\int_{\wp} \int_0^s r^{Q-1} (\psi_3(rw))^{1-p'} dr d\sigma(w) \right)^{q-1} \left(\int_s^\infty W_2(r) dr \right) \right) ds d\sigma(y).
\end{aligned}
$$

Here, using Hölder's inequality with three factors with indices $\frac{1}{p} + \frac{q-1}{p} + \frac{p-q}{p} = 1$, we can estimate

$$
\int_{\mathbb{G}} \phi_3(x) \left(\int_{B(0,|x|)} h(z)(\psi_3(z))^{1-p'} dz \right)^q dx \leq q K_1 K_2 K_3, \tag{5.31}
$$

where

$$
\begin{aligned}
K_1 &= \left(\int_{\wp} \int_0^\infty s^{Q-1} (h(sy))^p (\psi_3(sy))^{1-p'} ds d\sigma(y) \right)^{1/p} \\
&= \left(\int_{\mathbb{G}} (h(x))^p (\psi_3(x))^{1-p'} dx \right)^{1/p},
\end{aligned} \tag{5.32}
$$

$$
\begin{aligned}
K_2 = &\left(\int_{\wp} \int_0^\infty s^{Q-1} (\psi_3(sy))^{1-p'} \right. \\
&\times \left. \left(\frac{\int_{\wp} \int_0^s r^{Q-1} h(rw)(\psi_3(rw))^{1-p'} dr d\sigma(w)}{\int_{\wp} \int_0^s r^{Q-1} (\psi_3(rw))^{1-p'} dr d\sigma(w)} \right)^p ds d\sigma(y) \right)^{\frac{q-1}{p}}
\end{aligned} \tag{5.33}
$$

and

$$
\begin{aligned}
K_3 = &\left(\int_{\wp} \int_0^\infty s^{Q-1} (\psi_3(sy))^{1-p'} \left(\int_{\wp} \int_0^s r^{Q-1} (\psi_3(rw))^{1-p'} dr d\sigma(w) \right)^{\frac{(q-1)p}{p-q}} \right. \\
&\times \left. \left(\int_s^\infty W_2(r) dr \right)^{\frac{p}{p-q}} ds d\sigma(y) \right)^{\frac{p-q}{p}}.
\end{aligned} \tag{5.34}
$$

Leaving K_1 as it is, we will estimate K_2 and K_3. We rewrite K_2 as

$$
K_2 = \left(\int_{\mathbb{G}} \frac{(\psi_3(x))^{1-p'}}{(\int_{B(0,|x|)} (\psi_3(z))^{1-p'} dz)^p} \left(\int_{B(0,|x|)} (\psi_3(z))^{1-p'} h(z) dz \right)^p dx \right)^{(q-1)/p}.
$$

We want to apply (5.2) to K_2 with indices $p = q$, and with functions $f(x) = (\psi_3(x))^{1-p'} h(x)$ and

$$
\phi_1(x) = \frac{(\psi_3(x))^{1-p'}}{(\int_{B(0,|x|)} (\psi_3(z))^{1-p'} dz)^p}, \quad \psi_1(x) = (\psi_3(x))^{(1-p')(1-p)}.
$$

For that, we will check the condition that

$$\begin{aligned} A_1(R) = &\left(\int_{|x|\geq R}(\psi_3(x))^{1-p'}\left(\int_{B(0,|x|)}(\psi_3(z))^{1-p'}dz\right)^{-p}dx\right)^{1/p} \\ &\times\left(\int_{|x|\leq R}(\psi_3(x))^{1-p'}dx\right)^{1/p'} < \infty \end{aligned} \tag{5.35}$$

holds uniformly for all $R > 0$. Assuming (5.35) uniformly in $R > 0$ for a moment, the inequality (5.2) would imply that

$$\begin{aligned} K_2 &\leq C\left(\int_{\mathbb{G}}(\psi_3(x))^{(1-p')(1-p+p)}(h(x))^p dx\right)^{(q-1)/p} \\ &= C\left(\int_{\mathbb{G}}(h(x))^p(\psi_3(x))^{1-p'}dx\right)^{(q-1)/p}, \end{aligned} \tag{5.36}$$

which is something we will use later. So, let us check (5.35). For this, we denote

$$S(s) := \int_{\wp} s^{Q-1}(\psi_3(sw))^{1-p'}d\sigma(w).$$

Using integration by parts we have

$$\begin{aligned} A_1(R) &= \left(\int_{\wp}\int_R^{\infty} r^{Q-1}(\psi_3(rw))^{1-p'}\left(\int_0^r S(s)ds\right)^{-p}drd\sigma(w)\right)^{\frac{1}{p}}\left(\int_0^R S(s)ds\right)^{\frac{1}{p'}} \\ &= \left(\int_R^{\infty}\left(\int_0^r S(s)ds\right)^{-p}S(r)dr\right)^{1/p}\left(\int_0^R S(s)ds\right)^{1/p'} \\ &\leq \left(\frac{1}{p-1}\left(\int_0^R S(s)ds\right)^{1-p}\right)^{1/p}\left(\int_0^R S(s)ds\right)^{1/p'} = (p-1)^{-1/p} < \infty, \end{aligned}$$

so that (5.36) is confirmed. Next, for K_3, taking into account

$$\frac{1}{\delta} = \frac{1}{q} - \frac{1}{p} = \frac{p-q}{pq}$$

and using (5.26), we have

$$\begin{aligned} K_3 = &\left(\int_0^{\infty}\int_{\wp}\left(\int_s^{\infty}W_2(r)dr\right)^{\delta/q}\left(\int_{\wp}\int_0^r r^{Q-1}(\psi_3(rw))^{1-p'}drd\sigma(w)\right)^{\delta/q'}\right. \\ &\left.\times s^{Q-1}(\psi_3(sy))^{1-p'}d\sigma(y)ds\right)^{(p-q)/p} \end{aligned}$$

$$= \left(\int_{\mathbb{G}} \left(\int_{\mathbb{G}\setminus B(0,|x|)} \phi_3(z)dz \right)^{\frac{\delta}{q}} \left(\int_{B(0,|x|)} (\psi_3(z))^{1-p'}dz \right)^{\frac{\delta}{q'}} (\psi_3(x))^{1-p'}dx \right)^{\frac{p-q}{p}}$$

$$= A_3^{\frac{p-q}{p}} < \infty. \tag{5.37}$$

Now, plugging (5.32), (5.36) and (5.37) into (5.31), we obtain

$$\int_{\mathbb{G}} \phi_3(x) \left(\int_{B(0,|x|)} h(z)(\psi_3(z))^{1-p'}dz \right)^q dx$$
$$\leq C A_3^{\frac{p-q}{p}} \left(\int_{\mathbb{G}} (h(x))^p (\psi_3(x))^{1-p'}dx \right)^{\frac{1}{p}+\frac{q-1}{p}},$$

which implies (5.25) after taking $h := f\psi_3^{p'-1}$.

Let us now show the converse, namely, that (5.25) implies (5.26). For this, we consider a sequence of functions

$$f_k(x) := \left(\int_{|y|\geq|x|} \phi_3(z)dz \right)^{\delta/(pq)} \left(\int_{\alpha_k\leq|z|\leq|x|} (\psi_3(z))^{1-p'}dz \right)^{\delta/(pq')}$$
$$\times (\psi_3(x))^{1-p'} \chi_{(\alpha_k,\beta_k)}(|x|), \ k = 1, 2, \ldots.$$

Inserting these functions in the place of $f(x)$ in (5.25), we obtain (5.26), if we take $0 < \alpha_k < \beta_k$ with $\alpha_k \searrow 0$ and $\beta_k \nearrow \infty$ for $k \to \infty$. □

5.2 Convolution Hardy inequalities

In this section we discuss integral Hardy inequalities in the convolution form. Such inequalities are particularly useful if we make particular choices of the convolution kernels. For example, by taking the Riesz kernels of hypoelliptic differential operators on graded groups, such inequalities can be used to derive a number of hypoelliptic versions of Hardy inequalities. While this topic falls outside the scope of this book, we refer to [RY18a] for such applications. The inequalities that we will present here have been established in [RY18a] and we follow the proofs there in our exposition.

Theorem 5.2.1 (Convolution Hardy inequality). *Let $\mathbb{G}$ be a homogeneous Lie group of homogeneous dimension Q and with a homogeneous quasi-norm $|\cdot|$. Let $1 < p \leq q < \infty$, $0 < a < Q/p$, $0 \leq b < Q$ and $\frac{a}{Q} = \frac{1}{p} - \frac{1}{q} + \frac{b}{qQ}$. Assume that there is $C_2 = C_2(a, Q) > 0$ such that*

$$|T_a^{(1)}(x)| \leq C_2|x|^{a-Q} \tag{5.38}$$

holds for all $x \neq 0$. Then there exists a positive constant $C_1 = C_1(p, q, a, b) > 0$ such that

$$\left\| \frac{f * T_a^{(1)}}{|x|^{\frac{b}{q}}} \right\|_{L^q(\mathbb{G})} \leq C_1 \|f\|_{L^p(\mathbb{G})} \tag{5.39}$$

holds for all $f \in L^p(\mathbb{G})$.

The critical case $b = Q$ of Theorem 5.2.1 will be shown in Theorem 5.2.5.

Remark 5.2.2 (Riesz kernels). Let us briefly describe a typical situation when condition (5.38) is satisfied. Without going much into detail, let us assume that $\mathscr{R}$ is a positive homogeneous left invariant hypoelliptic differential operator on $\mathbb{G}$ of homogeneous degree ν. The existence of such an operator implies that the group $\mathbb{G}$ is graded, see [FR16, Section 4.1]. The operators $\mathscr{R}$ satisfying the above properties are called *Rockland operators.*

Let h_t denote the heat kernel associated to the operator $\mathscr{R}$, see [FR16, Section 4.3.4] for a thorough treatment of this and for the proof of the following notes. This heat kernel satisfies the following properties, see [FR16, Theorem 4.2.7 and Lemma 4.3.8]:

Theorem 5.2.3 (Heat kernels). *Let $\mathscr{R}$ be homogeneous left invariant hypoelliptic differential operator on $\mathbb{G}$ of homogeneous degree ν and let h_t be the associated heat kernel. Then each h_t is Schwartz and we have*

$$\forall s, t > 0 \;\; h_t * h_s = h_{t+s}, \tag{5.40}$$

$$\forall x \in \mathbb{G}, r, t > 0 \;\; h_{r^\nu t}(rx) = r^{-Q} h_t(x), \tag{5.41}$$

$$\forall x \in \mathbb{G} \;\; h_t(x) = \overline{h_t(x^{-1})}, \tag{5.42}$$

$$\int_{\mathbb{G}} h_t(x) dx = 1. \tag{5.43}$$

Moreover, we have

$$\exists C = C_{\alpha,N,\ell} > 0 \;\; \forall t \in (0,1] \;\; \sup_{|x|=1} |\partial_t^\ell X^\alpha h_t(x)| \leq C_{\alpha,N} t^N \tag{5.44}$$

for any $N \in \mathbb{N}_0$, $\alpha \in \mathbb{N}_0^n$ and $\ell \in \mathbb{N}_0$.

Furthermore, for any multi-index $\alpha \in \mathbb{N}_0^n$ and any real number a with $0 < a < (Q + [\alpha])/\nu$ there exists a positive constant $C > 0$ such that

$$\int_0^\infty t^{a-1} |X^\alpha h_t(x)| dt \leq C|x|^{-Q-[\alpha]+\nu a}. \tag{5.45}$$

The fractional powers $\mathscr{R}^{-a/\nu}$ for $\{a \in \mathbb{R},\ 0 < a < Q\}$ and $(I + \mathscr{R})^{-a/\nu}$ for $a \in \mathbb{R}_+$ are called *Riesz and Bessel potentials*, respectively, and they are well defined, see [FR16, Chapter 4.2]. Let us denote their respective kernels by $\mathcal{I}_a$ and $\mathcal{B}_a$. Then we have the relations

$$\mathcal{I}_a(x) = \frac{1}{\Gamma\left(\frac{a}{\nu}\right)} \int_0^\infty t^{\frac{a}{\nu}-1} h_t(x) dt \tag{5.46}$$

for $0 < a < Q$ with $a \in \mathbb{R}$, and

$$\mathcal{B}_a(x) = \frac{1}{\Gamma\left(\frac{a}{\nu}\right)} \int_0^\infty t^{\frac{a}{\nu}-1} e^{-t} h_t(x) dt \tag{5.47}$$

for $a > 0$, where Γ denotes the Gamma function. Consequently, it can be shown (see [FR16, Section 4.3.4]) that for any $0 < a < Q$ there exists a positive constant $C = C(Q, a)$ such that

$$|\mathcal{I}_a(x)| \leq C|x|^{-(Q-a)} \tag{5.48}$$

holds for all $x \neq 0$. Therefore, the Riesz kernel $\mathcal{I}_a$ gives a typical example of an operator satisfying condition (5.38).

Proof of Theorem 5.2.1. We split the integral in the left-hand side of (5.39) into three parts:

$$\int_{\mathbb{G}} |(f * T_a^{(1)})(x)|^q \frac{dx}{|x|^b} \leq 3^q (M_1 + M_2 + M_3), \tag{5.49}$$

with

$$M_1 := \int_{\mathbb{G}} \left(\int_{\{2|y|<|x|\}} |T_a^{(1)}(y^{-1}x) f(y)| dy \right)^q \frac{dx}{|x|^b},$$

$$M_2 := \int_{\mathbb{G}} \left(\int_{\{|x|\leq 2|y|<4|x|\}} |T_a^{(1)}(y^{-1}x) f(y)| dy \right)^q \frac{dx}{|x|^b}$$

and

$$M_3 := \int_{\mathbb{G}} \left(\int_{\{|y|>2|x|\}} |T_a^{(1)}(y^{-1}x) f(y)| dy \right)^q \frac{dx}{|x|^b}.$$

First, let us *estimate* M_1. We can assume without loss of generality that $|\cdot|$ is a norm (such a norm always exists, see Proposition 1.2.4, Part (2)) since replacing the seminorm by an equivalent one only changes the appearing constants.

Observe that by the reverse triangle inequality and the assumption $2|y| < |x|$ we have

$$|y^{-1}x| \geq |x| - |y| > |x| - \frac{|x|}{2} = \frac{|x|}{2}, \tag{5.50}$$

which is $|x| < 2|y^{-1}x|$. Taking into account this and that $T_a^{(1)}(x)$ is bounded by a radial function which is non-increasing with respect to $|x|$, we can estimate

$$\begin{aligned} M_1 &\leq \int_{\mathbb{G}} \left(\int_{\{2|y|<|x|\}} |f(y)| dy \right)^q \left(\sup_{\{|x|<2|z|\}} |T_a^{(1)}(z)| \right)^q \frac{dx}{|x|^b} \\ &\leq C \int_{\mathbb{G}} \left(\int_{\{2|y|<|x|\}} |f(y)| dy \right)^q \left(\frac{|x|}{2} \right)^{(a-Q)q} \frac{dx}{|x|^b}. \end{aligned} \tag{5.51}$$

We will now apply Theorem 5.1.1, Part (1), to estimate M_1. For this we need to check condition (5.3), that is, that

$$\sup_{R>0} \left(\int_{\{2R<|x|\}} \left(\frac{|x|}{2} \right)^{(a-Q)q} \frac{dx}{|x|^b} \right)^{\frac{1}{q}} \left(\int_{\{|x|<R\}} dx \right)^{\frac{1}{p'}} < \infty. \tag{5.52}$$

To check this, we consider two cases: $R \geq 1$ and $0 < R < 1$. For $R \geq 1$, we can estimate

$$\begin{aligned} &\left(\int_{\{2R<|x|\}} \left(\frac{|x|}{2} \right)^{(a-Q)q} \frac{dx}{|x|^b} \right)^{\frac{1}{q}} \left(\int_{\{|x|<R\}} dx \right)^{\frac{1}{p'}} \\ &\qquad \leq CR^{\frac{Q}{p'}} \left(\int_{\{2R<|x|\}} \left(\frac{|x|}{2} \right)^{(a-Q)q} \frac{dx}{|x|^b} \right)^{\frac{1}{q}} \\ &\qquad \leq CR^{\frac{Q}{p'}} \left(\int_{\{2R<|x|\}} |x|^{(a-Q)q-b} dx \right)^{\frac{1}{q}} \\ &\qquad \leq CR^{\frac{Q}{p'}} R^{\frac{(a-Q)q-b+Q}{q}} \\ &\qquad \leq C, \end{aligned} \tag{5.53}$$

which is uniformly bounded since $\frac{a}{Q} = \frac{1}{p} - \frac{1}{q} + \frac{b}{qQ}$ and $(a-Q)q-b+Q = -\frac{Qq}{p'} \neq 0$. Now let us check the condition (5.52) for $0 < R < 1$. Here, taking into account that $(a-Q)q-b+Q = -\frac{Qq}{p'} \neq 0$ we have

$$\int_{\{2R<|x|\}} \left(\frac{|x|}{2} \right)^{(a-Q)q} \frac{dx}{|x|^b} \leq CR^{(a-Q)q-b+Q}. \tag{5.54}$$

It follows with $\frac{a}{Q} = \frac{1}{p} - \frac{1}{q} + \frac{b}{qQ}$ that

$$\left(\int_{\{2R<|x|\}} \left(\frac{|x|}{2} \right)^{(a-Q)q} \frac{dx}{|x|^b} \right)^{\frac{1}{q}} \left(\int_{\{|x|<R\}} dx \right)^{\frac{1}{p'}} \leq CR^{a-Q-\frac{b}{q}+\frac{Q}{q}} R^{Q/p'} \leq C \tag{5.55}$$

holds for any $0 < R < 1$. Thus, we have checked (5.52). Applying Theorem 5.1.1, Part (1), we obtain

$$M_1^{\frac{1}{q}} \leq (p')^{\frac{1}{p'}} p^{\frac{1}{q}} A_1 \|f\|_{L^p(\mathbb{G})}. \tag{5.56}$$

Let us now *estimate* M_2. For this, we decompose M_2 as

$$M_2 = \sum_{k\in\mathbb{Z}} \int_{\{2^k \leqslant |x| < 2^{k+1}\}} \left(\int_{\{|x| \leqslant 2|y| \leqslant 4|x|\}} |T_a^{(1)}(y^{-1}x) f(y)| dy \right)^q \frac{dx}{|x|^b}.$$

Since $|x| \leqslant 2|y| \leqslant 4|x|$ and $2^k \leqslant |x| < 2^{k+1}$, we have $2^{k-1} \leqslant |y| < 2^{k+2}$. As in (5.50), assuming $|\cdot|$ is the norm and using the triangle inequality, we have

$$3|x| = |x| + 2|x| \geq |x| + |y| \geq |y^{-1}x|, \tag{5.57}$$

which implies $0 \leq |y^{-1}x| \leq 3|x| < 3 \cdot 2^{k+1}$. If we denote

$$\widetilde{I}_a(x) := C_2 |x|^{a-Q},$$

then by the assumption we have

$$|T_a^{(1)}(x)| \leq \widetilde{I}_a(x).$$

Taking into account these observations and applying Young's inequality in Proposition 1.2.13 with $1 + \frac{1}{q} = \frac{1}{r} + \frac{1}{p}$, $r \in [1, \infty]$, we can estimate M_2 by

$$\begin{aligned}
M_2 &\leq \sum_{k\in\mathbb{Z}} 2^{-kb} \int_{\mathbb{G}} (([f \cdot \chi_{\{2^{k-1} \leqslant |\cdot| < 2^{k+2}\}}] * \widetilde{I}_a)(x))^q dx \\
&= \sum_{k\in\mathbb{Z}} 2^{-kb} \|[f \cdot \chi_{\{2^{k-1} \leqslant |\cdot| < 2^{k+2}\}}] * \widetilde{I}_a\|^q_{L^q(\mathbb{G})} \\
&\leq \sum_{k\in\mathbb{Z}} 2^{-kb} \|\widetilde{I}_a \cdot \chi_{\{0 \leqslant |\cdot| < 3\cdot 2^{k+1}\}}\|^q_{L^r(\mathbb{G})} \|f \cdot \chi_{\{2^{k-1} \leqslant |\cdot| < 2^{k+2}\}}\|^q_{L^p(\mathbb{G})} \\
&= C_2 \sum_{k\in\mathbb{Z}} 2^{-kb} \left(\int_{|x| < 3\cdot 2^{k+1}} |x|^{(a-Q)r} dx \right)^{\frac{q}{r}} \|f \cdot \chi_{\{2^{k-1} \leqslant |x| < 2^{k+2}\}}\|^q_{L^p(\mathbb{G})} \qquad (5.58) \\
&\leq C \sum_{k\in\mathbb{Z}} 2^{-kb} (3 \cdot 2^{k+1})^{\left(\frac{(a-Q)pq}{pq+p-q} + Q\right)\frac{pq+p-q}{p}} \|f \cdot \chi_{\{2^{k-1} \leqslant |x| < 2^{k+2}\}}\|^q_{L^p(\mathbb{G})} \\
&= C \sum_{k\in\mathbb{Z}} 2^{-kb} (3 \cdot 2^{k+1})^b \|f \cdot \chi_{\{2^{k-1} \leqslant |x| < 2^{k+2}\}}\|^q_{L^p(\mathbb{G})} \\
&\leq C \sum_{k\in\mathbb{Z}} \|f \cdot \chi_{\{2^{k-1} \leqslant |x| < 2^{k+2}\}}\|^q_{L^p(\mathbb{G})} \\
&\leq C \|f\|^q_{L^p(\mathbb{G})},
\end{aligned}$$

since $\frac{(a-Q)pq}{pq+p-q} + Q = \frac{bp}{pq+p-q} > 0$ and $q \geq p$.

Now let us *estimate* M_3. Without loss of generality, we may assume again that $|\cdot|$ is a norm. Then, similarly to (5.50) we note that $2|x| < |y|$ implies $|y| < 2|y^{-1}x|$. Consequently we can estimate M_3 as

$$M_3 \leq \int_{\mathbb{G}} \left(\int_{\{|y|>2|x|\}} \left(\frac{|y|}{2}\right)^{(a-Q)} |f(y)|dy \right)^q \frac{dx}{|x|^b}.$$

We will apply Theorem 5.1.1, Part (2), to estimate M_3. For this, we need to check that

$$\sup_{R>0} \left(\int_{\{|x|<R\}} \frac{dx}{|x|^b} \right)^{\frac{1}{q}} \left(\int_{\{2R<|x|\}} \left(\frac{|x|}{2}\right)^{(a-Q)p'} dx \right)^{\frac{1}{p'}} < \infty. \tag{5.59}$$

To verify this, we consider two cases: $R \geq 1$ and $0 < R < 1$. First, for $R \geq 1$, using the assumption $|T_a^{(1)}(x)| \leq C|x|^{a-Q}$ and that $Q \neq ap$, one gets

$$\left(\int_{\{2R<|x|\}} \left(\frac{|x|}{2}\right)^{(a-Q)p'} dx \right)^{\frac{1}{p'}} \leq C \left(\int_{\{2R<|x|\}} |x|^{(a-Q)p'} dx \right)^{\frac{1}{p'}} \leq CR^{a-\frac{Q}{p}}. \tag{5.60}$$

Since $R \geq 1$ we can estimate

$$\left(\int_{\{|x|<R\}} \frac{dx}{|x|^b} \right)^{\frac{1}{q}} \left(\int_{\{2R<|x|\}} \left(\frac{|x|}{2}\right)^{(a-Q)p'} dx \right)^{\frac{1}{p'}} \leq CR^{a-\frac{Q}{p}+\frac{Q-b}{q}} \leq C,$$

since $b < Q$ and $\frac{a}{Q} = \frac{1}{p} - \frac{1}{q} + \frac{b}{qQ}$. Now let us check the condition (5.59) for the range $0 < R < 1$. In this case, noting that $ap - Q < 0$ we have

$$\int_{\{2R<|x|\}} \left(\frac{|x|}{2}\right)^{(a-Q)p'} dx \leq C \int_{\{2R<|x|\}} |x|^{(a-Q)p'} dx \leq CR^{(a-Q)p'+Q}. \tag{5.61}$$

Since

$$\int_{\{|x|<R\}} \frac{dx}{|x|^b} \leqslant CR^{Q-b},$$

we have

$$\left(\int_{\{|x|<R\}} \frac{dx}{|x|^b} \right)^{\frac{1}{q}} \left(\int_{\{2R<|x|\}} \left(\frac{|x|}{2}\right)^{(a-Q)p'} dx \right)^{\frac{1}{p'}} \leq CR^{\frac{Q-b}{q}} R^{\frac{(a-Q)p'+Q}{p'}} \leq C, \tag{5.62}$$

since $Q > b$ and $\frac{a}{Q} = \frac{1}{p} - \frac{1}{q} + \frac{b}{qQ}$. Thus, we have checked (5.59). Consequently, the application of Theorem 5.1.1, Part (2), to M_3 yields

$$M_3^{\frac{1}{q}} \leq (p')^{\frac{1}{p'}} p^{\frac{1}{q}} A_2 \|f\|_{L^p(\mathbb{G})}. \tag{5.63}$$

Thus, (5.56), (5.63) and (5.58) complete the proof of Theorem 5.2.1. □

Remark 5.2.4 (Schur test argument). In the case $p = q$, we can also prove Thecrem 5.2.1 by using Schur's test ([FR75]) as in the proof of Theorem 4.7.1. For the Riesz kernels of the sub-Laplacian on stratified groups such an argument was used in [CCR15], and the argument below was given in [RY18a].

For $p = q$, the condition $\frac{a}{Q} = \frac{1}{p} - \frac{1}{q} + \frac{b}{qQ}$ in Theorem 5.2.1 implies that $b = ap$, so we are interested in the L^p-boundedness of the operator

$$S_a f := |x|^{-b/p}(f * |x|^{a-Q}) = |x|^{-a}(f * |x|^{a-Q}).$$

The adjoint, defined by $(f, S_a^* g) = (S_a f, g)$, is given by $S_a^* g := (|x|^{-a} g) * |x|^{a-Q}$.

We now recall again the *Schur test*:

Assume that the integral operator S has a positive integral kernel, and that there exist a positive function h and constants A_p and B_p such that

$$S(h^{p'})(x) \le A_p(h(x))^{p'} \quad \text{and} \quad S^*(h^p)(x) \le B_p(h(x))^p$$

hold for almost all $x \in \mathbb{G}$. Then we have

$$\|S_a f\|_{L^p(\mathbb{G})} \le A_p^{1/p'} B_p^{1/p} \|f\|_{L^p(\mathbb{G})}$$

for all $f \in L^p(\mathbb{G})$.

Let us now take $h_c(x) := |x|^{c-Q}$ with $c > 0$ and consider the convolution integrals

$$h_c^{p'} * |x|^{a-Q} \quad \text{and} \quad (|x|^{-b/p} h_c^p) * |x|^{a-Q},$$

which arise in the computation of $S_a(h_c^{p'})$ and $S_a^*(h_c^p)$. We see that the homogeneity orders of $h_c^{p'}$ and $|x|^{-b/p} h_c^p$ are $(c-Q)p'$ and $(c-Q)p - b/p$, respectively. Then, the homogeneity orders of $h_c^{p'} * |x|^{a-Q}$ and $(|x|^{-b/p} h_c^p) * |x|^{a-Q}$ are $a - Q + (c-Q)p'$ and $a - Q + (c-Q)p - b/p$, respectively. Therefore, these convolution integrals converge absolutely in $\mathbb{G}\backslash\{0\}$ if and only if $0 < (c-Q)p' + Q < Q - a$ and $0 < (c-Q)p - b/p + Q < Q - a$, that is, if

$$\max\left(\frac{Q}{p}, \frac{a}{p} + \frac{Q}{p'}\right) < c < Q - \frac{a}{p'}$$

since $b = ap$. This condition is true if $0 < a < Q/p$.

Thus, it follows from Schur's test that

$$\||x|^{-a}(f * |x|^{a-Q})\|_{L^p(\mathbb{G})} \le A_{a,p}^{1/p'} B_{a,p}^{1/p} \|f\|_{L^p(\mathbb{G})},$$

where $0 < a < Q/p$, $1 < p < \infty$, and $f \in L^p(\mathbb{G})$.

Taking into account this and $|T_a^{(1)}(x)| \le C|x|^{a-Q}$, we obtain

$$\begin{aligned} \left\| \frac{f * T_a^{(1)}}{|x|^{\frac{b}{p}}} \right\|_{L^p(\mathbb{G})} &\le C \left\| \frac{|f| * |T_a^{(1)}|}{|x|^{\frac{b}{p}}} \right\|_{L^p(\mathbb{G})} \\ &\le C \||x|^{-a}(|f| * |x|^{a-Q})\|_{L^p(\mathbb{G})} \le C\|f\|_{L^p(\mathbb{G})}, \end{aligned} \tag{5.64}$$

which proves Theorem 5.2.1 in the case $p = q$.

Let us now show the critical case $b = Q$ of Theorem 5.2.1.

Theorem 5.2.5 (Critical convolution Hardy inequality). *Let $\mathbb{G}$ be a homogeneous Lie group of homogeneous dimension Q with a homogeneous quasi-norm $|\cdot|$. Let $1 < p < r < \infty$ and $p < q < (r-1)p'$, where $1/p + 1/p' = 1$. Assume that for $a = Q/p$ we have*

$$|T_a^{(2)}(x)| \le C_2 \begin{cases} |x|^{a-Q}, & \text{for } x \in \mathbb{G}\backslash\{0\}, \\ |x|^{-Q}, & \text{for } x \in \mathbb{G} \ \text{ with } |x| \ge 1, \end{cases} \tag{5.65}$$

for some positive $C_2 = C_2(a, Q)$. Then there exists a positive constant $C_1 = C_1(p, q, r, Q) > 0$ such that

$$\left\| \frac{f * T_{Q/p}^{(2)}}{\left(\log\left(e + \frac{1}{|x|}\right)\right)^{\frac{r}{q}} |x|^{\frac{Q}{q}}} \right\|_{L^q(\mathbb{G})} \le C_1 \|f\|_{L^p(\mathbb{G})} \tag{5.66}$$

holds for all $f \in L^p(\mathbb{G})$.

Remark 5.2.6. We note that compared to the condition (5.38), the decay assumption in (5.65) for large x is stronger. Continuing with the notation of Remark 5.2.2, we observe that the Bessel kernel (5.47) of the operator $(I + \mathscr{R})^{-a/\nu}$ for $0 < a < Q$ satisfies (5.65): there exists a positive constant $C = C(Q, a) > 0$ such that we have, in particular,

$$|\mathcal{B}_a(x)| \le \begin{cases} C|x|^{-(Q-a)}, & \text{for } x \in \mathbb{G}\backslash\{0\}, \\ C|x|^{-Q}, & \text{for } x \in \mathbb{G} \ \text{ with } |x| \ge 1. \end{cases} \tag{5.67}$$

We refer to [RY18a] for further details, as well as to the original proof of Theorem 5.2.5 that we follow here.

Proof of Theorem 5.2.5. Let us split the integral in the left-hand side of (5.66) into three parts,

$$\int_{\mathbb{G}} |(f * T_{Q/p}^{(2)})(x)|^q \frac{dx}{\left|\log\left(e + \frac{1}{|x|}\right)\right|^r |x|^Q} \le 3^q (N_1 + N_2 + N_3), \tag{5.68}$$

where

$$N_1 := \int_{\mathbb{G}} \left(\int_{\{2|y|<|x|\}} |T_{Q/p}^{(2)}(y^{-1}x) f(y)| dy \right)^q \frac{dx}{\left|\log\left(e + \frac{1}{|x|}\right)\right|^r |x|^Q},$$

$$N_2 := \int_{\mathbb{G}} \left(\int_{\{|x| \le 2|y| < 4|x|\}} |T_{Q/p}^{(2)}(y^{-1}x) f(y)| dy \right)^q \frac{dx}{\left|\log\left(e + \frac{1}{|x|}\right)\right|^r |x|^Q}$$

and

$$N_3 := \int_{\mathbb{G}} \left(\int_{\{|y|>2|x|\}} |T^{(2)}_{Q/p}(y^{-1}x) f(y)| dy \right)^q \frac{dx}{\left|\log\left(e + \frac{1}{|x|}\right)\right|^r |x|^Q}.$$

We begin by *estimating* N_1. Similar to the argument in (5.50), in the region $2|y| < |x|$ we have

$$|y^{-1}x| \geq |x| - |y| > |x| - \frac{|x|}{2} = \frac{|x|}{2}, \tag{5.69}$$

which is $|x| < 2|y^{-1}x|$. Denote

$$|T^{(2)}_a(x)| \leq \widetilde{B}_a(x) := C_2 \begin{cases} |x|^{a-Q}, & \text{for } x \in \mathbb{G}\backslash\{0\}, \\ |x|^{-Q}, & \text{for } x \in \mathbb{G} \text{ with } |x| \geq 1. \end{cases} \tag{5.70}$$

Since $T^{(2)}_{Q/p}(x)$ is bounded by $\widetilde{B}_{Q/p}(x)$ which is non-increasing with respect to $|x|$, then using (5.69) we get

$$\begin{aligned} N_1 &\leq \int_{\mathbb{G}} \left(\int_{\{2|y|<|x|\}} |f(y)| dy \right)^q \left(\sup_{\{|x|<2|z|\}} |T^{(2)}_{Q/p}(z)| \right)^q \frac{dx}{\left|\log\left(e + \frac{1}{|x|}\right)\right|^r |x|^Q} \\ &\leq \int_{\mathbb{G}} \left(\int_{\{2|y|<|x|\}} |f(y)| dy \right)^q \left(\widetilde{B}_{Q/p}\left(\frac{x}{2}\right) \right)^q \frac{dx}{\left|\log\left(e + \frac{1}{|x|}\right)\right|^r |x|^Q}. \end{aligned}$$

We will now apply Theorem 5.1.1, Part (1), to estimate N_1. For this we have to check the condition (5.3), that is, that

$$\left(\int_{\{2R<|x|\}} \left(\widetilde{B}_{Q/p}\left(\frac{x}{2}\right) \right)^q \frac{dx}{\left|\log\left(e + \frac{1}{|x|}\right)\right|^r |x|^Q} \right)^{\frac{1}{q}} \left(\int_{\{|x|<R\}} dx \right)^{\frac{1}{p'}} \leq A_1 \tag{5.71}$$

holds uniformly for all $R > 0$. To verify this uniform boundedness, we consider two cases: $R \geq 1$ and $0 < R < 1$. First, for $R \geq 1$, using the second equality in (5.70), we can estimate

$$\begin{aligned} &\left(\int_{\{2R<|x|\}} \left(\widetilde{B}_{Q/p}\left(\frac{x}{2}\right) \right)^q \frac{dx}{\left|\log\left(e + \frac{1}{|x|}\right)\right|^r |x|^Q} \right)^{\frac{1}{q}} \left(\int_{\{|x|<R\}} dx \right)^{\frac{1}{p'}} \\ &\leq CR^{\frac{Q}{p'}} \left(\int_{\{2R<|x|\}} \left(\widetilde{B}_{Q/p}\left(\frac{x}{2}\right) \right)^q \frac{dx}{|x|^Q} \right)^{\frac{1}{q}} = CR^{\frac{Q}{p'}} \left(\int_{\{2R<|x|\}} |x|^{-Qq-Q} dx \right)^{\frac{1}{q}} \\ &\leq CR^{-Q} R^{\frac{Q}{p'}} \leq C. \end{aligned} \tag{5.72}$$

Next, let us check (5.71) for $0 < R < 1$. We split the integral into two terms,

$$\begin{aligned}\int_{\{2R<|x|\}} \left(\widetilde{B}_{Q/p}\left(\frac{x}{2}\right)\right)^q \frac{dx}{\left|\log\left(e+\frac{1}{|x|}\right)\right|^r |x|^Q} &= \int_{\{2R<|x|<2\}} \left(\widetilde{B}_{Q/p}\left(\frac{x}{2}\right)\right)^q \frac{dx}{\left|\log\left(e+\frac{1}{|x|}\right)\right|^r |x|^Q} \\ &+ \int_{\{|x|\geqslant 2\}} \left(\widetilde{B}_{Q/p}\left(\frac{x}{2}\right)\right)^q \frac{dx}{\left|\log\left(e+\frac{1}{|x|}\right)\right|^r |x|^Q}. \end{aligned} \tag{5.73}$$

We note that the second integral in the right-hand side of (5.73) is finite by the second equality in (5.70). For the first integral, using the first equality in (5.70), we can estimate

$$\begin{aligned}\int_{\{2R<|x|<2\}} \left|\widetilde{B}_{Q/p}\left(\frac{x}{2}\right)\right|^q \frac{dx}{\left|\log\left(e+\frac{1}{|x|}\right)\right|^r |x|^Q} &\leq \int_{\{2R<|x|<2\}} \left|\widetilde{B}_{Q/p}\left(\frac{x}{2}\right)\right|^q \frac{dx}{|x|^Q} \\ &\leq C\int_{\{2R<|x|<2\}} |x|^{-Qq/p'-Q} dx \\ &\leq CR^{-Qq/p'}.\end{aligned}$$

Combining this with (5.73), we obtain

$$\begin{aligned}\left(\int_{\{2R<|x|\}} \left|\widetilde{B}_{Q/p}\left(\frac{x}{2}\right)\right|^q \frac{dx}{\left|\log\left(e+\frac{1}{|x|}\right)\right|^r |x|^Q}\right)^{\frac{1}{q}} \left(\int_{\{|x|<R\}} dx\right)^{\frac{1}{p'}} \\ \leq C(R^{-Q/p'}+1)R^{Q/p'} \leq C\end{aligned}$$

uniformly for all $0 < R < 1$. Thus, we have verified (5.71), so that applying Theorem 5.1.1, Part (1), to N_1 we obtain

$$N_1^{\frac{1}{q}} \leq (p')^{\frac{1}{p'}} p^{\frac{1}{q}} A_1 \|f\|_{L^p(\mathbb{G})}. \tag{5.74}$$

Now let us *estimate* N_2. We decompose it into the sum

$$\begin{aligned}N_2 = \sum_{k\in\mathbb{Z}} \int_{\{2^k\leqslant|x|<2^{k+1}\}} \left(\int_{\{|x|\leqslant 2|y|\leqslant 4|x|\}} |T^{(2)}_{Q/p}(y^{-1}x)f(y)|dy\right)^q \\ \times \frac{dx}{\left|\log\left(e+\frac{1}{|x|}\right)\right|^r |x|^Q}.\end{aligned}$$

Since the function $\left(\log\left(\frac{1}{|x|}\right)\right)^r |x|^Q$ is non-decreasing with respect to $|x|$ near the origin, there exists an integer $k_0 \in \mathbb{Z}$ with $k_0 \leqslant -3$ such that this function is non-decreasing in $|x| \in (0, 2^{k_0+1})$. Fixing this k_0, we decompose N_2 further as

$$N_2 = N_{21} + N_{22}, \tag{5.75}$$

where

$$N_{21} := \sum_{k=-\infty}^{k_0} \int_{\{2^k \leqslant |x| < 2^{k+1}\}} \left(\int_{\{|x| \leqslant 2|y| \leqslant 4|x|\}} |T^{(2)}_{Q/p}(y^{-1}x) f(y)| dy \right)^q \times \frac{dx}{\left|\log\left(e + \frac{1}{|x|}\right)\right|^r |x|^Q}$$

and

$$N_{22} := \sum_{k=k_0+1}^{\infty} \int_{\{2^k \leqslant |x| < 2^{k+1}\}} \left(\int_{\{|x| \leqslant 2|y| \leqslant 4|x|\}} |T^{(2)}_{Q/p}(y^{-1}x) f(y)| dy \right)^q \times \frac{dx}{\left|\log\left(e + \frac{1}{|x|}\right)\right|^r |x|^Q}.$$

Let us first estimate N_{22}. Since $|x| \leqslant 2|y| \leqslant 4|x|$ and $2^k \leqslant |x| < 2^{k+1}$, we must also have $2^{k-1} \leqslant |y| < 2^{k+2}$. Before starting to estimate N_{22}, using (5.65) and $q > p$, let us show that

$$\begin{aligned} \int_{\mathbb{G}} |T^{(2)}_{Q/p}(x)|^{\tilde{r}} dx &= \int_{|x|<1} |T^{(2)}_{Q/p}(x)|^{\tilde{r}} dx + \int_{|x|\geq 1} |T^{(2)}_{Q/p}(x)|^{\tilde{r}} dx \\ &\leq C_2 \left(\int_{|x|<1} |x|^{-\frac{Qq(p-1)}{pq+p-q}} dx + \int_{|x|\geq 1} |x|^{-\frac{Qpq}{pq+p-q}} dx \right) < \infty, \end{aligned} \tag{5.76}$$

where $\tilde{r} \in [1, \infty]$ is such that $1 + \frac{1}{q} = \frac{1}{\tilde{r}} + \frac{1}{p}$.

Then, (5.76) and Young's inequality in Proposition 1.2.13 with $1 + \frac{1}{q} = \frac{1}{\tilde{r}} + \frac{1}{p}$ and $\tilde{r} \in [1, \infty]$ imply that

$$\begin{aligned} N_{22} &\leqslant C \sum_{k=k_0+1}^{\infty} \int_{\{2^k \leqslant |x| < 2^{k+1}\}} \left(\int_{\{|x| \leqslant 2|y| \leqslant 4|x|\}} |T^{(2)}_{Q/p}(y^{-1}x) f(y)| dy \right)^q dx \\ &\leqslant C \|[f \cdot \chi_{\{2^{k-1} \leqslant |\cdot| < 2^{k+2}\}}] * T^{(2)}_{Q/p}\|^q_{L^q(\mathbb{G})} \\ &\leqslant C \|T^{(2)}_{Q/p}\|^q_{L^{\tilde{r}}(\mathbb{G})} \sum_{k=k_0+1}^{\infty} \|f \cdot \chi_{\{2^{k-1} \leqslant |\cdot| < 2^{k+2}\}}\|^q_{L^p(\mathbb{G})} \\ &= C \sum_{k=k_0+1}^{\infty} \left(\int_{\{2^k \leqslant |x| < 2^{k+1}\}} |f(x)|^p dx \right)^{q/p} \end{aligned}$$

$$\leqslant C\left(\sum_{k\in\mathbb{Z}}\int_{\{2^k\leqslant|x|<2^{k+1}\}}|f(x)|^p dx\right)^{q/p}$$
$$= C\|f\|^q_{L^p(\mathbb{G})}. \tag{5.77}$$

Next, let us estimate N_{21}. As in (5.69), assuming $|\cdot|$ is the norm and using the triangle inequality and $|y|\leqslant 2|x|$, we can estimate

$$3|x| = |x| + 2|x| \geq |x| + |y| \geq |y^{-1}x|. \tag{5.78}$$

Since $\left(\log\left(\frac{1}{|x|}\right)\right)^r |x|^Q$ is non-decreasing in $|x|\in(0,2^{k_0+1})$ and $3|x|\geqslant|y^{-1}x|$, we have

$$\left(\log\left(\frac{1}{|x|}\right)\right)^r |x|^Q \geq \left(\log\left(\frac{1}{|y^{-1}x/3|}\right)\right)^r \left|\frac{y^{-1}x}{3}\right|^Q.$$

Consequently, this and (5.65) yield

$$N_{21} \leq C\sum_{k=-\infty}^{k_0}\int_{\{2^k\leq|x|<2^{k+1}\}}\left(\int_{\{|x|\leq 2|y|\leq 4|x|\}}|y^{-1}x|^{-\frac{Q}{p'}}|f(y)|dy\right)^q$$
$$\times\frac{dx}{\left(\log\left(\frac{1}{|x|}\right)\right)^r|x|^Q}$$
$$= C\sum_{k=-\infty}^{k_0}\int_{\{2^k\leq|x|<2^{k+1}\}}\left(\int_{\{|x|\leq 2|y|\leq 4|x|\}}\frac{|y^{-1}x|^{-\frac{Q}{p'}}|f(y)|}{\left(\left(\log\left(\frac{1}{|x|}\right)\right)^r|x|^Q\right)^{\frac{1}{q}}}dy\right)^q dx$$
$$\leq C\sum_{k=-\infty}^{k_0}\int_{\{2^k\leq|x|<2^{k+1}\}}\left(\int_{\{|x|\leq 2|y|\leq 4|x|\}}\frac{|y^{-1}x|^{-\frac{Q}{p'}}|f(y)|}{\left(\left(\log\left(\frac{1}{|(y^{-1}x)/3|}\right)\right)^r|(y^{-1}x)/3|^Q\right)^{\frac{1}{q}}}dy\right)^q dx.$$

Since $|x|\leq 2|y|\leq 4|x|$ and $2^k\leq|x|<2^{k+1}$ with $k\leq k_0$, we must also have $2^{k-1}\leq|y|<2^{k+2}$ and $|y^{-1}x|\leq 3|x|<3\cdot 2^{k_0+1}\leq 3/4$, using (5.78) and $k_0\leq -3$. Taking into account these and setting

$$g(x) := \frac{\chi_{B_{\frac{3}{4}}(0)}(x)}{\left(\log\left(\frac{1}{|x|}\right)\right)^{\frac{r}{q}}|x|^{\frac{Q}{q}+\frac{Q}{p'}}},$$

we have for N_{21} that

$$N_{21}\leq C\sum_{k=-\infty}^{k_0}\int_{\{2^k\leq|x|<2^{k+1}\}}\left(\int_{\{|x|\leq 2|y|\leq 4|x|\}}\frac{|f(y)|}{\left(\log\left(\frac{1}{|y^{-1}x|}\right)\right)^{\frac{r}{q}}|y^{-1}x|^{\frac{Q}{q}+\frac{Q}{p'}}}dy\right)^q dx$$
$$\leq C\sum_{k=-\infty}^{k_0}\|[f\cdot\chi_{\{2^{k-1}\leq|\cdot|<2^{k+2}\}}]*g\|^q_{L^q(\mathbb{G})}.$$

Since $p < q < (r-1)p'$, we use Young's inequality in Proposition 1.2.13 with $1+\frac{1}{q}=\frac{1}{\tilde{r}}+\frac{1}{p}$ and $\tilde{r}\in[1,\infty)$, to get

$$N_{21} \leq C\|g\|^q_{L^{\tilde{r}}(\mathbb{G})} \sum_{k=-\infty}^{k_0} \|f\cdot\chi_{\{2^{k-1}\leq|\cdot|<2^{k+2}\}}\|^q_{L^p(\mathbb{G})} \leq C\|f\|^q_{L^p(\mathbb{G})}, \tag{5.79}$$

provided that $g\in L^{\tilde{r}}(\mathbb{G})$. Since $\left(\frac{Q}{q}+\frac{Q}{p'}\right)\tilde{r}=Q$, $\frac{r\tilde{r}}{q}=\frac{rp'}{p'+q}$ and $q<(r-1)p'$, then changing variables, we obtain

$$\|g\|^{\tilde{r}}_{L^{\tilde{r}}(\mathbb{G})} = \int_{B(0,3/4)} \frac{dx}{\left(\log\left(\frac{1}{x}\right)\right)^{\frac{rp'}{p'+q}}|x|^Q} = C\int_{\log\left(\frac{4}{3}\right)}^{\infty} \frac{dt}{t^{\frac{rp'}{p'+q}}} < \infty.$$

Let us *estimate* N_3 now. Without loss of generality, we may assume again that $|\cdot|$ is the norm. Similarly to (5.69) we obtain $|y|<2|y^{-1}x|$ from $2|x|<|y|$. Then, we have for N_3 that

$$N_3 \leq \int_{\mathbb{G}} \left(\int_{\{|y|>2|x|\}} \left|\widetilde{B}_{Q/p}\left(\frac{y}{2}\right)\right| |f(y)|dy\right)^q \frac{dx}{\left|\log\left(e+\frac{1}{|x|}\right)\right|^r |x|^Q}.$$

We will apply Theorem 5.1.1, Part (2), for the required estimate of N_3. For this we have to check the following condition:

$$\left(\int_{\{|x|<R\}} \frac{dx}{\left|\log\left(e+\frac{1}{|x|}\right)\right|^r |x|^Q}\right)^{\frac{1}{q}} \left(\int_{\{2R<|x|\}} \left|\widetilde{B}_{Q/p}\left(\frac{x}{2}\right)\right|^{p'} dx\right)^{\frac{1}{p'}} \leq A_2. \tag{5.80}$$

To check this, let us consider the cases: $R\geq1$ and $0<R<1$. Then, for $R\geq1$ by the second equality in (5.70), we get

$$\left(\int_{\{2R<|x|\}} \left|\widetilde{B}_{Q/p}\left(\frac{x}{2}\right)\right|^{p'} dx\right)^{\frac{1}{p'}} \leq C\left(\int_{\{2R<|x|\}} |x|^{-Qp'}dx\right)^{\frac{1}{p'}} \leq CR^{-\frac{Q}{p}}. \tag{5.81}$$

Moreover, we have

$$\int_{\{|x|<R\}} \frac{dx}{\left|\log\left(e+\frac{1}{|x|}\right)\right|^r |x|^Q} = \int_{\{|x|<\frac{1}{2}\}} \frac{dx}{\left|\log\left(e+\frac{1}{|x|}\right)\right|^r |x|^Q} + \int_{\{\frac{1}{2}\leqslant|x|<R\}} \frac{dx}{\left|\log\left(e+\frac{1}{|x|}\right)\right|^r |x|^Q},$$

and we note that the first summand in the right-hand side of above is finite since $r>1$. For the second term, we get

$$\int_{\{\frac{1}{2}\leq|x|<R\}} \frac{dx}{\left|\log\left(e+\frac{1}{|x|}\right)\right|^r |x|^Q} \leq \int_{\{\frac{1}{2}\leq|x|<R\}} \frac{dx}{|x|^Q} \leq C(1+\log R). \tag{5.82}$$

Combining (5.81) and (5.82), we have for $R \geq 1$ that

$$\left(\int_{\{|x|<R\}} \frac{dx}{\left|\log\left(e+\frac{1}{|x|}\right)\right|^{r}|x|^{Q}}\right)^{\frac{1}{q}}\left(\int_{\{2R<|x|\}}\left|\widetilde{B}_{Q/p}\left(\frac{x}{2}\right)\right|^{p'}dx\right)^{\frac{1}{p'}} \leq CR^{-\frac{Q}{p}}(1+\log R)^{\frac{1}{q}} \leq C.$$

Now let us check the condition (5.80) for $0 < R < 1$. We split the integral into two terms:

$$\begin{aligned}\int_{\{2R<|x|\}}&\left|\widetilde{B}_{Q/p}\left(\frac{x}{2}\right)\right|^{p'}dx\\ &=\int_{\{2R<|x|<2\}}\left|\widetilde{B}_{Q/p}\left(\frac{x}{2}\right)\right|^{p'}dx+\int_{\{|x|\geqslant 2\}}\left|\widetilde{B}_{Q/p}\left(\frac{x}{2}\right)\right|^{p'}dx.\end{aligned} \tag{5.83}$$

We note that the second integral in the right-hand side of above is finite by the second equality in (5.70). Then, using the first equality in (5.70) we get for the first integral that

$$\int_{\{2R<|x|<2\}}\left|\widetilde{B}_{Q/p}\left(\frac{x}{2}\right)\right|^{p'}dx \leq C\int_{\{2R<|x|<2\}}|x|^{-Q}dx \leq C\log\left(\frac{1}{R}\right).$$

Combined with (5.83), it follows that

$$\int_{\{2R<|x|\}}\left|\widetilde{B}_{Q/p}\left(\frac{x}{2}\right)\right|^{p'}dx \leq C\left(1+\log\left(\frac{1}{R}\right)\right). \tag{5.84}$$

Since

$$\int_{\{|x|<R\}}\frac{dx}{\left|\log\left(e+\frac{1}{|x|}\right)\right|^{r}|x|^{Q}} \leqslant C\left(\log\left(e+\frac{1}{R}\right)\right)^{-(r-1)},$$

and (5.84), and taking into account $r > 1$ and $q < (r-1)p'$ we obtain that

$$\begin{aligned}&\left(\int_{\{|x|<R\}}\frac{dx}{\left|\log\left(e+\frac{1}{|x|}\right)\right|^{r}|x|^{Q}}\right)^{\frac{1}{q}}\left(\int_{\{2R<|x|\}}\left|\widetilde{B}_{Q/p}\left(\frac{x}{2}\right)\right|^{p'}dx\right)^{\frac{1}{p'}}\\ &\leqslant C\left(\log\left(e+\frac{1}{R}\right)\right)^{-\frac{r-1}{q}}\left(1+\left(\log\left(\frac{1}{R}\right)\right)^{\frac{1}{p'}}\right) \leqslant C.\end{aligned} \tag{5.85}$$

Thus, we have checked (5.80). Consequently, applying Theorem 5.1.1, Part (2), for the term N_3, we obtain

$$N_3^{\frac{1}{q}} \leq (p')^{\frac{1}{p'}}p^{\frac{1}{q}}A_2\|f\|_{L^p(\mathbb{G})}. \tag{5.86}$$

Finally, a combination of (5.74), (5.86), (5.75), (5.77), (5.79) and (5.68) completes the proof of Theorem 5.2.5. □

5.3 Hardy–Littlewood–Sobolev inequalities on homogeneous groups

In this section we discuss the Hardy–Littlewood–Sobolev inequality on homogeneous groups. We show that it can be obtained as a simple consequence of the convolution Hardy inequality in Theorem 5.2.1. In fact, the argument implies a little more.

Theorem 5.3.1 (Hardy–Littlewood–Sobolev inequality). *Let $\mathbb{G}$ be a homogeneous Lie group of homogeneous dimension Q with a homogeneous quasi-norm $|\cdot|$. Let $0<\lambda<Q$ and $1<p,q<\infty$ be such that*

$$1/p+1/q+(\alpha+\lambda)/Q=2$$

with $0\leq\alpha<Q/p'$ and $\alpha+\lambda\leq Q$, where $1/p+1/p'=1$. Then there exists a positive constant $C=C(Q,\lambda,p,\alpha)>0$ such that

$$\left|\int_{\mathbb{G}}\int_{\mathbb{G}}\frac{\overline{f(x)}g(y)}{|x|^{\alpha}|y^{-1}x|^{\lambda}}dxdy\right|\leq C\|f\|_{L^p(\mathbb{G})}\|g\|_{L^q(\mathbb{G})} \tag{5.87}$$

holds for all $f\in L^p(\mathbb{G})$ and $g\in L^q(\mathbb{G})$.

Remark 5.3.2 (Stein–Weiss inequality).

1. The original Hardy–Littlewood–Sobolev inequality goes back to the work of Hardy–Littlewood [HL27], [HL30] and Sobolev [Sob38]. More specifically, in [HL27], Hardy and Littlewood considered the one-dimensional fractional operator on $(0,\infty)$, given by

$$T_\lambda f(x)=\int_0^\infty\frac{f(y)}{|x-y|^\lambda}dy,\quad 0<\lambda<1, \tag{5.88}$$

and proved that if $1<p<q<\infty$ and $\frac{1}{q}=\frac{1}{p}+\lambda-1$, then there is $C>0$ such that

$$\|T_\lambda f\|_{L^q(0,\infty)}\leq C\|f\|_{L^p(0,\infty)},$$

holds for all $f\in L^p(0,\infty)$. The N-dimensional analogue of (5.88) can be written by the formula

$$I_\lambda f(x)=\int_{\mathbb{R}^N}\frac{f(y)}{|x-y|^\lambda}dy,\quad 0<\lambda<N. \tag{5.89}$$

Consequently, if was shown by Sobolev in [Sob38] that if $1<p<q<\infty$ and $\frac{1}{q}=\frac{1}{p}+\frac{\lambda}{N}-1$, then there is $C>0$ such that

$$\|I_\lambda f\|_{L^q(\mathbb{R}^N)}\leq C\|f\|_{L^p(\mathbb{R}^N)},$$

holds for all $f\in L^p(\mathbb{R}^N)$. In [SW58], Stein and Weiss obtained the following two-weight extension of the Hardy–Littlewood–Sobolev inequality, which is nowadays called the *Stein–Weiss inequality.* More specifically, they have shown that if $0<\lambda<N$, $1<p<\infty$, $\alpha<\frac{N(p-1)}{p}$, $\beta<\frac{N}{q}$, $\alpha+\beta\geq 0$,

$\frac{1}{q} = \frac{1}{p} + \frac{\lambda+\alpha+\beta}{N} - 1$, and $1 < p \leq q < \infty$, then there is $C > 0$ such that we have

$$\||x|^{-\beta} I_\lambda f\|_{L^q(\mathbb{R}^N)} \leq C \||x|^\alpha f\|_{L^p(\mathbb{R}^N)}. \tag{5.90}$$

2. An extension of the Hardy–Littlewood–Sobolev inequality to the Heisenberg groups was considered in [FS74]. The sharp constants in the Hardy–Littlewood–Sobolev inequality in the cases of $\mathbb{R}^N$ and the Heisenberg group were obtained in [Lie83] and [FL12], respectively.
3. In [GMS10] the analogues of the Stein–Weiss inequality were obtained on Carnot groups. Note that in [HLZ12] the authors also proved an analogue of the Stein–Weiss inequality on the Heisenberg groups.
4. On general homogeneous groups the statement of Theorem 5.3.1 will be here obtained as a consequence of the integral Hardy inequalities. The estimate (5.87) contains the Hardy–Littlewood–Sobolev inequality and half of the Stein–Weiss inequality. The full Stein–Weiss inequality on homogeneous groups was obtained in [KRS18b]: *Let*

$$I_\lambda u(x) := \int_{\mathbb{G}} \frac{u(y)}{|y^{-1}x|^\lambda} dy, \quad 0 < \lambda < Q. \tag{5.91}$$

Let $0 < \lambda < Q$, $1 < p < \infty$, $\alpha < \frac{Q}{p'}$, $\beta < \frac{Q}{q}$, $\alpha + \beta \geq 0$, $\frac{1}{q} = \frac{1}{p} + \frac{\alpha+\beta+\lambda}{Q} - 1$, *where* $\frac{1}{p} + \frac{1}{p'} = 1$ *and* $\frac{1}{q} + \frac{1}{q'} = 1$. *Then for* $1 < p \leq q < \infty$ *we have*

$$\||x|^{-\beta} I_\lambda u\|_{L^q(\mathbb{G})} \leq C \||x|^\alpha u\|_{L^p(\mathbb{G})}. \tag{5.92}$$

5. **(Differential Stein–Weiss inequality on graded groups)**. Continuing with the notation of Remark 5.2.2, let $\dot{L}^p_a(\mathbb{G})$ be the homogeneous Sobolev space over L^p of order a associated to a Rockland operator. Such spaces are well defined on graded groups and do not depend on a particular choice of a Rockland operator, we refer to [FR17] or to [FR16, Section 4.4] for the extensive analysis and exposition of their properties. The following differential version of the Stein–Weiss inequality was obtained in [RY18a]:

Theorem 5.3.3 (Differential Stein–Weiss inequality). *Let* $\mathbb{G}$ *be a graded group of homogeneous dimension* Q *and let* $|\cdot|$ *be a quasi-norm on* $\mathbb{G}$. *Let* $1 < p, q < \infty$, $0 \leq a < Q/p$ *and* $0 \leq b < Q/q$. *Let* $0 < \lambda < Q$, $0 \leq \alpha < a + Q/p'$ *and* $0 \leq \beta \leq b$ *be such that* $(Q - ap)/(pQ) + (Q - q(b-\beta))/(qQ) + (\alpha+\lambda)/Q = 2$ *and* $\alpha + \lambda \leq Q$, *where* $1/p + 1/p' = 1$. *Then there exists a positive constant* $C = C(Q, \lambda, p, \alpha, \beta, a, b)$ *such that*

$$\left| \int_{\mathbb{G}} \int_{\mathbb{G}} \frac{\overline{f(x)} g(y)}{|x|^\alpha |y^{-1}x|^\lambda |y|^\beta} dx dy \right| \leq C \|f\|_{\dot{L}^p_a(\mathbb{G})} \|g\|_{\dot{L}^q_b(\mathbb{G})} \tag{5.93}$$

holds for all $f \in \dot{L}^p_a(\mathbb{G})$ *and* $g \in \dot{L}^q_b(\mathbb{G})$.

While the setting of graded groups falls outside the scope of this book, we follow [RY18a] in the proof of Theorem 5.3.1 below.

Proof of Theorem 5.3.1. Let $T_a(x) := |x|^{a-Q}$ with $0 < a < Q/r$ for some $1 < r < \infty$. Then, using Hölder's inequality we have

$$\begin{aligned}\left|\int_{\mathbb{G}}\int_{\mathbb{G}} \frac{\overline{f(x)}g(y)}{|x|^{\alpha}|y^{-1}x|^{\lambda}}dxdy\right| &= \left|\int_{\mathbb{G}} \overline{f(x)}\frac{(g * T_{Q-\lambda})(x)}{|x|^{\alpha}}dx\right| \\ &\leq \|f\|_{L^p(\mathbb{G})}\left\|\frac{g * T_{Q-\lambda}}{|x|^{\alpha}}\right\|_{L^{p'}(\mathbb{G})}.\end{aligned} \tag{5.94}$$

Note that the conditions $\alpha + \lambda \leq Q$ and $1/p + 1/q + (\alpha+\lambda)/Q = 2$ imply $q \leq p'$, while $0 < \lambda < Q$, $\alpha < Q/p'$ and $1/p + 1/q + (\alpha+\lambda)/Q = 2$ give

$$0 < Q - \lambda = Q - Q\left(2 - \frac{1}{p} - \frac{1}{q}\right) + \alpha < Q - Q\left(2 - \frac{1}{p} - \frac{1}{q}\right) + \frac{Q}{p'} = Q/q.$$

Since we have $1 < q \leq p' < \infty$, $0 \leq \alpha p' < Q$, $0 < Q - \lambda < Q/q$ and $(Q-\lambda)/Q = 1/q - 1/p' + \alpha/Q$, using Theorem 5.2.1 in (5.94) we obtain (5.87). □

Remark 5.3.4 (Reversed Hardy–Littlewood–Sobolev inequality)**.** Let us make some remarks concerning the reversed Hardy–Littlewood–Sobolev inequality on homogeneous groups. Namely, consider the inequality

$$\int_{\mathbb{G}}\int_{\mathbb{G}} f(x)|y^{-1}x|^{\lambda}f(y)dxdy \geq C_{Q,\lambda,p}\|f\|^{\theta}_{L^1(\mathbb{G})}\|f\|^{2-\theta}_{L^p(\mathbb{G})} \tag{5.95}$$

for any $0 \leq f \in L^1 \cap L^p(\mathbb{G})$ with $f \not\equiv 0$ and $0 < p < 1$, where $\lambda > 0$ and $\theta := (2Q - p(2Q+\lambda))/(Q(1-p))$.

In the Euclidean case $\mathbb{G} = (\mathbb{R}^n, +)$, i.e., with $Q = n$, the case $p = 2n/(2n+\lambda)$ was investigated in [DZ15] and [NN17], while the case $p > n/(n+\lambda)$ was studied in [DFH18].

Following [RY18a], let us briefly recapture the argument in the setting of general homogeneous groups. Namely, let us show that in the case $0 < p \leq Q/(Q+\lambda)$ the inequality (5.95) is not valid, namely, (5.95) fails for any $C_{Q,\lambda,p} > 0$. In the Euclidean case this was shown in [CDP18] when $p < n/(n+\lambda)$ and in [DFH18] when $p \leq n/(n+\lambda)$.

Let f be a non-negative function with compact support, and let h be a non-negative smooth function such that $\int_{\mathbb{G}} h(x)dx = 1$. Then, for some $A > 0$, consider the function

$$f_\varepsilon(x) := f(x) + A\varepsilon^{-Q}h(x/\varepsilon).$$

Suppose now that (5.95) holds for some $C_{Q,\lambda,p} > 0$. Putting this f_ε in the inequality (5.95), we obtain

$$\begin{aligned}C_{Q,\lambda,p} &\leq \frac{\int_{\mathbb{G}}\int_{\mathbb{G}} f_\varepsilon(x)|y^{-1}x|^{\lambda}f_\varepsilon(y)dxdy}{\|f_\varepsilon\|^{\theta}_{L^1(\mathbb{G})}\|f_\varepsilon\|^{2-\theta}_{L^p(\mathbb{G})}} \\ &\to \frac{\int_{\mathbb{G}}\int_{\mathbb{G}} f(x)|y^{-1}x|^{\lambda}f(y)dxdy + 2A\int_{\mathbb{G}}|x|^{\lambda}f(x)dx}{(\int_{\mathbb{G}} f(x)dx + A)^{\theta}(\int_{\mathbb{G}}(f(x))^p dx)^{(2-\theta)/p}}\end{aligned} \tag{5.96}$$

as $\varepsilon \to 0_+$, where we have used that

- $\int_{\mathbb{G}} f_\varepsilon(x)dx = \int_{\mathbb{G}} f(x)dx + A$;
- when $\varepsilon \to 0_+$, we have

$$\int_{\mathbb{G}} (f_\varepsilon(x))^p dx \to \int_{\mathbb{G}} (f(x))^p dx;$$

- when $\varepsilon \to 0_+$, we have

$$\begin{aligned}
&\int_{\mathbb{G}} \int_{\mathbb{G}} f_\varepsilon(x)|y^{-1}x|^\lambda f_\varepsilon(y)dxdy \\
&\quad = \int_{\mathbb{G}} \int_{\mathbb{G}} f(x)|y^{-1}x|^\lambda f(y)dxdy + 2A \int_{\mathbb{G}} \int_{\mathbb{G}} f(x)|(\varepsilon^{-1}y)^{-1}x|^\lambda h(y)dxdy \\
&\quad + A^2\varepsilon^{-2Q} \int_{\mathbb{G}} \int_{\mathbb{G}} h\left(\frac{x}{\varepsilon}\right) h\left(\frac{y}{\varepsilon}\right) dxdy \\
&\quad \to \int_{\mathbb{G}} \int_{\mathbb{G}} f(x)|y^{-1}x|^\lambda f(y)dxdy + 2A \int_{\mathbb{G}} |x|^\lambda f(x)dx,
\end{aligned}$$

since $\int_{\mathbb{G}} h(x)dx = 1$.

Note that in (5.96) we can take also the limit as $A \to +\infty$ since it is valid for all $A > 0$. Then, when $\theta > 1$, that is, for $p < Q/(Q+\lambda)$, taking $A \to +\infty$ in (5.96) we see that $C_{Q,\lambda,p} = 0$. In the case $\theta = 1$, that is, for $p = Q/(Q+\lambda)$, taking the limit as $A \to +\infty$ in (5.96) we get

$$C_{Q,\lambda,p} \leq \frac{2\int_{\mathbb{G}} |x|^\lambda f(x)dx}{\left(\int_{\mathbb{G}} (f(x))^p dx\right)^{1/p}}. \tag{5.97}$$

Finally, we show that the right-hand side of (5.97) goes to zero as $R \to \infty$ if we insert the function

$$f_R(x) = \begin{cases} |x|^{-(Q+\lambda)}, & \text{for } 1 \leq |x| \leq R, \\ 0, & \text{otherwise}, \end{cases} \tag{5.98}$$

for any $R > 1$. Indeed, in this case $p = Q/(Q+\lambda)$, and from (5.97) we obtain that

$$C_{Q,\lambda,p} \leq \frac{2\int_{\mathbb{G}} |x|^\lambda f_R(x)dx}{\left(\int_{\mathbb{G}} (f_R(x))^p dx\right)^{1/p}} = 2(|\wp| \log R)^{-\lambda/Q} \to 0 \tag{5.99}$$

as $R \to \infty$, where $|\wp|$ is a $Q-1$-dimensional surface measure of the unit quasi-sphere in $\mathbb{G}$.

Summarizing, we conclude that for $0 < p \leq Q/(Q+\lambda)$ the reversed Hardy–Littlewood–Sobolev inequality (5.95) is not valid with any constant $C_{Q,\lambda,p} > 0$.

5.4 Maximal weighted integral Hardy inequality

Here we present a maximal Hardy inequality in the integral form, involving the maximal function

$$(\mathcal{M}f)(x) := \frac{1}{|B(0,|x|)|}\int_{B(0,|x|)} f(z)dz.$$

Theorem 5.4.1 (Maximal integral weighted Hardy inequality). *Let $\mathbb{G}$ be a homogeneous group of homogeneous dimension Q with a homogeneous quasi-norm $|\cdot|$. Let ϕ and ψ be positive functions defined on $\mathbb{G}$. Then there exists a constant $C > 0$ such that*

$$\int_{\mathbb{G}} \phi(x)\exp(\mathcal{M}\log f)(x)dx \leq C\int_{\mathbb{G}} \psi(x)f(x)dx \tag{5.100}$$

holds for all positive $f \geq 0$ if and only if

$$A := \sup_{R>0} R^Q \int_{|x|\geq R} \frac{\phi(x)\exp\left(\mathcal{M}\log\frac{1}{\psi}\right)(x)}{|x|^{2Q}}dx < \infty. \tag{5.101}$$

Remark 5.4.2. Inequalities of the type of those in Theorem 5.4.1 in the Abelian case $\mathbb{G} = (\mathbb{R}^n, +)$ were studied in [HKK01] for the one-dimensional case $n = 1$, and in [DHK97] for the multidimensional case $n \geq 1$. Theorem 5.4.1 was proved in [RSY18a] and we follow the presentation there.

Proof of Theorem 5.4.1. Let us first show (5.101) implies (5.100) for all $f \geq 0$. Denoting

$$W_3(x) := \phi(x)\exp\left(\mathcal{M}\log\frac{1}{\psi}\right)(x), \tag{5.102}$$

as well as $u(x) := f(x)\psi(x)$, and $z = |x|\xi$, we have

$$\begin{aligned}
&\int_{\mathbb{G}} \phi(x)\exp(\mathcal{M}\log f)(x)dx\\
&= \int_{\mathbb{G}} \phi(x)\exp\left(\frac{1}{|B(0,|x|)|}\left(\int_{B(0,|x|)}\log\left(\frac{1}{\phi}\right)(z)dz + \int_{|z|\leq|x|}\log(\phi f)(z)dz\right)\right)dx\\
&= \int_{\mathbb{G}} \phi(x)\exp\left(\mathcal{M}\log\frac{1}{\phi}\right)(x)\exp\left(\frac{1}{|B(0,|x|)|}\int_{|z|\leq|x|}\log(\phi(z)f(z))dz\right)dx\\
&= \int_{\mathbb{G}} W_3(x)\exp\left(\frac{1}{|B(0,|x|)|}\int_{|z|\leq|x|}\log(u(z))dz\right)dx\\
&= \int_{\mathbb{G}} W_3(x)\exp\left(\frac{1}{|x|^Q|B(0,1)|}\int_{B(0,1)}\log(u(|x|\xi))|x|^Q d\xi\right)dx.
\end{aligned} \tag{5.103}$$

Since

$$\int_{B(0,1)}\log(|\xi|^Q)d\xi = Q\int_{\wp}\int_0^1 r^{Q-1}\log r\,dr\,d\sigma(y) = -|B(0,1)|,$$

and by using Jensen's inequality, we obtain

$$\begin{aligned}
&\int_{\mathbb{G}} \phi(x)\exp(\mathcal{M}\log f)(x)dx \\
&= \int_{\mathbb{G}} W_3(x)\exp\left(\frac{1}{|B(0,1)|}\left(\int_{B(0,1)}\log(|\xi|^Q u(|x|\xi))d\xi - \int_{B(0,1)}\log(|\xi|^Q)d\xi\right)\right)dx \\
&= \int_{\mathbb{G}} W_3(x)\exp\left(\frac{1}{|B(0,1)|}\int_{B(0,1)}\log(|\xi|^Q u(|x|\xi))d\xi + 1\right)dx \\
&= e\int_{\mathbb{G}} W_3(x)\exp\left(\frac{1}{|B(0,1)|}\int_{B(0,1)}\log(|\xi|^Q u(|x|\xi))d\xi\right)dx \\
&\leq \frac{e}{|B(0,1)|}\int_{\mathbb{G}} W_3(x)\int_{B(0,1)}|\xi|^Q u(|x|\xi)d\xi dx \\
&= \frac{e}{|B(0,1)|}\int_{\wp}\int_{\wp}\int_0^\infty r^{Q-1}W_3(rw)\int_0^1 s^{2Q-1}u(rsy)dsdrd\sigma(y)d\sigma(w),
\end{aligned}$$

where $|\xi| = s$ and $|x| = r$. Furthermore, with $t = rs$ we get

$$\begin{aligned}
&\int_{\mathbb{G}} \phi(x)\exp(\mathcal{M}\log f)(x)dx \\
&\quad \leq \frac{e}{|B(0,1)|}\int_{\wp}\int_{\wp}\int_0^\infty r^{Q-1}W_3(rw)\int_0^1 s^{2Q-1}u(rsy)dsdrd\sigma(y)d\sigma(w) \\
&\quad = \frac{e}{|B(0,1)|}\int_{\wp}\int_{\wp}\int_0^1 s^{2Q-1}\int_0^\infty W_3\left(\frac{t}{s}w\right)\left(\frac{t}{s}\right)^{Q-1}u(ty)\frac{dt}{s}dsd\sigma(y)d\sigma(w) \\
&\quad = \frac{e}{|B(0,1)|}\int_{\wp}\int_{\wp}\int_0^1 s^{Q-1}\int_0^\infty W_3\left(\frac{t}{s}w\right)t^{Q-1}u(ty)dtdsd\sigma(y)d\sigma(w) \\
&\quad = \frac{e}{|B(0,1)|}\int_{\wp}\int_{\wp}\int_0^\infty t^{Q-1}u(ty)\left(\int_0^1 s^{Q-1}W_3\left(\frac{tw}{s}\right)ds\right)dtd\sigma(y)d\sigma(w) \\
&\quad = \frac{e}{|B(0,1)|}\int_{\wp}\int_{\wp}\int_0^\infty t^{Q-1}u(ty)\left(\int_t^\infty \left(\frac{t}{r}\right)^{Q-1}W_3(rw)t\frac{dr}{r^2}\right)dtd\sigma(y)d\sigma(w) \\
&\quad = \frac{e}{|B(0,1)|}\int_{\wp}\int_{\wp}\int_0^\infty t^{2Q-1}u(ty)\left(\int_t^\infty r^{-Q-1}W_3(rw)dr\right)dtd\sigma(y)d\sigma(w) \\
&\quad \leq \frac{e}{|B(0,1)|}\int_{\wp}\int_{\wp}\int_0^\infty t^{Q-1}u(ty)t^Q\left(\int_t^\infty \frac{r^{Q-1}W_3(rw)dr}{r^{2Q}}\right)dtd\sigma(w)d\sigma(y) \\
&\quad = \frac{e}{|B(0,1)|}\int_{\mathbb{G}} u(x)\left(|x|^Q\int_{|z|\geq|x|}\frac{W_3(z)}{|z|^{2Q}}dz\right)dx \leq A\frac{e}{|B(0,1)|}\int_{\mathbb{G}} u(x)dx,
\end{aligned}$$

yielding (5.100), where we have used (5.101) in the last line.

We now show that (5.100) implies (5.101). From (5.103) we notice that (5.100) is equivalent to

$$\int_{\mathbb{G}} W_3(x)\exp\left(\frac{1}{|B(0,|x|)|}\int_{|z|\le|x|}\log(u(z))dz\right)dx \le C\int_{\mathbb{G}} u(x)dx. \tag{5.104}$$

Furthermore, for a function

$$u(x) = R^{-Q}\chi_{(0,R)}(|x|) + e^{-2Q}|x|^{-2Q}R^{Q}\chi_{(R,\infty)}(|x|),\ x\in\mathbb{G},\ R>0, \tag{5.105}$$

we have

$$\begin{aligned}
&\int_{\mathbb{G}} W_3(x)\exp\left(\frac{1}{|B(0,|x|)|}\int_{|z|\le|x|}\log(u(z))dz\right)dx \\
&\le C\int_{\mathbb{G}} u(x)dx = C\int_{\wp}\int_0^\infty s^{Q-1}u(s)dsd\sigma(y) \\
&= C|\wp|\left(\int_0^R s^{Q-1}R^{-Q}ds + \int_R^\infty e^{-2Q}s^{Q-1}R^{Q}s^{-2Q}ds\right) \\
&= C|\wp|\left(\frac{1}{Q}+\frac{e^{-2Q}}{Q}\right) =: C(Q) < \infty,
\end{aligned}$$

since χ is the cut-off function. Thus, from this, by plugging (5.105) into the left-hand side of (5.104) we calculate

$$\begin{aligned}
\infty > C(Q) &\ge \int_{\mathbb{G}} W_3(x)\exp\left(\frac{1}{|B(0,|x|)|}\int_{|z|\le|x|}\log(u(z))dz\right)dx \\
&= \int_{\wp}\int_0^\infty s^{Q-1}W_3(sy)\exp\left(\frac{1}{|B(0,s)|}\int_{\wp}\int_0^s r^{Q-1}\log(u(r))drd\sigma(w)\right)dsd\sigma(y) \\
&= \int_{\wp}\int_0^\infty s^{Q-1}W_3(sy)\exp\left(\frac{|\wp|}{s^Q|B(0,1)|}\int_0^s r^{Q-1}\log(u(r))dr\right)dsd\sigma(y) \\
&= \int_{\wp}\left(\int_0^R s^{Q-1}W_3(sy)\exp\left(\frac{|\wp|}{s^Q|B(0,1)|}\int_0^s r^{Q-1}\log(u(r))dr\right)ds\right)d\sigma(y) \\
&\quad+\int_{\wp}\left(\int_R^\infty s^{Q-1}W_3(sy)\exp\left(\frac{|\wp|}{s^Q|B(0,1)|}\left(\int_0^R r^{Q-1}\log(R^{-Q})dr\right.\right.\right. \\
&\qquad\left.\left.\left.+\int_R^s r^{Q-1}\log(e^{-2Q}r^{-2Q}R^{Q})dr\right)\right)ds\right)d\sigma(y) \\
&\ge \int_{\wp}\left(\int_R^\infty s^{Q-1}W_3(sy)\exp\left(\frac{|\wp|}{s^Q|B(0,1)|}\left(\int_0^R r^{Q-1}\log(R^{-Q})dr\right.\right.\right. \\
&\quad\left.\left.\left.+\int_R^s r^{Q-1}\log(e^{-2Q}r^{-2Q}R^{Q})dr\right)\right)ds\right)d\sigma(y)
\end{aligned}$$

$$
\begin{aligned}
&= \int_{\wp} \left(\int_R^\infty s^{Q-1} W_3(sy) \exp \left(\frac{|\wp|}{s^Q |B(0,1)|} \left(\int_0^R r^{Q-1} \log(R^{-Q}) dr \right.\right.\right. \\
&\quad + \int_R^s r^{Q-1} \log(e^{-2Q}) dr - 2Q \int_R^s r^{Q-1} (\log r) dr \\
&\quad \left.\left.\left. + \int_R^s r^{Q-1} \log(R^Q) dr \right) \right) ds \right) d\sigma(y) \\
&= \int_{\wp} \left(\int_R^\infty s^{Q-1} W_3(sy) \exp \left(\frac{|\wp|}{s^Q |B(0,1)|} \left(\frac{R^Q \log(R^{-Q})}{Q} - 2Q \frac{s^Q - R^Q}{Q} \right.\right.\right. \\
&\quad \left.\left.\left. -2Q \left[\frac{r^Q \log r}{Q} - \frac{r^Q}{Q^2} \right]_R^s + \frac{s^Q - R^Q}{Q} \log(R^Q) \right) \right) ds \right) d\sigma(y) \\
&= \int_{\wp} \int_R^\infty s^{Q-1} W_3(sy) \\
&\quad \exp \left(\frac{|\wp|}{|B(0,1)|} \left(-2 + \frac{2R^Q}{s^Q} - 2\log s + \frac{2}{Q} - \frac{2R^Q}{Q s^Q} + \log R \right) \right) ds d\sigma(y) \\
&\geq e^{(2-2Q)} \int_{\wp} \int_R^\infty s^{Q-1} W_3(sy) \frac{R^{\frac{|\wp|}{|B(0,1)|}}}{s^{\frac{2|\wp|}{|B(0,1)|}}} ds d\sigma(y) \\
&= e^{2-2Q} R^Q \int_{|x| \geq R} \frac{W_3(x)}{|x|^{2Q}} dx = e^{2-2Q} R^Q \int_{|x| \geq R} \frac{\phi(x) \exp\left(\mathcal{M} \log \frac{1}{\psi} \right)(x)}{|x|^{2Q}} dx,
\end{aligned}
$$

which implies (5.101), where we have used $\frac{|\wp|}{|B(0,1)|} = Q$, $\frac{2R^Q}{s^Q} - \frac{2R^Q}{Qs^Q} > 0$, and (5.102) in the last two lines. □

Permissions

All chapters in this book were first published in HIHG, by Springer by Michael Ruzhansky and Durvudkhan Suragan; hereby published with permission under the Creative Commons Attribution License or equivalent. Every chapter published in this book has been scrutinized by our experts. Their significance has been extensively debated. The topics covered herein carry significant information for a comprehensive understanding. They may even be implemented as practical applications or may be referred to as a beginning point for further studies.

We would like to thank the editorial team for lending their expertise to make the book truly unique. They have played a crucial role in the development of this book. Without their invaluable contributions this book wouldn't have been possible. They have made vital efforts to compile up to date information on the varied aspects of this subject to make this book a valuable addition to the collection of many professionals and students.

This book was conceptualized with the vision of imparting up-to-date and integrated information in this field. To ensure the same, a matchless editorial board was set up. Every individual on the board went through rigorous rounds of assessment to prove their worth. After which they invested a large part of their time researching and compiling the most relevant data for our readers.

The editorial board has been involved in producing this book since its inception. They have spent rigorous hours researching and exploring the diverse topics which have resulted in the successful publishing of this book. They have passed on their knowledge of decades through this book. To expedite this challenging task, the publisher supported the team at every step. A small team of assistant editors was also appointed to further simplify the editing procedure and attain best results for the readers.

Apart from the editorial board, the designing team has also invested a significant amount of their time in understanding the subject and creating the most relevant covers. They scrutinized every image to scout for the most suitable representation of the subject and create an appropriate cover for the book.

The publishing team has been an ardent support to the editorial, designing and production team. Their endless efforts to recruit the best for this project, has resulted in the accomplishment of this book. They are a veteran in the field of academics and their pool of knowledge is as vast as their experience in printing. Their expertise and guidance has proved useful at every step. Their uncompromising quality standards have made this book an exceptional effort. Their encouragement from time to time has been an inspiration for everyone.

The publisher and the editorial board hope that this book will prove to be a valuable piece of knowledge for students, practitioners and scholars across the globe.

Index

www.ingramcontent.com/pod-product-compliance
Lightning Source LLC
LaVergne TN
LVHW051048220926
848485LV00001BA/47
* 9 7 8 1 6 3 9 8 7 3 0 7 4 *